THE ANCIENT ENGINEERS

Books by L. Sprague de Camp

Historical Fiction:

An Elephant for Aristotle
The Bronze God of Rhodes
The Dragon of the Ishtar Gate

Science Fiction:

Lest Darkness Fall
Divide and Rule
The Wheels of If
Genus Homo (with P. Schuyler Miller)
Rogue Queen
The Continent Makers
Sprague de Camp's New Anthology
Cosmic Manhunt
The Tower of Zanid
The Glory That Was
The Search for Zei
A Gun for Dinosaur

Fantasy:

The Incomplete Enchanter (with Fletcher Pratt)
The Land of Unreason (with Fletcher Pratt)
The Carnelian Cube (with Fletcher Pratt)
The Castle of Iron (with Fletcher Pratt)
The Undesired Princess
Tales from Gavagan's Bar (with Fletcher Pratt)
The Tritonian Ring
Tales of Conan (with Robert E. Howard)
The Return of Conan (with Björn Nyberg)
Solomon's Stone
Wall of Serpents (with Fletcher Pratt)

Non-fiction:

Inventions and their Management (with Alf K. Berle)
The Evolution of Naval Weapons
Lands Beyond (with Willy Ley)
Science-Fiction Handbook
Lost Continents
Inventions, Patents, and their Management (with Alf K. Berle)
The Heroic Age of American Invention
The Ancient Engineers

Juvenile:

Engines
Man and Power
Energy and Power
Man and Life

THE ANCIENT ENGINEERS

L. SPRAGUE DE CAMP

BARNES & NOBLE BOOKS
NEW YORK

Several sections of this book have appeared as articles in *Fate, Isis,* and *Science Digest,* for which permission to use is gratefully acknowledged. Permission also is gratefully acknowledged from the British Museum, the Deutsches Museum of Munich, the *Illustrated London News,* the Metropolitan Museum of Art (New York), Guido Ucelli, the University Museum of the University of Pennsylvania, and the Vorderasiatisches Museum of the Staatliche Museen zu Berlin for permission to reproduce the photographs and paintings duplicated in the plates. Plate XIX, an illustration by John Christiansen from *Heavenly Clockwork* by Joseph Needham, Wang Ling, and Derek J. de Solla Price, is reprinted by permission of Cambridge University Press.

We are grateful to the following publishers for permission to use excerpts from their copyrighted publications: Cambridge University Press for material from *Heavenly Clockwork* by Joseph Needham, Wang Ling, and Derek J. de Solla Price; The Clarendon Press, Oxford, for material from *The Legacy of Egypt* by S. R. K. Glanville and Aristotle's *Mechanics, On Marvelous Things Heard,* and *Politics;* Dodd, Mead & Company for six lines from "Lepanto" from *The Collected Poems of G. K. Chesterton,* copyright 1932 by Dodd, Mead & Company; Harvard University Press for excerpts from *The Loeb Classical Library;* Charles Scribner's Sons for material from *Technology* by Friedrich Klemm; Springer-Verlag for material from *The Engineering Techniques in Ancient History* by Curt Merckel; The Williams & Wilkins Company for material from *Engineers and Engineering in the Renaissance* by William B. Parsons.

This edition published by Barnes & Noble, Inc.,
by arrangement with Spectrum Literary Agency

1993 Barnes & Noble Books

ISBN 0-88029-456-6

Printed and bound in the United States of America

M 9 8 7 6

To my wife and collaborator,

Catherine Crook de Camp

PREFACE

The system of indicating dates in this book is based upon those used by the late George Sarton in his *History of Science* and by Joseph Needham in his *Science and Civilisation in China*. Centuries are indicated by Roman numerals preceded by + or – according to whether they are centuries of the Christian era or B.C.; hence –VIII means eighth century B.C. Years are treated likewise, with Arabic instead of Roman numerals; for instance, +412 = A.D. 412. The plus sign is, however, omitted from years after +1000, because the meaning of the numeral is obvious in such cases.

In the text, most Greek names are spelled in the Greek manner, instead of the Latin (hence Keraunos instead of Ceraunus) because I like it better and think it will in time prevail. But in the notes and bibliography, most names of Greek writers are given in Latinized or Anglicized form to make it easier to find standard editions and translations.

For help in one way or another with this work–procuring books for me, answering questions, checking my translations, and criticizing parts of the text–I am grateful to Allen T. Bonnell, Lionel Casson, Jack Coggins, Bern Dibner, Caroline Gordon Dosker, A. G. Drachmann, I. E. S. Edwards, R. J. Forbes, Umberto Forti, Samuel Freiha, Samuel N. Kramer, Willy Ley, William McDermott, Robert P. Multhauf, Derek J. de Solla Price, Pellegrino Claudio Sestieri, Guido Ucelli, Donald N. Wilbur, Howard H. Williams, and Conway Zirkle; and to the Burndy Library (Norwalk, Conn.), the Swarthmore College Library, the Union Library Catalogue, and the University of Pennsylvania Library. Finally, my wife's work of editing the manuscript has gone far beyond the call of duty.

L. Sprague de Camp

CONTENTS

PLATES

ILLUSTRATIONS IN TEXT

THE ANCIENT ENGINEERS

THE COMING OF THE ENGINEERS

ONE

Civilization, as we know it today, owes its existence to the engineers. These are the men who, down the long centuries, have learned to exploit the properties of matter and the sources of power for the benefit of mankind. By an organized, rational effort to use the material world around them, engineers devised the myriad comforts and conveniences that mark the difference between our lives and those of our forefathers thousands of years ago.

The story of civilization is, in a sense, the story of engineering—that long and arduous struggle to make the forces of nature work for man's good. The story of engineering, pieced together from dusty manuscripts and crumbling relics, explains as well the state of the world today as all the accounts of kings and philosophers, generals and politicians.

To appreciate the accomplishments of the engineers, we must understand the changes that have taken place in human life during the last million years. A million years ago, at the beginning of the Pleistocene Period, our ancestors were small, apelike primates, much like the man-apes whose fossil remains have been found in Africa.

Two things distinguished our ancestors from modern apes, such as the gorilla and chimpanzee. First, they lived mostly on the ground and regularly walked upright, so that their limbs were proportioned much like ours. They did not have the long hooklike arms, the short bowed legs, and handlike feet of modern apes. Their brains were essentially the same as those of modern apes.

Probably as early as 100,000 years ago, before the last advance of the Pleistocene glaciers, and certainly by 10,000 years ago, the forces of evolution had caused these man-apes to evolve into men, every bit as human in form and as intelligent as we are. Differences in climate in different parts of the world had split the human stock into three major and several minor races.

These men, like all the men who had gone before them, lived by food-gathering. They sought a precarious livelihood by hunting, fishing, picking berries, and digging up edible roots and tubers. They greedily gobbled lizards, insects, and carrion. Today only small bands of African Bushmen and Pygmies, a few Australian aborigines, and a handful of Eskimos —a tiny fraction of 1 per cent of humanity—subsist in this manner.

Because of the difficulty of getting food, in Pleistocene times only a few hundred thousand people existed on the entire face of the globe. But there is no reason to think that we today are one bit cleverer than the men of –8000, at the time of the great Neolithic agricultural revolution that turned hunters into peasants. For one thing, 10,000 years is too short a time for evolution to have had a measurable effect. For another, many geneticists believe that civilization causes the human stock slowly to degenerate, by enabling persons with unfavorable mutations to live and breed, when in a wild state they would quickly perish.

However that may be, man has spent about 99 per cent of his history, since he first learned to make tools, as a hunting and food-gathering tribesman. Civilization has arisen only during the remaining 1 per cent of this time, since 9,000 to 10,000 years ago, when men discovered how to raise crops and tame animals. These discoveries enabled a square mile of fertile land to support 20 to 200 times as many people as before and freed some of these people for other, specialized occupations.

This revolution seems to have first taken place in the hills that curve around to the north of Iraq and Syria. From Iraq and Syria the Agricultural Revolution quickly spread to the valleys of the Nile and the Indus, which in their turn became centers of cultural radiation.

The Agricultural Revolution brought about changes fully as drastic in people's lives as those caused by the Industrial Revolution of the last two centuries. Permanent villages took the place of temporary campsites. One theory holds that men were first persuaded to give up their wandering life by the discovery that mashed grass-seeds could be used to make beer, since they had to stay put long enough for the mash to ferment.

In another three or four thousand years, some of the farming villages of the Near and Middle East grew into cities. Then with a rush came metals, writing, large-scale government, science, and all the other features of civilization.

As farmers learned to raise more food than they themselves needed, other men were able to spend all their time in making useful things, which they exchanged for surplus foods. Thus specialization arose.

Human society had long known a couple of specialists: the tribal priest or wizard and the tribal chief or war leader. As specialization increased, merchants, physicians, poets, smiths, and craftsmen of many kinds came into being. Instead of making their own houses, carts, wells, and boats, men began to buy them from workmen skilled in these arts. Soon the arts advanced to the point where even a wise and experienced workman could not know all that had to be known about his craft.

As the chiefs evolved into kings and the wizards into high priests, they waxed rich and powerful. They acquired helpers, messengers, bodyguards, and other servants, who outranked the simple peasants. Slavery—at first a humane invention, which made it no longer necessary to slaughter one's prisoners of war—introduced still another class. Thus society became seamed and fissured into a multitude of specialized occupations.

Wealth and experience piled up. Men undertook projects too large for a single craftsman, even with the help of his sons and apprentices. These projects called for the work of hundreds or even thousands of men, organized and directed towards a common goal. Hence arose a new class of men: the technicians or engineers, who could negotiate with a king or a priesthood for building a public work, plan the details, and direct the workmen. These men combined practical experience with knowledge of general, theoretical principles. Sometimes they were inventors as well as contractors, designers, and foremen, but all were men who could imagine something new and transform a mental picture into physical reality.

Invention has been going on ever since our apish ancestors learned to feed a fire and flake a flint. But the conditions under which invention takes place, and the pace of invention, have changed greatly since the beginning of historic times.

Some primitive inventions, like the manioc squeezer of the South American Indians, the Australian boomerang, and the Eskimo toggle-joint harpoon, are extremely ingenious. They point to inventive talents as keen as anything the civilized world can show.

Nevertheless, during nearly all of the last million years, invention progressed with glacial slowness. Men chopped with ax heads held in the fist for hundreds of thousands of years before they learned to fasten handles to their axes. During the earlier part of the Pleistocene Period, it is possible that men were too stupid to be very inventive. By 100,000 years ago, however, men had probably become quite as intelligent as we are–but still technology advanced at a crawl.

The reasons for the sloth of invention in primitive societies are not hard to understand. For one thing, primitive peoples live a hand-to-mouth existence. Most of their foods cannot be stored, so that they have no economic surplus. Therefore they can less well afford to risk experiment than more advanced peoples. If an experiment fails, they die.

As a result, primitive societies are very conservative. Tribal customs prescribe exactly how everything shall be done, on pain of the gods' displeasure. An inventor is likely to be liquidated as a dangerous deviationist.

Peasant farmers are almost equally conservative. Man's inventive faculties are stimulated by the breakdown of established custom that takes place in the urban environment; hence most inventions have been made by city dwellers.

Another cause of the slowness of primitive invention is the scarcity of inventors. A hunting and food-gathering technology can support only a very small population for a given area. Thus the few hundred thousand members of the human species living at any time before the Agricultural Revolution were divided into many isolated little hunting bands.

Such a band seldom exceeds fifty or a hundred people, counting the many but short-lived children. Because the radius of action of the hunters is limited to the distance they can walk to kill their game and carry it back to camp, an increase in numbers does not enlarge the area that can be hunted at one time. It merely causes the same area to be hunted more intensively. So, if the band grows too large, game in the neighborhood becomes scarce; and the band must migrate or starve. Eventually it will have to split up. Perhaps human factiousness–our tendency to divide up into factions on almost any pretext (racial, religious, cultural, political, economic, or sporting) and fight it out–is a survival mechanism evolved during man's hunting phase, to insure that hunting bands split up before they grew too large to feed themselves.

Now, in any society, only a few human beings ever have original ideas or make inventions. Of these inventors, only a fraction have the courage, stubbornness, and energy to keep on bettering their inventions until they really work and to keep on promoting them until they persuade others to take them up.

A rough idea of the percentage of inventors among modern Americans can be obtained from the statistics of the United States Patent Office. The Patent Office issues about 40,000 patents every year. So we can estimate that the mid-twentieth-century American population of 180,000,000 people produces about one patentable invention each year for every 4,500 citizens.

Suppose, now, that all Americans were wiped out except one band of forty-five people. If this group continued to produce inventions at the same rate, it would turn out only one invention every century! This is of course a gross oversimplification. But it does indicate why a small tribal society, no matter how clever the tribesmen, cannot be expected to produce inventions rapidly.

In actual fact, the rate of inventions among Stone Age hunters was enormously slower than among our imaginary band of forty-five Americans. For modern Americans are encouraged to invent in ways that primitive folk are not. We are used to the thought that men can improve their lot by inventing things, and that invention is a worthy act. On the contrary, primitive people, who have all they can do to keep alive and who cannot afford to support a fellow tribesman in idleness while he dreams up new ideas, regard inventors with glowering suspicion.

Suppose now that there are two bands of forty-five Americans. If they are isolated from each other, each band will produce one invention a century, so that each progresses at the same rate as before. Their cultures will diverge somewhat, as they will hit upon the same inventions only rarely, by chance. But each group will plod along at the same old rate of one invention a century.

However, if they meet and join forces, then all ninety persons will take advantage of the inventions produced by any one of them. The combined group will produce inventions twice a century instead of once. In other words, they will progress technologically twice as fast.

To sum up: Progress in civilization depends upon invention, and a rapid rate of invention in turn depends upon the sizable populations that are only possible under civilization. The crucial inventions that made such progress possible–knowledge of raising domesticated, edible animals and plants–took place in Syria and Iraq about –8000.

Once the Agricultural Revolution had taken place, much denser and more numerous populations than had ever before existed could and did live in the valleys of the Nile, the Euphrates, and the Indus. As the

Reverend Thomas Malthus pointed out a hundred and sixty years ago, people quickly breed up to the greatest density the land will support at the current technological level. At that point the population levels off, because excess people are destroyed by starvation, pestilence, or war.

The mere fact of having large interconnected populations, then, meant that inventions took place at a faster rate than before, and these inventions in turn made denser and more widely interconnected populations possible. Hence civilized men tended to draw farther and farther ahead of their primitive fellows.

Moreover, the inventions on which civilization was founded tended to spread. These inventions did not spread out evenly in all directions. They spread along trade routes, and they spread to lands where these ideas could be profitably applied. They were stopped by strong natural barriers, such as deserts and oceans; and they died out where conditions made them useless.

Thus the idea of raising cotton or dates could not spread to Europe, because the cotton tree and the date palm will not grow there. The wheel failed to spread from Iraq to neighboring Arabia, because there was no place in the wastes of the Arabian desert where wheeled vehicles would have been very useful.

As a result of this speed-up and spread of technology, a high level of civilization had been achieved a thousand years before Christ in a broad belt stretching from the lands around the Mediterranean through the Middle East, India, and Southeast Asia to China. Any new invention, originating at one end of this Main Civilized Belt, traveled in a few centuries to the other. China, partly isolated at one end of the Belt by the Mongolian deserts, the Tibetan mountains, and the jungles of Southeast Asia, was a thousand years late in getting started but soon became as civilized as the rest.

Some of these advances in technics spread to Central Asia and Central Europe as well. Civilization had little effect on northern Europe and northern Asia, however, because the population of these lands was very thinly scattered and conditions of life were so different from those of the Belt that most inventions made in warmer lands were of little use there.

Civilization also failed to penetrate Negro Africa, being stopped by the barrier of the Sahara Desert, the swamps of the White Nile, and the mountains of Abyssinia. This barrier isolated sub-Saharan Africa as effectively as if it had been an island. Furthermore, Old World civilization failed to leap the watery barriers to reach the Pacific Islands, Australia, or the Americas. In another millennium, however, the peoples of Central and South America began independently to develop their own civilizations.

It would seem, then, that the main factor in determining whether any particular people took part in the technological adventure that followed the Agricultural Revolution was neither race, nor climate, nor local resources. The main factor was simply a matter of geography—where the people lived with respect to the river valleys in which this revolution took place. Those lucky enough to dwell along the cultural highways from China to Spain received the benefits of the speed-up; those who lived elsewhere did not, or did so only tardily.

I have spoken of the spread of inventions through the Main Civilized Belt and into lands outside this area. A few decades ago, a tremendous dispute on the spread of inventions arose among anthropologists. This dispute is called the Diffusionist Controversy.

The basis of the argument is this: If you find the same culture trait —such as a blowgun or a flood legend—in two widely separated groups of people, and the intermediate peoples lack this trait altogether, did the two groups invent it independently, or did they somehow get it from the same source?

Certain Britons—the psychologist Rivers, the anatomist Elliot Smith, and the anthropologist W. J. Perry—developed the extreme diffusionist or dispersionist theory. According to this hypothesis, all civilization came from one (or at most a few) Old World centers. The diffusionists deemed invention so rare that the same invention could *never* have been made independently by different peoples. Wherever close similarity was found, even on opposite sides of the globe, they averred that the trait had been spread by trade or migration.

Hence the diffusionists inferred, for instance, that the Mayas and Aztecs must have learned to build pyramids from the ancient Egyptians —despite the fact that, when the Mayas and Aztecs began to erect these structures, Egypt was already thousands of years old and had long since stopped building pyramids. They argued that all human civilization must have originated in one spot on the earth. Elliot Smith named Egypt, but others found their source of illumination in Brazil, the Ohio Valley, India, the Arctic, or Plato's fictional Atlantis.

Diffusionism became a cult. This cult attracted people of the sort who seek arcane wisdom in the measurements of King Khufu's pyramid or hunt for the Lost Ten Tribes of Israel among the Irish, the Iroquois, the Japanese, or the Zulus. By insisting that the same invention could never have been made twice over, the cult appealed to people who, never having had an original idea themselves, find it impossible to imagine anybody's else having one.

In later years this nonsense declined as sane anthropologists pointed out, over and over, that every invention contains some borrowing and every borrowing some invention. Where you draw the line between diffusion and original invention, then, is a matter of convenience.

Furthermore, there are many well-known cases of independent invention. As we shall see, the crossbow was independently invented in the Far East and in the Mediterranean. In civilized countries, simultaneous invention occurs all the time. That is why the United States Patent Office has a special procedure called an "interference" to find out who in such a case is legally entitled to the patent.

On the other hand, there are many cases of worldwide diffusion of an invention. Thus the bow reached the Americas from Asia, and later the tobacco pipe traveled around the world during the Age of Exploration. It is often hard to decide whether an invention traveled from one land to another or was independently created. Each case must be judged on its merits.

A specimen or a working diagram of an invention need not make the journey. A man may hear a rumor of an invention practiced in some foreign land, and the mere idea is enough to set him to thinking and tinkering in order to develop a similar invention on his own. Several systems of writing, devised by West African natives in +XIX, furnish examples of this "stimulus diffusion" as the anthropologists call it.

The first engineers were irrigators, architects, and military engineers. The same man was usually expected to be an expert at all three kinds of work. This was still the case thousands of years later, in the Renaissance, when Leonardo, Michelangelo, and Dürer were not only all-round engineers but outstanding artists as well. Specialization within the engineering profession has developed only in the last two or three centuries.

Irrigators laid out the canal systems on which the early river-valley civilizations depended. The Babylonian *gugallu* or irrigation inspector was such an expert. Irrigation enabled farmers to raise so much more food that an increasing number of specialists, relieved of peasant's chores, were able to gather in cities to practice their specialities. Today's city is still essentially a place where specialists live and work, even though the farming class, once almost the whole population, has dwindled in industrial lands to a small minority.

Soon the kings who ruled these early cities desired houses larger and more comfortable than the huts of stone, clay, and reeds wherein they had been living. So they called upon architects to build them palaces.

Next, priests insisted that the gods would be offended if they were not housed at least as splendidly as the kings. So the architects put up temples, containing statues of the gods and other works of art.

To protect the wealth of the gods and the kings, military engineers built walls and dug moats around cities. In the lower Euphrates Valley, where there is practically no stone, walls were made of brick. Elsewhere they were made of stone—preferably the largest stones that could be moved.

Even before mortar was invented, men could build a good solid wall of small stones, which would stand up to the weather for years. However, all an enemy had to do to such a wall was to pry out a few stones with his spear, and the wall collapsed.

Therefore, many early fortifiers made their walls of very large stones, trimmed to fit roughly together. The sheer weight of these stones prevented the foe from pulling them out, especially if defenders atop the wall were raining missiles upon him. Such walls are called "cyclopean" because the ancient Greeks, seeing the ruins of walls of that kind built several centuries earlier, thought they must have been made by the mythical one-eyed giants called Cyclopes.

The hoards of metals, jewels, fine raiment, and foodstuffs in the temples and palaces also required men and means to keep track of them. Thus came about the invention of arithmetic and writing. Writing was done on the surfaces of some local material: in Egypt, on paper made of strips of papyrus reed; in Mesopotamia, on slabs of clay; in India, on paper made from palm fronds; in China, on strips of bamboo. Stone, wood and leather were also used as writing materials. In Mesopotamia, writing originated in the little clay tokens—spheres, disks, cones, and pyramids—used to keep accounts of property. Then it was found easier to draw pictures of the spheres and so forth on wet clay than to model them.

Many ancient writings on stone and clay have survived; but those on perishable materials have disappeared, save where people were interested enough to copy and recopy them.

As a result, the high school student of ancient history gets the curious impression that during the Golden Age of Greece, the Greeks were the only people in the world who were really alive. It seems as though the folk of all the other lands were standing around like waxen dummies in a state of suspended animation.

Of course that is not true. During the Golden Age of Greece, all along the Main Civilized Belt from Spain to China, teeming multitudes toiled. Everywhere princes preened; politicians plotted; priests prayed; merchants haggled; warriors clashed; thinkers pondered; lovers sighed; drunkards reeled; poets declaimed; prophets ranted; sorcerers conjured; charlatans beguiled; slaves shirked; thieves filched; and people joked, quarreled, sang, wept, lusted, blundered, yearned, schemed, and carried on the business of living in quite as lively a fashion as the Greeks were doing.

But, because the Greeks put their experiences down in writing, and because good luck has saved a small part of these writings for us, we know a lot about them. We know much, for instance, of the little up-country brawls of tiny Greek city-states. On the other hand, we know almost nothing about the score of thunderous battles by which Darius the Great and his generals defeated the many rival claimants to the Persian throne, although these battlefields may have seen quite as brilliant feats of generalship and as gallant deeds of dought as the fields of Koronea and Leuktra.

For the same reason, we know quite a lot about Greek and Roman engineering, but very little about ancient Iranian, Indian, and Chinese engineering. In Iran, India, and China either the subject was not written about, or the writings have perished; or, where records have come down, many have never been published in European languages.

Even today, numbers of ancient manuscripts lie in the great libraries of Asia and North Africa, unread, uncatalogued, and untranslated. Many might shed additional light on medieval oriental science and engineering. Some may even be translations of supposedly lost Greek works on these subjects. One of the most urgent tasks of scholarship is the publication and translation of these works before the originals are vaporized in another war. A few scholars work at this task as time and chance permit, but the number of workers is small for the size of the job.

As nearly as we can reconstruct the evidence, the earliest civilizations were patchworks of little independent city-states, ever fighting one another. Government varied as power shuffled back and forth among the dominant groups: the king and his cronies, the priesthood, the senate (a gathering of the heads of the richest families), and the assembly (a meeting of the fighting men of the group). Women, poor men, and slaves, having neither wealth, arms, nor magical powers, did not count.

The government—whether a theocracy, a monarchy, or a republic—controlled not only the dwellers in the city but also as many of the peasants of the neighboring countryside as could be persuaded or coerced into accepting the city's "protection." In return for military service and taxes, the peasants, willy-nilly, got centralized control of their irrigation systems, defense against foreign invaders, and some rough-and-ready law and justice.

In time, the march of technology made the city-state obsolete. Where a river system forms a single large watershed, an irrigation system works better when it is ruled by one central administration. Thus, in the valleys of the Nile, the Tigris and Euphrates, the Indus, and the Hwang-ho, conditions favored the extension of one state's rule over all the others in the watershed. Historians argue whether empire came first and made possible large-scale irrigation, or whether large-scale irrigation came first and encouraged the growth of empire. Probably the former is more nearly right, but there was also a mutual effect. Each institution fostered and strengthened the other as it grew.

In the large watersheds of wet countries, such as the valleys of the Ganges and the Mekong, irrigation was less important. But here the need to protect the valley dwellers from floods promoted the centralization of government.

Because of the benefits of large-scale government in such a river valley, a city-state or a king who had conquered half of a watershed could easily gobble up the remaining half. The conqueror's subjects accepted him, however grudgingly, because of these economic advantages. And, once established, he was hard to get rid of.

Under the conditions of early river-valley civilization, even a bad emperor might be better than none at all. While men feared cruel and rapacious rulers, even more they feared a time of anarchy. The Indians called it "the way of the fishes," when the strong devoured the weak without hindrance. Their poets chanted:

> A river without water,
> A forest without grass,
> A herd of cattle without a herdsman,
> Is the land without a king.[1]

So important was the distribution of water in such a polity that the German-American scholar Wittfogel refers to a watershed empire of the type we have discussed as a "hydraulic state." While the government of city-states took various forms, such as limited monarchy, aristocratic republic, and popular dictatorship, ancient empires tended to be absolute monarchies of the most despotic kind. The king was deemed a god, or the son of a god, or at least the special agent of a god. His word was law. Government was a centralized, authoritarian despotism of–it would seem to us–the most tyrannical and oppressive sort.

Moreover, nobody seems to have seriously considered a large-scale government of any other kind. In ancient republics the voters, who were only a fraction of the total population, had to gather together to vote in person. Although such a scheme shares power to some extent and works fairly well in a small city-state, it is impractical in a large nation.

There were plenty of revolts, revolutions, and civil wars in the ancient empires. It was a rare king whose death did not result in a war among his would-be successors, and provinces that had once been separate nations repeatedly sought to regain their independence. But, while many kings were overthrown or murdered, the sole result was to replace one despot by another who, his supporters hoped, would prove a better king.

Sometimes a watershed empire broke up into parts as a result of domestic disorder or foreign conquest. But, after a few decades of the joys and sorrows of anarchy and incessant strife, the people of the watershed were once more prepared to submit to the rule of an all-powerful emperor.

From the rise of the first watershed empires down to the achievement of temporary world mastery by Europe after 1600, man's history largely consists of the story of the mighty empires that rose in the Main Civilized Belt, spread far beyond the confines of a single watershed, flourished for a time, and withered away. Sometimes they lasted for centuries, sometimes for a few years only.

Thus the Assyrian Empire gave way to the Median, and that to the Persian, and that to the Macedonian, and that to the Roman, and that to the Arab, and that to the Turkish. A long succession of other empires, in Iran, India, China, and Central Asia, flourished beside these westerly realms. And many of the rulers of these domains–however good or bad in other respects–were among the world's greatest builders of public works and, therefore, the greatest patrons of the engineering profession.

For, whatever their sins and oppressions, some early despots did much for those they ruled. A king with any brains tries to make his people prosper, if only so that he can tax them. Rulers of ancient empires built roads, which fostered commerce and communication. But the principal purpose of these roads, as of the governmental postal systems that operated over them, was to keep a swift stream of commands and inquiries flowing out from the capital to all parts of the realm, and an equally lively stream of information and tribute flowing back, for the benefit of the ruler. However they might disagree on other matters, a king and his subjects had a common interest in keeping up roads and canals, suppressing brigandage and piracy, and maintaining order.

Nowadays we draw fine distinctions among the meanings of such words as craftsman, engineer, technician, and inventor. The United States Patent Office has elaborate rules for deciding whether an invention is original, or whether it is merely "an improvement obvious to one skilled in the art," such as a change in size, strength, speed, proportions, or materials.

In speaking of ancient technical men, however, there is no point in observing such delicate differences. Every time an ancient craftsman made something that was not a close copy of a previous article, he invented, even though his invention might not be patentable according to modern laws.

We think of an engineer as a man who designs some structure or machine, or who directs the building of it, or who operates and maintains it. In practice most ancient engineers were inventors; while most ancient inventors, at least after the rise of civilization, could also be classed as engineers. So let us lump all these ancient innovators and designers together as "engineers."

Despite the enormous importance of engineers and inventors in making our daily life what it is, history does not tell much about them. The earliest historical records were made by priests praising their gods and poets flattering their kings. Neither cared much about such mundane matters as technology.

As a result, ancient legend and history are one-sided. We hear much about mighty kings and heroic warriors, somewhat less about priests, philosophers, and artists, and very little about the engineers who built the stages on which these players performed their parts. The warriors Achilles and Hector were celebrated in song and story—but the forgotten genius who, about the time of the siege of Troy, invented the safety pin, lies wholly forgotten. Everybody has heard of Julius Caesar —but who knows about his contemporary Sergius Orata, the Roman building contractor who invented central indirect house heating? Yet Orata has affected our daily lives far more than Caesar ever did.

Nevertheless, of all the phases of civilized life, the advance of technology gives the best ground for belief in progress. If there is any consistent pattern of evolution in politics and government, it is not easy to discern. Great soldiers and statesmen have built up empires—but a few generations later these empires faded away as though they had never been. In the field of government, many people thought half a century ago that there was a natural evolutionary trend towards the democratic republic–but then many parts of the world turned in the other direction, towards authoritarian despotism. It is mere soothsaying to predict what form of government, if any, will finally prevail.

Likewise, great world religions like Buddhism, Judaism, Christianity, Islam, and Hinduism, with their tightly organized priesthoods and their closely reasoned theologies, have in the last two thousand years won most of the world away from the unorganized pagan and tribal cults. But the world religions differ basically among themselves and are no nearer to scientific proof of their discordant claims about the nature of man and the gods than when they were founded. Today, in many lands, they are losing ground to the pseudo-scientific philosophy of Marxism.

Pure science has advanced enormously in the last three centuries. But, looked at over the whole stretch of recorded history, the advance of science has been erratic. It has leaped ahead in sudden spurts, shot off on pseudo-scientific tangents like astrology and alchemy, become embroiled in religious and political conflicts, and sometimes been repudiated by whole nations.

In the arts, people's tastes have changed from age to age, but in a capricious and faddish manner. People have often abandoned some canon of beauty in painting, sculpture, architecture, music, or poetry and embraced another simply because they were bored with the old and eager to try something new.

But through all the ages of history, one human institution–technology –has plodded ahead. While empires rose and fell, forms of government went through their erratic cycles, science flared up and guttered out, men burned each other over differences of creed, and the masses pursued bizarre fads and fashions, the engineers went ahead with raising their city walls, erecting their temples and palaces, paving their roads, digging their canals, tinkering with their machines, and soberly and rationally building upon the discoveries of those who had gone before.

So, if there is any one progressive, consistent movement in human history, it is neither political, nor religious, nor aesthetic. Until recent centuries it was not even scientific. It is the growth of technology, under the guidance of the engineers.

Technology has progressed continuously from the time of the Agricultural Revolution 10,000 years ago, slowly and hesitantly at first, then with increasing sureness and speed. The sixteenth century marked the beginning of modern engineering because, from that time on, professional societies were formed, treatises on engineering subjects were printed in quantity, engineering schools sprang up, specialization within the profession began, and engineers began to take advantage of the brilliant scientific discoveries of the time. The Industrial Revolution, which started two centuries ago and is still going on, was a surge in the growth of technology. Barring nuclear war, the end of this fruition of engineering is nowhere in sight.

Today, in technologically advanced lands, men live very similar lives in spite of geographical, religious, and political differences. The daily lives of a Christian bank clerk in Chicago, a Buddhist bank clerk in Tokyo, and a Communist bank clerk in Moscow are far more alike than the life of any one of them is like that of any single man who lived a thousand years ago. These resemblances are the result of a common technology, and this technology is what many generations of engineers have built up, with the greatest skill and diligence of which human beings are capable, and handed down to us.

Many readers already know of the doings of the engineers and inventors of recent times. They have heard of James Watt and his steam engine, of John Augustus Roebling and his Brooklyn Bridge, or of George W. Goethals and the Panama Canal. But few know about the remote predecessors of these modern engineers–about the men who laid the foundations on which their modern colleagues have built. Therefore, this book will be devoted to all these neglected early engineers who, much more than the soldiers, politicians, prophets, and priests, have built civilization.

THE EGYPTIAN ENGINEERS

TWO

Serious archeological work began in Egypt and Mesopotamia only about a hundred years ago, but since then much has been learned about the early civilizations of these lands. Although no definite date can be given to the beginning of either civilization, most scholars now believe that the civilization of the Euphrates Valley is several centuries older than that of the Nile.

The monuments of early Egypt, however, are far better preserved and much more impressive than those of its sister civilization of Iraq. The Egyptians had abundant supplies of good limestone and granite in the bluffs that paralleled their river for hundreds of miles. And, as most of the country gets hardly any rain, some of the monuments that the Egyptians built of these stones have lasted with but little weathering for thousands of years.

On the other hand, the Euphratean plain has no stone, and its date palms do not furnish good timber. Any timber the Mesopotamians used had to be brought down the Tigris from the Assyrian hills. Moreover, kiln-dried or burnt brick, which stands up to wet weather, was costly because of the scarcity of fuel for the kilns. Therefore it was used only to face the most important buildings. The interiors of the walls of these buildings, and the whole of ordinary dwellings, were made of sun-dried or mud brick.

Now, although mud brick can be made fairly strong by drying it in the sun for two to five years before use, it still softens and crumbles when wet. When a crack developed in the burnt-brick facing of a Mesopotamian temple or palace and was not at once repaired, the sharp winter rains dissolved the mud brick within, and the building crumbled into ruin. Hence the upper parts of the walls of public buildings in ancient Mesopotamia have almost entirely disappeared. All we know of these buildings is what we can discover by digging around the foundations, which have been protected from complete dissolution by the piled-up débris of the upper stories.

Therefore, to Egypt we must go to find great engineering works of earliest historic times still in recognizable condition and, as it happens, to learn about the most ancient engineer whom we know by name. This is the man who invented the pyramids, the most famous monuments of the ancient world. Of all the Seven Wonders of the World, only the pyramids survive to this day.

What were the Seven Wonders? Several Greek writers, beginning with Antipatros of Sidon (about –100) drew up lists of the seven most wonderful engineering feats they knew about. The usual list of Wonders comprised: 1. The Pyramids of Egypt. 2. The Hanging Gardens of Babylon. 3. The statue of Zeus by Pheidias at Olympia. 4. The temple of Artemis at Ephesos. 5. The tomb of King Mausolos of Karia at Halikarnassos. 6. The Colossus of Rhodes. 7. The Pharos or lighthouse of Alexandria.

Subsequent writers drew up their own lists of Wonders, sometimes substituting other structures, such as the walls of Babylon or the Temple of Jupiter in Rome, for some of those on the original list. Of course, classical writers could only list things they had heard of. They did not know about the Great Wall of China, or the huge dam at Ma'rib in Arabia, or the enormous Buddhist stupas of Ceylon. If they had, their lists might have been different.

The first recorded engineering work of early Egypt was the wall of the city of Memphis. This capital of the Old Kingdom stood at the point of the Delta, on the western bank of the Nile twelve miles above modern Cairo. Here the Nile, winding like a vast blue serpent athwart the North African desert belt, fissions into a dozen branches, which writhe across the flat, fertile, fan-shaped Delta to the sea.

A visitor of classical times–let's say the Greek historian Herodotos (–V)–in crossing the Nile beheld a lofty wall of pearly limestone. Over this wall appeared the upper parts of a forest of huge stone statues, 30 to 75 feet tall. These colossi were the eidolons of the conquering kings of the New Empire, the Rameseses and Senuṣerts.

In the midst of the city rose the citadel, the White Castle. This was an artificial hill surrounded by 40-foot limestone walls and bearing palaces and barracks on its top. Beyond the city, for many miles along the western bank of the Nile, clumps of pyramids pierced the skyline with blunt triangular teeth of buff-colored limestone. In these gigantic tombs lay the Pharaohs of the Old Kingdom, already a fading memory in the minds of the teeming, swarthy folk of the land of Khem.

This was a city of many names. In Herodotos' time it was called Men-nofer, the Memphis of the Greeks. It was also known as the City of the White Castle and the Abode of the Soul of Ptaḥ. Memphis was as ancient to Herodotos as Herodotos is to us. The business of catering to tourists who had come from afar to view its antique wonders was already well in hand.

Now let us go yet farther back in time, to the very beginning of the Old Kingdom, as far as we can dimly discern the events of that distant day through the mist of centuries. About –3000 Mena, king of the South, conquered all of Egypt. At the boundary between the former separate kingdoms of Upper and Lower Egypt he built his new capital, Memphis, and surrounded it with a great white wall. This wall was probably made at first of brick with a coating of gypsum plaster. In later times a wall of stone took its place.

Three centuries after Mena, in the reign of King Joṣer,[1] lived the first engineer and architect known to us by name. This was Imḥotep, who built the first pyramid for his sovran. Imḥotep is mentioned, though not by name, in a history of Egypt written in Greek thousands of years later by an Egyptian priest, Manetho. In his book Manetho wrote: "Tosorthos, [that is to say Joṣer, who reigned] for twenty-nine years; who, because of his medical skill has the reputation of Asklepios among the Egyptians, and who was the inventor of the art of building in hewn stone. He also devoted his attention to writing."[2]

Later Greek and Egyptian allusions, however, show that Manetho was mistaken or that copyists dropped some words out of the text. For the man who was "styled Asklepios" and who built in stone and wrote was not Joṣer himself, but his minister Imḥotep.[3]

If we can trust our scanty sources, Imḥotep was born in Memphis, the son of the royal architect Kanofer. He held various posts and titles, including Royal Chancellor, Administrator of the Great Mansion, Hereditary Noble, and Heliopolitan High Priest.

Imḥotep left a son, Raḥotep, from whom descended a long line of architects. At least, so says the inscription of Khnumabra, Minister of Public Works under the Persian king Darius I about –490. Khnumabra claimed descent from Imḥotep and listed a line of twenty-five architects, beginning with Kanofer and ending with himself. Simple arithmetic shows that this number is much too small for the 2,000 years from Kanofer to Khnumabra; a complete pedigree covering that length of time would contain about three times as many generations. So Khnumabra either left out many ancestors or, like some modern folk who yearn for eminent ancestry, was faking his genealogy.

Otherwise there is no real history of Imḥotep and his royal master. A papyrus of Ptolemaic times relates how the kingdom was afflicted by famine for several years because the Nile failed to rise. Joṣer accordingly took counsel with Imḥotep, who explained that Khnum, the god of the Cataracts, was wroth. So the king deeded lands for temples to the god, and all was well. While there is no reason to think that this story has any historical basis, it provides the kernel of the biblical legend of Joseph and the seven lean years.

Although no trustworthy details of the lives of Joṣer and Imḥotep have come down, we can be sure that they were able men who worked long and effectively together. Probably Imḥotep was a universal genius like Archimedes and Leonardo da Vinci. Such was his repute as a physician, architect, writer, wizard, statesman, and all-round sage that in later times collections of wise sayings circulated under his name.

From the monuments that he and Joṣer built, we can tell something of how Imḥotep came to invent the pyramid.

Most peoples believe in life after death. This belief may have originally been based upon the dreams of primitive men about persons whom they knew to be dead. Most ancient peoples did not make much of this belief, thinking the afterlife a dim and shadowy affair.

The Egyptians, however, developed elaborate beliefs about life after death. One of these beliefs was that such afterlife could be enjoyed only so long as the body was kept intact. Hence arose the practices of mummifying corpses and of building massive tombs, designed to foil tomb robbers forever.

Tomb robbers were drawn by the jewels and precious metals buried with kings and nobles, who thought that in the afterworld the spirit of a dead man needed the spirits of the things he used in life to keep him happy. In the early days of Egypt, Mesopotamia, and China, in addition to stores of food, clothing, weapons, and ornaments, scores of attendants and guardsmen were killed and buried with the king to serve him in the afterworld.

Before King Joṣer, Egyptian kings and nobles were buried in a tomb called a mastaba.[4] This was a rectangular structure of brick, with inward-sloping walls, set over an underground chamber. The reason for the inward-sloping walls is that most Egyptian building of this time was in mud brick.

Although mud brick is one of the feeblest of structural materials, the Egyptians learned that, if they made their walls taper upward, these walls would not crumble away so quickly. When they began building walls of stone, they continued to taper their walls from bottom to top, although this batter was no longer needed. The Egyptians, after the first few dynasties, became the world's most conservative people—so conservative, in fact, that more than two thousand years later, in Ptolemaic times, they were still tapering stone walls upward!

After the burial of a king or noble in his mastaba, heavy slabs of stone were dropped down vertical shafts to block off the passage to the burial chamber. Kings of the Third Dynasty built larger mastabas and began to use stone instead of brick.

Then, when Joṣer came to the throne, he and Imḥotep experimented. First, west of Memphis at modern Ṣaqqâra, they built a stone mastaba of unusual size and shape. It was square instead of oblong like its predecessors, and it was over 200 feet on a side and 26 feet high.

Not yet satisfied, Joṣer and Imḥotep enlarged this mastaba twice by adding stone to the sides. Before the second of these enlargements was completed, the king changed his mind again. He decided not only to enlarge the structure still further, but also to make it into a step pyramid, resembling four square mastabas of decreasing size piled one atop the other.

Then Joṣer changed his mind once more. The tomb ended as a step pyramid of six stages, 200 feet high on a base 358 by 411 feet. The main body of the pyramid was made of blocks of limestone quarried from local outcrops. To the outside, Imḥotep added a facing of high-grade limestone—almost marble—from quarries across the Nile at Troyu.[5]

Under the pyramid lay a burial chamber, whence many corridors branched out, probably to hold the wealth that Joṣer hoped to take with him. Around the pyramid was built a walled inclosure, about 885 by 1,470 feet. This contained Joṣer's mortuary temple, where the priests of a permanent staff were supposed to perform rituals forever to promote the welfare of the king in the afterlife. The temple compound included living quarters for these priests, tombs for royal relatives, and other structures, all made of gleaming golden-buff limestone.

Imḥotep's reputation expanded after his death until he was said to have been a son of Ptaḥ, the god of property, the god of the arts and crafts, and the tutelary deity of Memphis. Imḥotep was worshiped as the god of medicine, with his own temple in Memphis. He appeared in dreams to people who slept in the courtyard of this temple to give them medical advice. When the Greeks settled in Egypt they identified him with their own Asklepios,[6] mentioned as a wise physician in Homer's *Iliad* and later, like Imḥotep, promoted to godhood.

Joṣer's successors began step pyramids like his. But these pyramids were abandoned at an early stage or else have been so plundered for stone that little is left of them. A few decades after Joṣer, however, three large pyramids arose: two at Dahshûr, a few miles south of Ṣaqqâra, and one at Maydûm, about twenty-five miles farther south.

The pyramid at Maydûm was begun as a step pyramid of the Saqqâra type with seven steps. Then it was enlarged to a step pyramid of eight steps. At last, the steps were filled in and the structure converted to a true, smooth-sided pyramid. Nowadays, perhaps as a result of a heavy rain, or an earthquake, or both, the last addition has fallen away from the upper part of the pyramid, leaving the top of the second stepped stage protruding from a pile of débris.

The southernmost of the two pyramids at Dahshûr was begun as a true pyramid. But, about halfway to the top, the angle of inclination of the sides decreases sharply, so that the sides appear folded in. Hence this pyramid is called the Bent or Blunted Pyramid. The likeliest reason for this odd change of shape is that the king for whom the pyramid was built expired before its completion, and his successor hurried and cheapened the work by finishing it off with a top lower than had been planned.

The other pyramid at Dahshûr, usually credited to King Ṣeneferu, was the first large true pyramid to reach completion. It still stands—huge, silent, and solitary—in the desert near the new road from Cairo to the Fayyûm, as impressive in its isolation as the Great Pyramid on the crowded hill west of Giza. Although the names of the kings who reigned when the Dahshûr and Maydûm pyramids were built are known, it is not certain which king built which tomb.

The second king of the Fourth Dynasty, Khufu (the Cheops of Herodotos) built the largest pyramid of all on a hill five miles west of Giza, a town on the west bank of the Nile just above Cairo. Khufu called his masterpiece *Khuit-Khufu,* "Khufu's Horizon." Although some cultists have denied Khufu's authorship of this monument, there is no doubt about it. Besides the testimony of Manetho and Herodotos,

Khufu's name was found in red paint on some of the stones of the interior.

This enormous pyramid measures 756 feet square. It originally rose to a height of about 480 feet, although the uppermost thirty feet are now missing because of the quantities of stone that have been stolen from the outside. The cathedrals of Florence, Milan, St. Peter's at Rome, St. Paul's in London, and Westminster Abbey could all be placed at once on an area the size of its base.

The Great Pyramid is made of about 2,300,000 blocks of stone, weighing an average of two and a half tons apiece.[7] Except for the Great Wall of China, it was the largest single human construction of antiquity.

Khufu's Great Pyramid is not only the largest of the pyramids; it is also in many ways the best built, despite Kipling's derisive verse:

> Who shall doubt the secret hid
> Under Cheops' pyramid
> Was that the contractor did
> Cheops out of several millions?[8]

The sides of the base come to within 7 inches of forming a perfect square. They are also oriented to within less than 6 minutes of arc—one-tenth of a degree—of the true north-south and east-west directions, and the south side is within 2 minutes of the true east-west direction. Such accuracy is amazing. None of the other pyramids is oriented so closely, albeit some approach the Great Pyramid in this respect.

Like his predecessors, Khufu used limestone from local outcrops for the bulk of his pyramid, while for casing he used fine limestone from Troyu and the Moqaṭṭam Hills east of Cairo. The capstone was probably gilded. But nearly all the fine stone was peeled off by the medieval Muslim rulers of Egypt to build bridges and houses in Cairo.

Khufu changed his mind twice during the construction. Perhaps the real secret of the Great Pyramid is that King Khufu was a claustrophobe and, after the building had begun, called in his architect and told him that the thought of all those tons of stone lying on top of his final resting place gave him the creeps.

In any case, Khufu made up his mind not to be buried in the usual underground chamber of rock. This chamber was therefore abandoned and a large room, misleadingly called the "Queen's Chamber," was built into the structure. This Queen's Chamber had been roofed but not completely floored when Khufu decided to go higher yet. Hence work was stopped on the Queen's Chamber and the architects changed their plans to allow for a third and higher room, the so-called "King's Chamber."

As the construction had already risen above the level of the Queen's Chamber, the passage to the new chamber was partly bored through the existing masonry. Moreover, lest an earthquake cause the King's Chamber to collapse, several small rooms, one above the other, were built into the structure above this chamber to reduce the weight on its roof.

The passage from the outside of the pyramid first slopes downwards towards the underground chamber. Then this passage forks, one branch continuing down to the underground chamber and the other, the Ascending Corridor, sloping up on its way to the Queen's Chamber. This corridor forks in its turn. One branch runs horizontally to the Queen's Chamber. The other, still rising, opens out into the Grand Gallery. This is a high, narrow, sloping tunnel in the form of a corbelled vault, leading to the vestibule of the King's Chamber.

The corbelled arch and vault were used in Mesopotamia and in Egypt before the invention of the true arch and vault. Corbelling is laying courses or layers of stone or brick so that each course overhangs the one below. When walls are corbelled out from two sides until they meet, a corbelled arch or vault results. Although a structure of this kind is neither so strong nor so roomy as a true arch or vault, it is easy to make and does not require centering–that is, a wooden scaffolding, shaped to match the inner surface of the arch or vault, which holds up the stones or bricks during construction.

The corbel is one of the four devices that builders have developed for holding up the roofs or upper stories of houses. The other three are the post-and-lintel, the arch-and-vault, and the truss. Each device is best carried out by certain building materials. Brick and stone are suitable for the corbel and the arch-and-vault. Stone and wood are both suitable for the post-and-lintel. But, for the truss, wood alone, of the materials the ancients had, was satisfactory.

Hence different parts of the world developed building styles best suited to the local material. In ancient times, these materials were mainly clay, stone, and wood. Nowadays, of course, with steel and reinforced concrete, we can build structures that the men of old never dreamed of.

Mesopotamia, having plenty of clay but no stone or wood to speak of, early favored the corbel and the arch-and-vault. Egypt, having stone and clay, and Greece and China, having stone, clay, and wood to choose from, long adhered to post-and-lintel construction. It remained for wood-rich Europe to develop the truss.

Around the base of the Great Pyramid was built the usual inclosure, with mortuary temples and a great stone causeway leading down to the Nile. Herodotos, who saw these structures in good condition, deemed them as impressive as the Great Pyramid itself. Now, however, they have almost entirely disappeared.

Herodotos also reported various stories told him by his guides. They said, for instance, that Khufu had prostituted his own daughter to help to pay for the Great Pyramid; that it took a hundred thousand laborers, working in three-month shifts, twenty years to build this pyramid; that Khufu's sarcophagus lay on an island in an underground lake beneath the pyramid; and that the hieroglyphics carved on the outer casing of the pyramid recorded the food consumed by the workers. All of these stories were untrue. But the guides, like some of their descendants today, told whatever tale they thought would send the tourist away happy.

When Khufu died, his attendants placed his mummy in a wooden coffin. They carried this coffin up the Ascending Corridor and the Grand Gallery to the King's Chamber. Here they put the coffin into a plain granite sarcophagus, which must have been installed during the building of the pyramid because it is a little too wide to go through the narrow passage to the King's Chamber. The sarcophagus had a heavy stone lid, so made that when it was slid into place, stone bolts dropped into recesses in the trough and secured the lid–it was hoped–for all time.

On their way out, the workmen knocked loose some props in the vestibule of the King's Chamber, allowing three huge portcullis blocks to fall to the floor of the vestibule, blocking it. Removal of more props in the Grand Gallery allowed three great granite plugs to slide from the Grand Gallery down into the Ascending Corridor, blocking it also.

Khufu's son and successor Dedefra began a pyramid at Abu Roâsh, five miles north of Khufu's pyramid. Nothing but its base remains. Perhaps Dedefra died before the tomb was built; or perhaps it was demolished in the course of a feud among Khufu's sons over the succession.

Dedefra was succeeded by Khafra,[9] probably another son of Khufu, though the relationships of these early kings are uncertain. At Giza, Khafra constructed the Sphinx, a colossal lion with Khafra's own head on its shoulders, partly carved from an outcrop and partly built up of limestone blocks. The rest of the outcrop was quarried away for pyramid stones, so that the Sphinx lies in a depression formed by this quarry. A Muslim fanatic battered off the nose of the Sphinx about 1400.

Besides the Sphinx, Khafra built a pyramid slightly smaller than Khufu's. But it looks even taller than the Great Pyramid because it stands upon higher ground. This pyramid has none of the complicated interior corridors and chambers of the Great Pyramid, only a single underground burial chamber with passages leading to it.

Khafra's successor Menkaura[10] built a much smaller pyramid on Pyramid Hill, and other kings continued the custom down to the Twelfth Dynasty. Amenemḥat III,[11] whose pyramid, much the worse for wear, stands near Hawwâra, built such an elaborate mortuary temple that, centuries later, Greek visitors called it the Labyrinth, after the underground maze supposedly made for the legendary King Minos of Crete by the engineer Daidalos.

The last Egyptian pyramids were built about –1600; some think the very last one was made by Ahmose I.[12] By this time, about seventy pyramids dotted the land of Khem. Most of the later ones, however, were filled with rubble instead of good cut stone. Hence they eroded away to mere mounds after subsequent builders stole their limestone facings.

In –VIII, when the rule of Egypt was divided amongst a multitude of quarreling local lords, the kings of Kush conquered the land of Khem. Kush was the Ethiopia of the Greeks, corresponding to the modern Sudan. Less than a century later, the Assyrians drove out the Kushites. When troubles at home recalled the Assyrian armies, Egypt recovered its unity and independence.

The Kushite kings, who copied Egyptian culture and customs, had already imitated the custom of burying kings under pyramids. Back in the Sudan, they continued to build small pyramids for themselves and their queens clear down to +350, when the Abyssinians overthrew the Kushite kingdom. Remains of sixty-odd Kushite pyramids still exist near the ancient Kushite capitals of Napata and Meroê.

Robbers broke into all the Egyptian pyramids, despite the granite plugs, false passages, and other elaborate precautions of their builders. The Great Pyramid held out until the Caliph al-Ma'mûn (+IX) got past the granite plugs by boring through the softer limestone around them. Caring nought for relics of the Days of Ignorance, as Muslims call the ages before Muhammad, he smashed the lid of the sarcophagus and tore Khufu's mummy to bits for the gold that decked it. However, some archeologists think this pyramid had been robbed long before, about –XXIII, and that the mummy that fell victim to al-Ma'mûn's greed was not Khufu's but that of a later intruder.

The pyramids have long been a fertile source of pseudo-scientific speculation. Many people have made wild guesses about the purpose of these structures: that they were ostentatious displays of royal power, vaults wherein the sages of old stored their archives, Joseph's granaries against the seven lean years, models of Noah's ark, astronomical observatories, phallic symbols, Masonic halls, and standards of measurement.

These notions can all be easily disposed of. For instance, the passages inside the pyramids were blocked up as soon as the kings were laid to rest within, so they could not have been used for granaries, stargazing, or Masonic meetings. Modern archeology agrees with Herodotos that these buildings were tombs pure and simple.

The modern pseudo-scientific cult of Pyramidology began when Colonel Howard Vyse blasted his way into Khufu's and Menkaura's pyramids with gunpowder in the 1830s. From Vyse's measurements, the London publisher John Taylor and the Scottish astronomer Charles Piazzi Smyth evolved the theory that the Great Pyramid had been built by Noah, Melchizedek, or some other Old Testament patriarch under divine guidance; and that it incorporated in its structure such cosmic wisdom as the true value of π (the ratio of the circumference of a circle to its diameter), the mass and circumference of the earth, and the distance of the sun. The sarcophagus was supposed to be a standard of measurement, as if anyone but a lunatic would take as a volumetric standard a vessel holding the awkward amount of a ton and a quarter of water and then shut it up in a man-made mountain so that it could not be used.

The measurements in the Grand Gallery were taken to prophesy the history of mankind. Smyth, a religious fanatic whose strongest passion was to discredit Egyptian "idolatry," inferred from these measurements that a miracle, comparable to the Second Coming of Christ or the Millennium, would occur in 1881. When no miracle took place in 1881, other Pyramidologists reshuffled the numbers to make other predictions, which likewise failed to come true. The last one was that the world would end in 1953. This nonsense can go on forever, because new cultist minds are always being born. As engineering achievements, the pyramids are quite remarkable enough without bedecking them with occult whimseys.

Some people think that the ancient Egyptians must have used powered machinery like ours to build the pyramids, or even that they called upon occult powers whose secret has been lost. As the modern Egyptian poet Hafiz Ibrahim put it:

For they had crafts beyond our ken
And sciences that lesser men
Lack wit to grasp; with dexterous hand
To rich invention wed, they planned
Fair idols men might be forgiven
For worshiping in hope of heaven . . .[13]

Herodotos is responsible for this picture of pyramids being built by modern construction machinery. He wrote:

The pyramid was built in steps, battlement-wise, as it is called, or, according to others, altar-wise. After laying the stones for the base, they raised the remaining stones to their places by means of machines formed of short wooden planks. The first machine raised them from the ground to the top of the first step. On this there was another machine, which received the stone upon its arrival, and conveyed it to the second step, whence a third machine advanced it still higher.[14]

As nobody has found any trace in Egyptian art, architecture, or literature of anything like these wooden hoisting machines, it is likely that they were merely the fantasy of some guide or priest, recounted to the eminent Greek tourist in the hope of extracting an extra obolos from him. When Herodotos wrote, pyramids had not been built in Egypt for more than a thousand years, and it is unlikely that his guides would have any clear idea of the engineering methods of their long-dead predecessors.

But from various sources–tool marks on stone, quarries with blocks half detached, ancient tools found in modern times, and tomb paintings that show Egyptians working–we know much of how the Egyptians built large constructions of stone. From these sources we learn that the Egyptians of Khufu's time used very simple methods indeed. They lacked tongs and pulleys. They had no tools of any metal but copper. They made but little use of the wheel.

It is not even certain whether they moved heavy stones on rollers. Later engineers used rollers–for instance, Domenico Fontana moved his obelisk to St. Peter's by this means in 1586. But an Egyptian picture from the end of the pyramid-building age (–XX) shows 172 men pulling the 60-ton statue of a nobleman[15] on a sled without rollers. To make the sled move more easily, a man poured a liquid–probably milk, the fat content of which makes it a better lubricant than water—on the ground before it.

The true secrets of the ancients' engineering triumphs were three: first, the intensive and careful use of such simple instruments and devices as they had; second, unlimited manpower and the ability to organize and command it; and lastly, no need for haste. The most important of these was the last—the infinite patience they applied to their projects. The ancients were perfectly capable of duplicating many of our large modern public works, provided they did not require structural steel; but it would take them many more man-hours to do so.

The early Egyptian structures that have survived are nearly all tombs and temples. Although the Egyptian kings built handsome palaces, practically nothing is left of these, because the palaces were made of mud brick while the temples and tombs were of stone. From the Egyptians' point of view, this was logical. Since they took the afterlife seriously, they built palaces of brick, meant to last through their own lifetimes only; but tombs and temples were for eternity.

King Joṣer was not the first to build in stone; his predecessor Khasekhemui had used stone for the inner part of his mastaba. Joṣer made his step pyramid and mortuary temple compound of comparatively small stones, because the workers did not yet know how to handle larger ones.

By Khufu's time, techniques had so improved that the crews could handle not only the 2.5-ton blocks of which most of the pyramid was composed, but also a number of granite slabs, weighing over 50 tons each, for roofing the chambers. By the time of Menkaura, they were building the king's mortuary temple of stones weighing as much as 220 tons; by the time of Rameses II they were moving 1,000-ton statues.

The pyramids and other Egyptian monuments were not, as is often thought, built by hordes of slaves. Although Egypt was a land of vast class differences, slavery in the strict sense never played much part in its history.

For that matter, while slavery was found everywhere in the ancient world, the actual number of slaves at any one time and place was small, save in certain exceptional cases. During the growth of an empire by rapid conquest, as in the case of the Roman Empire in the last centuries of the Republic, many thousands of persons were enslaved by the conquering armies. With this exception, such slaves as existed were mostly the house servants of rich men and officials.

On the other hand, forced labor was common. It was the standard method of building roads, canals, temples, and other public works, because tax-gathering machinery was not yet effective enough to make the hiring of voluntary workers practical. Moreover, the inefficiency of forced labor had not yet been realized. About +150, King Rudradaman of Ujjain, in India, proudly boasted in an inscription that he had rebuilt and enlarged an important irrigation dam without resorting either to forced labor or to special taxation.

Simple calculations show that Herodotos' tale of the building of the Great Pyramid by 100,000 men working for twenty years–two million man-years of labor–is much exaggerated. Even with the simple methods of the times, the pyramid could have been built with a fraction of that labor.

Probably there was a small permanent staff of skilled workmen. A set of barracks of rough stone and dried mud, whose ruins were found near Khafra's pyramid, may have housed this permanent staff. The barracks are thought to have held about 4,000 men.

In addition, the king conscripted tens of thousands of peasants to help with the heavy work during the season of the annual flood of the Nile, when these farmers would otherwise have been idle. It was forced labor, but the laborers were conscripts, not slaves. They were probably paid in food, because money did not yet exist. They were organized in gangs with such heartening names as "Vigorous Gang" and "Enduring Gang." The kings also freely pressed their soldiers into service for work on such monuments.

While it is not likely that the workers were constantly lashed with whips, as the slaves of legend are supposed to have been, we need not think that building such a monument was all sweetness and light, either. Egyptian tomb paintings show the foremen of gangs as carrying yard-long limber rods, and these were probably not mere symbols of office. An occasional whack with a stick has been a customary part of bossing a gang of Egyptian workers, slave or free, from ancient times almost to the present day. The architect Nekhebu, in boasting on his tomb of his many virtues and kindnesses, mentions the fact that he never struck a workman hard enough to knock him down.

With forced labor, such methods are to be expected, because the conscripted worker, like the slave, has nothing much to gain by working hard. He therefore does as little as he possibly can. It is no use to threaten him with dismissal, because he would like nothing better.

It has been suggested that the pyramid workers labored willingly, deeming it their pious duty to preserve the body of their god-king and perhaps believing that such preservation helped the masses in some magical way. Considering the amount of voluntary work done on cathedrals in medieval Europe as a pious duty, such an attitude is not impossible, at least among some of the workers. But there is no way to settle the question by interviewing Khufu's subjects to see what they thought.

Most of the stone for the pyramids, cut from local outcrops, could be dragged directly to the site on sleds. Fine limestone from Troyu had to be rafted across the Nile. Granite for the linings of chambers came from Swenet (modern Aswân) and were floated on barges down the Nile. For pyramids built on low ground, the kings had canals dug from the Nile partway to the pyramid, so that the stones could be brought near the site before being dragged overland.

The Egyptians had various methods of quarrying. One was to cut notches in the rock along the line of fracture, drive wooden wedges into these notches, and wet the wedges. When the wood swelled, the block split off. Another method was to drive copper wedges between thin copper feathers on the sides of the notches.

Still another way was to have a crew pound at the rock with balls of hard stone (diorite) held in both hands until they had bashed out a trench all around the stone to be detached. A modern experiment has shown that pounding granite with a similar ball, on an area a little over a square foot, lowers the level of the stone at a rate of one-fifth of an inch an hour. This laborious method seems to have been used for long pieces of granite, such as obelisks, perhaps because it created less risk of cracking the stone than the wedging methods.

The stones were moved by the lavish use of levers and ramps, first to get the stones on their sleds, then to bring the sleds to the sites where the stones were to be set. The Egyptians made enormous ropes of palm fiber or reed. If, as a tomb painting indicates, 172 men could move a 60-ton statue, 8 men should have been able to move an ordinary 2.5-ton pyramid block, at least on the level. Sometimes oxen were used instead of men.

While some men were quarrying the stones for a pyramid, others were clearing and leveling the site of the tomb. The sides of the base were measured off with cords to form a square. There were several possible ways to check the trueness of the square, such as measuring the diagonals. For leveling, a long narrow trough of clay, into which water was poured, served just as well as a modern spirit level.

It is not known for sure how the Egyptians found the true north so accurately. The star nearest the north celestial pole in Khufu's time was Alpha Draconis, and that may not have been close enough to be very helpful.[16] While it is possible to mark the directions of the sun's rising and setting and bisect the angle between these directions, the sun is too large an object to sight on accurately.

A likelier method is to build an artificial horizon–that is, a circular wall, high enough so that a person seated in the center cannot see any earthly objects over the top of the wall. The seated observer, with his head at the center of the circle, watches a star rise and directs another surveyor to mark the place on the wall where the star appeared. When the star sets, he causes another mark to be made.

By lowering a plumb bob from the marks on the wall, the places at the foot of the wall, inside, and directly below the marks are found. Lines are drawn to the center of the circle. By bisecting the angle between these lines with cords and markings, the true north is found. The surveyors probably made a number of tries to be sure of getting the correct direction.

In building a pyramid, the stones were sledded to the site, levered off their sleds, and shoved into place with much prying and grunting. Probably the masons spread a layer of thin mortar or mud on the rock over which the stone was to be slid, to make the job easier.

As the pyramid rose, the builders raised an earthen mound on all sides of it, with one or more long ramps for hauling up the stones. Remains of such ramps have been found near some of the pyramids and other monuments. As each course was laid, the mound and the ramp were raised to another level. When the job was done and the gilded capstone had been set in place, all this vast mass of earth had to be hauled away.

The core stones of common limestone were only roughly fitted together, but the fine limestone blocks of the casing were fitted so carefully that a knife blade could hardly be thrust between them. The joints between adjacent blocks are all more or less askew, which implies that each row of casing stones was lined up on the ground and trimmed to fit one another before being hauled up the ramp and pushed into place. And Ptah help the foreman who tried to insert Block Number 6 where Number 5 was supposed to go!

Lastly, during the removal of the mound and the ramp, masons standing on the mound trimmed away any irregularities left in the facing of each side.

Ancient and modern critics alike have berated the kings of Egypt for spending the kingdom's wealth and their subjects' labor on such useless monuments. Of course, the kings who built the pyramids did not regard them as useless, but as a sure means of gaining a pleasant and everlasting afterlife. As things turned out, the pyramids were not all wasted effort. In building them, Egyptian engineers learned much about quarrying, shaping, and moving heavy stones. This knowledge became a part of the world's general fund of technological wisdom.

Egypt was a land of vast social distances between noble and commoner, between king and subject. It was a land of rigid class lines not easily crossed. Still, engineers succeeded in crossing them, because engineering ability is not a common gift. The architect Nekhebu (–XXVI) told on his tomb the story of his rise from humble beginnings:

> His Majesty found me a common builder; and His Majesty conferred upon me the offices of Inspector of Builders, then Overseer of Builders, and Superintendent of a Guild. And His Majesty conferred upon me the offices of King's Architect and Builder, then Royal Architect and Builder under the King's Supervision. And His Majesty conferred upon me the offices of Sole Companion, King's Architect and Builder in the Two Houses.[17]

Nekhebu goes on to boast of how he built mortuary chapels for King Pepi I, dug two canals, and executed other royal commissions. For these labors, the king conferred additional titles upon him and rewarded him with gold, bread, and beer.

When Third Dynasty kings first began to build structures in stone, they proceeded with a kind of nervous caution. To be on the safe side, they adhered to architectural forms as much as possible like the earlier forms of brick, wood, and other materials.

Joṣer's pyramid and the other buildings of its temple compound are full of such imitations. Stone walls are carved to look like reed matting; a stone roof is carved on the underside to resemble a roof of rounded logs.

Columns imitate supporting members gathered from the plant world. They mimic the palm trunks, the bundles of reeds and saplings, the papyrus stems–singly and in bunches–with which Egyptians had been holding up their flimsy dwellings. The ribbing of columns indicates bunches of slender stems lashed together. Foliage of the appropriate kind is carved in stone at the tops of the columns.

After all, such slavish imitation of plant forms might not make any difference in the strength of the column. But then again it might, and why take chances? Stress analysis was undreamed of, and no architect wanted his temple to collapse because he had dared to break away from known shapes.

Fig. 1. Ancient capitals (*after Brown, Daremberg & Saglio, and Edwards*).

By the time of King Khafra, architects no longer thought that a stone supporting member had to look like a tree trunk. Khafra had his so-called "valley temple" near the Sphinx built with plain square columns, as starkly functional as anything in modern architecture.

Not long after Khafra's time, the architects also began to make the tops of their columns in the forms of lotus blossoms–that is, water lilies. This is certainly no imitation of any real wooden column, because the stem of a water lily has no more structural strength than a piece of spaghetti. Therefore, these lotus capitals must have been used for religious or artistic reasons.

Withal, the habit of carving imitation tree trunks in stone to hold up the roofs of public buildings persisted. The familiar three orders of classical Greek temple columns–Doric, Ionic, and Corinthian–probably evolved from Egyptian temple columns with their lotus, papyrus, and date-palm capitals. In other parts of the world, the capitals of columns were derived from other plant and animal forms. Just remember, next time you pass a bank with conventional Greek columns before it, that you are beholding an imitation in concrete of an imitation in stone of a simple wooden log.

After Aḥmose I, the kings of Egypt ceased to make artificial mountains for sepulchers. Perhaps costs of construction rose, as they seem to have been doing from that day to this. Perhaps the liberalization of Egyptian religious doctrines, which at last allowed common men to hope for immortality without benefit of pyramids, had an effect.

In any case, the subsequent kings drove tunnels into real mountains, mostly on the west bank of the Nile opposite the city of Opet (the Egyptian Thebes of the Greeks, and the modern Luxor), which was the capital of the New Empire. Although these Pharaohs tried to hide their tombs, grave robbers gained access to all but one: that of the boy-king Tutankhamon (–XIV). The entrance to this tomb was covered by débris thrown out in digging the tomb of the later Rameses VI (–XII); and thus it was buried, forgotten, and saved for posterity.

Relieved of the staggering costs of pyramids, the kings put more of their wealth into temples and monuments, like the obelisks. Obelisks were monuments to the sun god Amon-Ra, in the form of a tall tapering shaft of Aswân granite, surmounted by a pyramidion (little pyramid) originally plated with metal–copper, gold, or electrum–to reflect the rays of the divine sun. On these shafts the kings usually included, besides the dedicatory inscription, boasts of their own virtues and feats. A 90-footer still lies in the quarry at Aswân where it was abandoned after flaws appeared in the rock. To raft the obelisks down the Nile, Egyptian rulers built 200-foot barges, the largest ships that the world had yet seen.

At one time several score of these monuments dotted the land of Khem, but only five remain. Some were felled and broken by earthquakes or settling. Many were carried off by Roman emperors to decorate Rome and Constantinople.

Others were given away to foreigners by the nineteenth-century rulers of Egypt, especially Mehmet Ali, the crafty and energetic Albanian tyrant. Mehmet Ali cared nothing for monuments and everything for money and power. Having demolished several ancient temples to make factories of the stones, he proposed to take the Great Pyramid apart for its stone and was only dissuaded when told that it would be more efficient to use a quarry nearer Cairo. As a result of the Albanian's openhandedness with the relics of the Days of Ignorance, Paris, London, and New York have one obelisk apiece.

The largest obelisk of ancient Egypt was the 105-footer made by Thothmes III (–XV) now at the Church of San Giovanni in Laterano in Rome. The largest still standing in Egypt is one of four erected for the famous Queen Ḥatshepsut, the aunt of Thothmes III, by her architect and favorite Senmut.

It is likely that Senmut was named Royal Architect because he was Ḥatshepsut's favorite, not that he became her favorite because he was such a good architect. In lowering Ḥatshepsut's surviving obelisk into place, Senmut's crew missed the groove along the upper edge of one side of the pedestal. Such a groove was cut in all obelisk plinths to insure that the monument should settle into place squarely centered on the pedestal. As a result of Senmut's blunder, this obelisk is several inches off-center and slightly askew. Poor Senmut was probably liquidated, along with Ḥatshepsut's other supporters, when Thothmes III took the throne and tried to obliterate the memory of his hated aunt by chiseling her name off all her monuments.

For moving these shafts, the Egyptians used methods like those employed in moving pyramid blocks. To erect an obelisk, they probably hauled it up an earthen ramp and dug the earth away under the butt until the obelisk had tipped up to an angle of about 45°. When the lowest edge was seated in the groove along the upper edge of the pedestal, they hauled the obelisk upright with ropes and shear legs.[18] We can imagine their taking many precautions with guy ropes, braces, and cushions of brushwood to make sure that the stone did not get away from them.

In 1961 the archeologist Millet watched an Egyptian crew erect an obelisk of Rameses II in a park in the Gezira ("Island") of Cairo by much the same methods, except that two modern steel winches took the place of "the huge gangs of men who had erected the monolith in its original setting in the great temple at Tanis." Millet adds: "I went off home in the snug glow that affects every self-respecting archaeologist when he sees a survival of the past holding its own in the present. It proves what he has always secretly believed–that they did things better in those ancient days."[19]

Similar methods were probably used by the primitive Britons at about the time of Thothmes III in setting up the circle of upright twenty-six-ton stones for their outdoor temple at Stonehenge. In prehistoric times, thousands of similar stones were also set up in long parallel rows near Carnac, in Brittany.

Of the later Egyptian temples, the most impressive is that of Amon of Opet, usually called after the modern village of Karnak. Parts of it go back as far as –2000. Some of the early buildings were demolished to make room for grander structures, whereas others were allowed to stand while later buildings rose around them. In the course of the centuries, the temple acquired the form of a long rectangle, about 400 by 1,200 feet.

However, the symmetry of this rectangle was spoiled by kings who haphazardly added more buildings and monuments, until the vast temenos became a chaos of walls, columns, pylons, courts, temples, statues, obelisks, and shrines. Small detached temples stood here and there outside the main mass or inside the courts of the larger temples. The walls were covered with painted reliefs, showing the kings dispatching their foes and adoring their gods.

The most impressive part of the Temple of Amon is the Hypostyle Hall built by Rameses II (–XIII). Rameses was one of the ablest, most aggressive, and most self-conceited of all the kings–nearly 300 of them –who ruled the land of Khem before the Persians came. He had a mania for statues of himself, the bigger the better. After his time, Egypt was dotted with colossi displaying idealized versions of Rameses' lanky form and large hooked nose.

One colossus, dug up near Saqqâra, now stands in front of the railroad station in Cairo. Remains of another, fallen and broken into three pieces at Rameses' mortuary temple in the City of the Dead, across the Nile from Opet, inspired Shelley's *Ozymandias:*

I met a traveller from an antique land
Who said: Two vast and trunkless legs of stone
Stand in the desert . . . Near them, on the sand,
Half sunk, a shattered visage lies, whose frown
And wrinkled lip, and sneer of cold command,
Tell that its sculptor well those passions read
Which yet survive, stamped on these lifeless things,
The hand that mocked them, and the heart that fed:
And on the pedestal these words appear:
'My name is Ozymandias, king of kings:
Look on my works, ye Mighty, and despair!'
Nothing beside remains. Round the decay
Of that colossal wreck, boundless and bare
The lone and level sands stretch far away.[20]

The fact that the real Ramesseum and its fallen colossus do not answer at all well to Shelley's description does not detract from the beauty of the poem, which across the gulfs of time and space has awakened in the minds of millions a sense of the mighty stream of human history.

The most astonishing feature of the Hypostyle Hall is the forest of 134 immense columns. Of these, the columns of the two central rows, twelve in all, are 69 feet tall and almost 12 feet in diameter. They are topped by 11-foot foliage capitals. The remaining 122 columns are smaller. All are carved with reliefs and inscriptions. Whereas most ancient columns were built up of drums–cylindrical stones fitting one above the other–the columns of the Hypostyle Hall are made of half-drums, as whole drums would have been so large as to be awkward to handle.

Although the Hypostyle Hall is one of the world's most celebrated buildings, a closer look shows it to be more an expression of Rameses' megalomania than a worthy house for the gods. The construction is gimcracky. The columns had no foundations other than pavements of small stones. As a result, when the Nile rose to record heights in 1899, eleven columns fell over. Since then the archeologists have reset them, secured them, and erected a concrete roof over them. But the columns are so massive that a worshiper within could see practically nothing of what was going on at the far end of the temple.

Although to our skeptical age it might seem that the Egyptians devoted undue effort to tombs and temples, Egyptian engineers also worked at more mundane projects. A bas-relief on an ornamental stone mace shows an early king, known only as "Scorpion," grasping a hoe as he officiates at a canal-digging ceremony.

From Scorpion's time on, the construction of canals was a major concern of the Pharaohs and their servants. Among the first duties of a provincial governor were the digging and repair of canals. These canals were used to flood large tracts of the country during the Nile's high water, which occurs in the autumn as a result of summer rains in the lands to the south of Egypt.

The land to be flooded was cut up by dykes into a checkerboard pattern of small basins. When the basins were full, the dykes were closed up and the water was kept standing until, perhaps a month later, the ground was thoroughly soaked. Then the surplus water was drained off into the canals.

Although the Nile is the world's most reliable river, an excessively high or low rise of the Nile spelled disaster. If the river did not rise high enough, it failed to flood the tracts laid out for that purpose, and no crops grew. If it rose too high, it washed away the dykes and nearby villages, drowning thousands.

Where fields were not low enough to be flooded directly, the peasants drew water from the canals and from the Nile by means of the swape or *shadûf*. The swape consisted of a bucket on the end of a cord, which hung from the long end of a pivoted boom, counterweighted at the short end. Such swapes, mostly of very rough construction, are still used in Egypt. The posts bearing the axle on which the boom is pivoted are made of dried mud. The boom is a tree trunk with the branches lopped, and the counterweight is another mass of mud. A farmer in such a rainless land may spend up to half his working time irrigating his little plot.

Canal building continued down the centuries. Senuṣert III (–XXI) was such a vigorous canal builder that he became known as "the king who built the canals," and later storytellers ascribed to this one king the deeds of many. At length he became the legendary "Sesostris" of the Greek historians, whose mythical deeds included damming the Fayyûm depression west of the Nile and conquering the whole civilized world.

The Egyptians did in truth know how to build dams. At some time—perhaps as far back as King Khufu's reign—they built one in the Wadi Garâwi, twenty miles southeast of Cairo, to store water for the use of workers in the nearby quarries. This dam is of rough masonry 33 feet high, between 200 and 370 feet long, and between 150 and 270 feet thick. Later—perhaps about –1300—Egyptian engineers threw a much larger stone dam across the Orontes River in Syria. This dam, a mile and a quarter long, created the Lake of Ḥomṣ. Lake and dam are both still there and still in use for local irrigation.

As a result of Egyptian canal building, Herodotos noted:

. . . whereas Egypt had formerly been a region suited for horses and carriages, henceforth it became entirely unfit for either. Though a flat country throughout its whole extent, it is now unfit for either horse or carriage, being cut up by the canals, which are extremely numerous and run in all directions.[21]

This is an exaggeration; the many Egyptian pictures of chariots and the remains of the chariots themselves show that there were always some fairly good roads. Borrowing the idea of the chariot from Mesopotamia, Egyptian wainwrights developed a light, openwork chariot of refined design for sport and war.

The most far-sighted of all the canal projects, however, was begun by Nikau II[22] about –600. This was a ship canal to connect the Red Sea with the Mediterranean. But it did not run north and south through the Serbonian Swamp, between Egypt and Sinai, as does the present Suez Canal.

Instead, Nikau's canal ran east and west, from the easternmost branch of the Nile (near modern Zagazig) to Lake Timsâḥ, the mid-point of the present Suez Canal. Thence it turned south and followed more or less the course of the Suez Canal, skirting the Bitter Lakes to the head of the Red Sea.

Nikau gave up the project, we are told, when an oracle warned him that he was laboring for the benefit of "the foreigner." Three-quarters of a century later the foreigner arrived, under the standards of the conquering army of Cambyses[23] the Persian. After the conquest of Egypt, the great Darius I[24] completed the canal.

From then on the canal was alternately open and closed, as careless rulers let it fill up with sand and energetic ones dredged it out again. Ptolemaios II (–III) not only restored it but added some sort of lock or water gate. The Roman emperor Trajan restored it again; so did the Arab general 'Amr ibn-al-'Âṣ, who conquered Egypt in the +640s.

In +VIII, however, the canal went out of use for good. At this time, disorderly Arab rule allowed the whole Egyptian canal system to fall into disrepair, with the result that the teeming population of Egypt was halved by starvation. Communication between the Mediterranean and the Red Sea was not reopened until the completion of the Suez Canal in 1869, with bands, fireworks, and Empress Eugénie. Verdi composed *Aïda* for the event but was two years late.

So ends the story of engineering in the days of independent ancient Egypt. To learn more about the great public works of ancient times, we must turn to that other great watershed culture, the civilization of Mesopotamia. For, if the Egyptians surpassed all other pre-classical peoples in the art of building in stone, the Mesopotamians excelled in many other aspects of civilization. We derive much more of our science, religion, and commerce from ancient Iraq than from Egypt; and in engineering, too, the ancient Mesopotamians were second to none.

THE MESOPOTAMIAN ENGINEERS

THREE

Save in the extreme north, Iraq is an enormous flat plain, fading off into the Arabian deserts. It is barren and desolate except where irrigation brings in the waters of the Tigris and Euphrates rivers, which wander placidly over its level surface. The Greeks called this land Mesopotamia—"the land between the rivers."

So vital is irrigation to Mesopotamia that an ancient Babylonian curse was: "May your canal be filled with sand!"[1] Many of the ancient laws dealt with canals and water rights, like this one from about –VI:

The gentleman who opened his wall for irrigation purposes, but did not make his dyke strong and hence caused a flood and inundated a field adjoining his, shall give grain to the owner of the field on the basis of those adjoining.[2]

The great Khammurabi, called the Law-giver, who lived about –1700, was constantly writing his governors about the repair of canals:

Unto Governor Sid-iddinam say: Thus saith Khammurabi. Thou shalt call out the men who hold lands along the banks of the Damanum-canal that they may clear out the Damanum-canal. Within the present month shall they complete the work . . .[3]

Khammurabi also let the thirty-third year of his reign be known as the year in which he "redug the canal called 'Khammurabi-spells-abundance-for-the-people, the Beloved-of-Anu-and-Enlil,' thus he provided Nippur, Eridu, Ur, Larsa, Uruk, and Isin with a permanent and plentiful water supply . . ."[4] after these cities had been threatened with destruction by the drying up of the Euphrates.

In southern Mesopotamia, at the beginning of recorded history, the Sumerians—a people of unknown origin—built the city walls and temples and dug the canals that comprised the world's first engineering works. Here, for over two thousand years, little city-states bickered and fought over water rights.

The Sumerian element in the population was gradually swamped beneath a tide of Arabian nomads, who drifted in from the desert to take up the lives of farmers and city dwellers. The Semitic Akkadian tongue replaced Sumerian and spread all over the Near East as a trade language. Lugalannemundu of Sumer, Sargon of Akkad, Ur-Engur of Ur, Khammurabi of Babylon, Gandash the Kassite, and Nebuchadrezzar I of Babylon founded short-lived empires until the Assyrians overcame them all (–XII).

Before the rise of cities, the Sumerians lived in round huts built by the stud-and-mud or wattle-and-daub method. A number of slender rods —canes, saplings, or withes—were stuck in the ground in a circle. These were then plastered over with clay or mud to form a wall. Similar houses were built in Egypt and are still made in Iraq today.

When cities arose and wealth accumulated, city-dwellers changed from these circular huts to rectangular houses of brick. The Sumerian brick mold, still in use there, may be called the world's first mass-production device. Asphalt from the natural oil well of Id[5] was used for mortar. A house might be whitewashed or, if pretentious enough, might be coated with plaster.

Instead of hinges like ours, the door had a pair of vertical pins at the hinge corners. The bottom pin rested in a stone door socket, usually the only piece of stone in the house. The upper pin was held in place by a strap, and the door frame was painted red to scare away demons. Windows were often barred by a grille or shutter of brick.

Most of the roofs were made of palm logs laid in a row from wall to wall. Over the logs was spread a layer of palm fronds and, over that, a layer of earth, rolled flat with a stone roller. After every rain, the Mesopotamians climbed up to their flat roofs to re-roll them, as some still do today. Many householders even raised vegetable gardens in the earth of their roofs.

Some houses had, over one or more rooms, roofs in the form of corbelled domes. These domes, like clusters of magnified beehives, are still seen in Syrian villages.

As houses grew larger, they took on the shape that ever since has been popular in southwestern Asia and the Mediterranean region. The rooms formed a hollow square; the doors and windows faced the open court or garden in the middle of the square. The larger homes often had two stories, the upper one being of flimsy mud-and-reed construction for lightness. The outside of the house, except for the front door, presented a blank brick wall to the outer world.

There were several reasons for this outside-in construction. One was the intense summer heat of the region. A learned man who went east with Alexander the Great wrote that at Susa,[6] "when the sun is hottest, at noon, the lizards and snakes could not cross the streets in the city quickly enough to prevent their being burnt to death in the middle of the streets" and "barley spread out in the sun bounces like parched barley in ovens."[7] The hollow square allowed the house owner to sit outdoors in the shade at all times of day.

Moreover, such a house was less vulnerable to burglary than one with windows opening to the outside.

Finally, a blank wall shielded the householder from the eager eyes of tax gatherers and other royal agents. Some men living under a despotism managed to get rich, even though they were not members of the governmental apparatus. Such men—merchants for example—were careful to hide their wealth from the despot's agents, lest they be beggared by sudden tax claims or slain on some trumped-up charge to give the autocrat an excuse for taking all. An Iranian proverb expressed it: "If you are being fattened by someone, you may expect very quickly to be slaughtered by him."[8]

Because of the scarcity of fuel in Mesopotamia, kiln-dried bricks were rarely used in private houses. As a result, the householder was kept busy during the winter trying to patch up his house as fast as the rains dissolved it away. When, once in a generation, the task became hopeless, he brought out his movables, knocked down the walls, leveled and smoothed off the débris to make a new floor, and built another house on the ruins of the old.

As the early cities of the ancient watershed kingdoms had neither sewers, garbage disposal, nor trash collection, rubbish accumulated and constantly raised the level of the streets. Older houses could be distinguished by the fact that their entrances were below street level.

This was particularly true of temples. Most large temples stood in a sacred precinct, the temenos, surrounded by a wall. Because the temple was more substantially built than private houses, it lasted longer. Also, as the temenos was holy, it was kept free of rubbish, so that the level of its grounds did not rise with that of the street outside. In time a temple and its inclosure might be almost buried from sight as the street levels rose around them.

Thus cities originally built on the plain slowly rose on hills of their own débris. Today Iraq is dotted with these hills or tells,[9] scores of which still await the picks, shovels, and whisk brooms of archeologists.

When men first began to build large numbers of houses close together, two methods of arranging them grew up. If a city grew out of a village, the dwellers were likely to continue to let everybody put his house where he pleased.

The result was a city laid out like the oldest parts of Paris or Boston today. Narrow, winding alleys, hardly wide enough for two men to pass and unusable by large beasts of burden or vehicles, ran every which way. This was no great fault in a village, where too few people were abroad at any one time to constitute a traffic problem.

But, as the population grew, traffic congestion grew faster. Hence the ancient metropoleis were forced, like modern cities, to regulate traffic. The winding-alley village layout also aggravated the problems of waste disposal, fire protection, and law enforcement as the city waxed larger.

So when, as sometimes happened, a group of people came to a likely place and said: "Let us build a city here," their leaders often had the wit to plan the city from the start. They laid it out with straight streets, some of them wide avenues, in a regular pattern. They kept at least one area clear of buildings for a marketplace, and reserved another space for temples and palaces.

The usual pattern was a gridiron or checkerboard with streets crossing at right angles. The cities of the Indus Valley showed this plan, as did some Mesopotamian and Egyptian cities. Peoples much given to sending out colonies, such as the Phoenicians, the Greeks, and the Romans, had many occasions for laying out cities in this manner.

A ceremony marked the founding of a city. The head man traced out the line of the wall, often by plowing a furrow with gaps for the gates. This line then became invested with magical properties.

The simple gridiron plan is not always the best possible, according to modern city planners, who like to include a few diagonal streets to carry heavy traffic and a winding layout for residential sections. But the gridiron was the best that ancient city planners could envisage.

In locating a city, ancient town planners were often torn between two choices: to put the city at the bottom of a valley for nearness to water, or to place it on a hilltop to make it easy to defend. The choice depended upon the likelihood of attack.

The survival of the hilltop city or fortress depended on getting water when besieged. Some cities solved the problem thus: When a spring flowed from the hill below the walls, the builders drove an inclined tunnel from inside the walls down through the rock to the spring, which they then walled in so that it could be approached only by this tunnel. Sometimes another tunnel carried the water inwards, under the fortified place, where it filled an underground cistern.

The kings of Mycenae in Greece, reigning at the time of the legendary Trojan War (–XII), took this precaution. So did the Jebusites, who lived in Jerusalem before the Israelites captured it. The Jebusites ran a tunnel from the spring of Gihon, southeast of the city, to a natural cave beneath the city. Then they excavated a 40-foot vertical shaft from this cave up to the surface, so that women could lower their vessels from the surface down to the reservoir thus created.

When David attacked Jerusalem (about –1000) the warrior Joab led a party of Israelites up the shaft, captured the city, and thus made himself David's commander in chief. About 300 years later, King Hezekiah of Judah blocked off the Jebusite tunnels and made a tunnel of his own, leading to the Pool of Siloam at the southern end of the Valley of Cheesemongers. An inscribed tablet marked the place where the two tunneling gangs, starting from points a third of a mile apart and boring from opposite directions, met.

The two great Mesopotamian rivers meander southeastward across the Euphratean plain, approaching to within twenty miles of each other near Baghdad. Then they diverge for another 300 miles. At last they join and flow together for fifty miles into the Persian Gulf. According to one theory (with which not all students agree) this gulf in ancient times extended farther to the northwest than it now does, and the rivers entered it separately. Silt from these rivers has since filled up the head of the gulf.

Both rivers, like the Mississippi and other large flood-plain streams, have changed their courses many times, often leaving prosperous cities stranded, to decay and die in the midst of a desert. In the upper part of its course, where the slope is steep and the current swift, a river picks up silt from the bottom. Then farther down, as the river nears the sea, its slope becomes gentler and its current slower, so that it drops the silt it carried. Therefore the bottom in this part builds up higher and higher.

In time, during the high-water season, the river overflows its banks.

Sometimes it makes a whole new channel in some other part of the flood plain. Because men try to keep the river in one place by building up the banks with levees, the river rises higher and higher above the surrounding country, and a flood is more destructive when it does come.

Flood problems were more acute in Mesopotamia than in Egypt. For one thing, the Tigris and Euphrates carry about five times as much silt for a given volume as the Nile. Hence the bottoms of these rivers rise faster, and they change their courses more often.

For another, they are less regular than the Nile in the date and degree of the rise of their waters. Their high level occurs in the spring at an awkward time, too late to help with winter crops and too early for summer crops. Therefore, much storage of water is needed to raise good crops on this fertile flatland.

Although the Euphrates is much the longer of the two rivers, the Tigris carries over twice as much water as its sister. It is also swifter and more unpredictable. Moreover, being faster, it digs a deeper trench and so is less easily used for irrigation.

Because the leading civilizations of antiquity arose in broad river valleys, and because floods are the deadliest natural catastrophes in such valleys, these civilizations all developed flood legends as part of their mythology. The Sumerians had a legend about the pious Ziusudra, who by building an ark saved himself and his family from a great flood sent by the gods. The legend evolved down the centuries and passed from folk to folk, so that Ziusudra became the Utnapishtim of the Assyrians, the Noah of the Hebrews, and the Deukalion of the Greeks.

Across the yawning gulf of 5,000 years, we see the sun-browned Sumerians beginning the endless task of breaking the rivers and the plain to the use of man. As century followed century, Mesopotamia came to be damascened by an azure web of canals, which tamed the mighty Euphrates, clothed the desert in rippling fields of golden grain, and moistened the roots of date palms planted along their banks in endless rows.

In the third millennium B.C., for example, King Entemanna of Lagash built an especially large canal, which ran from the Tigris south along the 46th meridian to the Euphrates. This canal can still be traced by a line of lakes, streams, and marshes. Later an even larger canal, the Nahrwân, over 200 miles long and 400 feet wide, paralleled the Tigris along its left bank from Baghdad (then a mere village) to a place near modern Kut al-'Amâra.

Keeping up such a canal system in Mesopotamia presented special difficulties. Because of the lack of timber and stone, there was no easy way to reinforce the canal banks. When these banks were simply made of piled-up mud, they easily fell into disrepair.

An irrigation canal must be carefully planned and maintained. It must be a little above the surface to be irrigated, and it must have a slight but constant slope to keep the water flowing. If the slope is too steep, the water flows too fast and eats away the banks. If it is too gentle, weeds and silt block the channels. The domestic goat, that ancient scourge of the Near East, breaks down the banks by scrambling up and down them. Constant repairs are therefore needed. As soon as a section begins to silt up, it must be dredged out lest it ruin the circulation of the area.

Lacking stone and wood, the Mesopotamians used cane reeds, tied in bundles or woven into mats, as reinforcing material. They used these reeds not only in their canal work but also in house building. Therefore Mesopotamian towns maintained a curious institution: the municipal marsh, a patch of swamp deliberately kept as a wetland where the useful reeds could grow.

Mesopotamian irrigation was of the basin type, like that in the Egyptian Delta. As such basins do not have mechanical gates or sluices, they are opened by digging a gap in the surrounding embankment and closed by shoveling mud into this gap again. Hence Mesopotamian irrigation farming was a very laborious business.

For hoisting water, the Mesopotamians used a swape like that of Egypt. Sometimes they employed a battery of these swapes, the first one hoisting water to a certain height, the next hoisting it still farther, and so on. They also devised the first advances on the simple bucket hoist. One was the pulley, which appeared before –1500 and which made much easier the task of drawing water from a well.

A main form of engineering advance is the substitution of continuous for intermittent motion, and rotary for back-and-forth motion. The Mesopotamians attained this stage by about –1200. A legal document on clay orders a man to replace a water-raising treadwheel, 20 feet long with seventeen steps, which he had borrowed and lost.

Mesopotamian laws not only required farmers to keep their basins and feeder canals in repair but also called upon everybody in the kingdom to turn out with hoe and shovel in times of flood, or when a new canal had to be dug or an old one repaired. Times of trouble caused the canal system to decay, so that extra effort was required to put it back into good condition.

With the best of care, however, canals would last only about a thousand years. Then they were abandoned and others were built. Today, four or five thousand years later, Iraq is still ridged with the embankments of these abandoned canals, crisscrossing the country in parallel lines.

For four thousand years the Mesopotamian canal system supported a denser population than lives there today. Then in 1258 the Mongols of Hulagu Khan conquered Mesopotamia, sacked Baghdad, and killed the last of the Khalîfahs of Baghdad. Looking upon all sedentary peoples as vermin to be wiped out, the Mongols destroyed the irrigation system and allowed a famine to reduce the population to a fraction of its former size.

Iraq remained under Mongol rule for about a century. As nothing was done to rebuild the canals, the land went back to desert and swamp. Subsequently plundered by Arabs, Turks, and Kurds, the region was almost depopulated, as happened to many other lands in ancient times as the result of barbarian raids and conquests. Iraq never even began to recover until it became independent in +XX.

However, the ruin of Mesopotamian irrigation may not have been due solely to the Mongols. There is reason to think that agriculture in this region had been decaying for centuries before the Mongol invasion. The reason for this decay has to do with salt.

In the lower strata of the alluvial soil of Mesopotamia lie thick beds of salt. This salt may be a relic of a prehistoric time when the sea covered the whole Euphratean plain. When such a soil is irrigated again and again for thousands of years, capillary action draws salt water to the surface. As the water evaporates, the salt remains and little by little makes the land useless for farming. So perhaps, even with modern agricultural methods, Iraq will never again be the teeming farmland it was in ancient times.

Once the Mesopotamians had learned to irrigate their land and wall their cities, they could turn their attention to building temples and palaces—the only gay and handsome structures in this almost treeless flatland of brown mud-brick villages.

The modern ceremony of breaking ground for a public work, with some puffing politician turning the first spadeful of earth, goes back to ancient Mesopotamia. A relief from the third millennium B.C. shows the Sumerian king, Ur-Nanshe of Lagash, in the unkingly pose of bearing a basket on his head. Presumably the basket contains the first bricks for a temple or other public work, because a similar relief shows the dreaded Ashurbanipal of Assyria (–VII) carrying a similar basket for the rebuilding of the great temple of Marduk in Babylon.

The first temples of Mesopotamia were built in the fourth millennium B.C. with starkly rectangular lines. They had blank brick outer walls, pierced high up by a few small triangular windows. The austere aspect of these temples gives the impression that they were designed, by some trick of time travel, by twentieth-century architects of the functionalist school. Strict symmetry and right-angled corners, however, the Mesopotamians did not especially admire.

Later Mesopotamian temples were brightly decorated. This was done by pressing cones of colored brick, about the size of a finger, into the wet plaster that covered the walls and pillars, so that only the bases of the cones could be seen. The walls and pillars of the temples of Uruk[10] and Uqair were gay with mosaics formed by these cones, placed in gaudy patterns of colored polka-dots. Even richer effects were obtained by affixing wafers of copper plating or colored stone to the bases of the cones.

To make them more impressive, temples were sometimes raised upon pyramidal platforms of brick. These pyramids became larger and larger until two distinct types of sacred structure evolved. The first was the temple proper: a massive, pillared hall on the ground. The other was the ziggurat,[11] a lofty pyramid of brick, with setbacks, staircases, and a shrine on top. The only ziggurat that still survives in anything like its original form stands at Ur in the South, amid what is now a desolate wilderness.

There are several theories about the purpose of ziggurats. The theory that most persuades me is that these towers were used like the Palestinian "high places" mentioned in the Bible, or the large wooden pillars on the grounds of Syrian temples. In Syria, a priest would climb to the top of such a pillar. Another priest at the foot of the pillar collected offerings from the faithful as they asked him questions.

This priest shouted each question up to the priest on the pillar, who in turn shouted it up to the gods. The pillar brought the petitioning priest nearer to heaven so that the gods could hear him more plainly. So, I suspect that ziggurats likewise furnished Mesopotamian priests with an elevated platform whence to address the powers above with the needed audibility.

The most famous ziggurat was raised at Babylon in honor of Marduk, the Babylonian Jupiter. The Bible calls this ziggurat the Tower of Babel. To the Babylonians it was Etemenanki, the Cornerstone of the Universe, originally built by the gods themselves. After several destructions and rebuildings, it reached its final form under Nebuchadrezzar II, around –600. Then it towered skyward for nearly 300 feet and was covered with enameled bricks in colorful patterns, as if it were clothed in the scaly skin of some monstrous reptile. Could the legend of the Confusion of Tongues be an echo of labor troubles during the building of Etemenanki? It is a tempting speculation, but–alas! no evidence supports it.

Although the ziggurats of Mesopotamia look a little like the pyramids of Egypt, there is no reason to think that there is any real connection between the two types of structure. The ziggurats evolved from temple platforms, whereas the Egyptian pyramids were never anything but tombs.

The seeming similarity of Mesopotamian ziggurats to Egyptian pyramids, and for that matter to the pyramids of Central America, is explained by the state of engineering in these lands when these structures were made. If you set out to build an edifice several hundred feet high when architecture is in its infancy, the arch and vault are practically unknown, and metal reinforcement is undreamed-of, you have to adopt a pyramidal form for the sake of stability.

In –XII, this teeming farmland with its web of blue canals and its looming ziggurats resounded to the tramp of the dreaded soldiery of Assyria–burly, bearded, hooknosed men in heavy boots and crested bronzen helms. For 500 years, one of the most ferociously militaristic governments known to history held the Land Between the Rivers in its merciless grip. Assyrian kings were always putting up monuments boasting:

> I destroyed them, tore down the wall, and burned the town with fire; I caught the survivors and impaled them on stakes in front of their towns . . . Pillars of skulls I erected in front of the town . . . I fed their corpses, cut into small pieces, to dogs, pigs, vultures . . . I slowly tore off his skin . . . Of some I cut off the hands and limbs; of others the noses, ears, and arms; of many soldiers I put out the eyes . . . I flayed them and covered with their skins the wall of the town . . .[12]

Other ancient kings played rough, too, but without quite such fiendish gusto. Most Egyptian and Babylonian kings preferred to boast of their justice, piety, and public works, rather than of their cruelties and atrocities.

There is, however, another side to the Assyrians. They were gifted and energetic inventors and engineers. Some of their inventions stemmed from their militarism. For, despite the harm that it does in other directions, war certainly stimulates technology.

Thus the Assyrians were the first to equip armies with weapons of iron (–VIII). Iron had been known for seven or eight centuries. According to tradition it was discovered by a tribe in Asia Minor, the Chalybes, but for several centuries it was too precious for mass armament.

The Assyrians also developed remarkable new wheeled war machines. The wheel had already been known for several thousand years; there is a picture of a chariot on a Mesopotamian vase that may have been painted before –4000. In the 1920s Sir Leonard Woolley excavated the grave of the Sumerian king, Abargi of Ur (about –2500). In the tomb Woolley found, along with the skeletons of sixty-five attendants slain to serve the king in the next world, the remains of two wagons.

Each wagon was drawn by three oxen, which had been killed with the attendants. Each wagon had four wheels, as did the war chariots shown on an inlaid panel found in the same tomb. As the horse had not yet been tamed, each chariot was drawn by a pair of onagers–that is, Asiatic asses. This animal, however, did not prove well suited to domestication. Hence the modern domestic ass is descended from the smaller, African species.

The wheels of these vehicles were solid, built up of two semicircles of wood fastened together. To protect the rim from wear, they probably had tires of leather straps, through which large-headed copper nails were driven. Much later, in Assyrian times, tires of copper or bronze came into use, and later still iron tires.

Historians of technology suppose that the wheel evolved from the roller and that the first wheels were rigidly fastened to the axle, which turned with them. This seems logical, although there is little real evidence. The revolving axle, however, had the disadvantage that both wheels, forming an integral part of the axle, had to turn at the same speed. Therefore, when the vehicle rounded a corner, one wheel was bound to skid or drag. By King Abargi's time, the axle was fixed to the vehicle and the wheels rotated loosely and independently on the axle.

In Abargi's four-wheeled wagons and chariots, both axles were rigidly fastened to the body. Neither axle was pivoted, as is the front axle of any modern wagon so that it can follow the team around corners. Therefore these vehicles had to be manhandled around turns. Moreover, as anybody who has pushed a baby carriage knows, a four-wheeled vehicle with fixed axles tends to zigzag.

To make a wagon turn easily as the draft animals draw it around a bend, the front axle must be pivoted on a king bolt. This improvement, however, took a long, long time–perhaps twenty or thirty centuries. In the days of the Persian Empire, a four-wheeled carriage called a *harmanaxa* came into use, it may possibly have had a pivoted front axle. But, even after this mechanism was known, four-wheeled vehicles long remained rare. The simpler two-wheeled carts and chariots were more commonly used throughout ancient times. A shortcoming of ancient four-wheelers was that, even after they had pivoted front axles, their front wheels were made as large as their rear ones. Therefore they could not turn sharply because, if they tried to do so, their front wheels scraped against the sides of their bodies.

The Assyrians also exploited the wheel by inventing the belfry, or helepolis, or movable siege tower. Fierce fighters though the Assyrians were, climbing a scaling ladder was always a desperate business. There was an excellent chance that the defenders would either drop a heavy stone or beam on the attacker or push the ladder over backwards.

Therefore some clever Assyrian engineer fixed the ladder to a wooden framework, too heavy to be pushed over, and put wheels under the framework. As the defenders could still pepper the attackers with arrows and other missiles, the next step was to board up the sides and front. Untanned hides were nailed to the structure to keep it from being set afire. The resulting wheeled tower or belfry remained a standard siege engine for over 2,000 years, until cannon made lofty walls useless for defense.

A further Assyrian improvement was to combine the belfry with a battering ram. A small ram was simply a log carried by a number of men. But, for breaching the walls of a real city, something larger was needed. Hence a tree trunk was shod on one end with a mass of metal and hung by chains from the roof of a shed on wheels called a tortoise.

The fully developed Assyrian belfry had six wheels and a ram working through a hole in front. Over the forward part of the tortoise rose the tower, as high as the wall to be attacked. During the approach, archers on the tower tried to clear the wall of defenders so that the attackers could scale it in safety. In several lively bas-reliefs, the defenders shower the belfry with torches, while a man in the tower hastily puts the fires out by pouring water on them from a dipper.

Sometimes, instead of a ram, the belfry was equipped with a bore. This was like the ram except that it had a sharp spearlike head. In attacking a wall of soft mud brick, such a tool might be more useful than a ram.

The Assyrians also exploited a new military arm, horse cavalry. Around –2000 or earlier, on the great grassy plain that stretches from Poland to Turkestan, a wandering tribe of cattle-raising nomads tamed the most important of man's working animals. This was the horse, which then ran wild from the forests of Germany to the deserts of Mongolia.

This feat of domesticating the horse had momentous results. The horse tamers set out in their rattling chariots and easily conquered their neighbors. They imposed their language upon their subjects and intermarried with them. These mixed peoples set out in their turn and conquered more neighboring nations. Thus the horsemen spread their speech from Portugal to Bengal. The family of tongues derived from that of the original horsemen is called Indo-European. The horsemen who conquered Iran and India called themselves *Arya,* "noble ones." Therefore, the original conquerors and their descendants are sometimes called Indo-Europeans and sometimes Aryans.

However, there is no "Aryan race." Whatever the race of the first horsemen, it has long since disappeared by intermarriage and dilution. The words for certain plants and animals, common to widely separated Indo-European languages, suggest that the original point of dispersion was south of the Baltic Sea–that is, on the plains of Poland.

For over 2,000 years, waves of Aryan barbarians–Kassites, Hittites, Cimmerians, Scythians, Medes, Persians, Dorians, Thracians, Celts, Germans, and Sarmatians–washed over the more cultured lands to the east, south, and west of their northern homes. Sometimes they were driven back; sometimes they set up mighty empires.

Their horsemen both rode and drove horses. For fighting, they drove chariots. Although they rode to carry messages or to travel fast, they did not usually fight on horseback. For one thing, the early horses were not big enough to carry an armored man very far. For another, having no stirrups and being topheavy from the weight of their armor, the riders were in constant danger of falling off.

One Indo-European tribe, the Medes, settled in western Iran. Here, on the Nisaean Plain, they bred horses for size until the horses were as big as modern riding horses. These could bear a man even when wearing armor. For centuries the Medes, the Persians, and the other Iranians had the world's best cavalry, not only because they were skillful riders but also because they had the largest horses.

The kings of Assyria, however, soon copied the Medes. Again and again they sent marauding armies over the mountains into Media to capture some of the famous horses for their own breeding stock.

In the ancient world, the chariot was built in many sizes and shapes and was pulled by one to four horses, mules, or asses. The maximum load was limited by the fact that the animals were always harnessed abreast, never in tandem. Therefore the number that could be hitched to a given vehicle was restricted by the width of roads and gates. Thus, although four-horse chariots were sometimes used for war and racing, cars for ordinary transportation were usually drawn by one or two beasts.

The Persians and the ancient Britons fastened scythe blades to the hubs of their war chariots to mow down foot soldiers. The people of the south and east of the Mediterranean preferred a chariot high in front and open in back, while those to the north and west made chariots high in back and low in front, with a seat across the back. In later centuries, the open-rear chariot, in which the driver stood up, was preferred for fighting, hunting, and racing, whereas the open-front chariot was preferred for pleasure riding and simple transportation.

As a chariot needed smooth ground to run on, its military use was limited. Once men had the fine Median riding horse, chariots became less effective than the same number of men mounted on horses. After the time of Alexander the Great, chariots went out of use for fighting, albeit they were kept for private carriages, governmental mail carts, and racing vehicles.

In addition to developing engines of war and cavalry, the Assyrian kings found time for peaceful public works. When Sargon II invaded Armenia in −714, he saw an irrigation system not yet known in Mesopotamia. This system, also used in Iran, can be called either by the Arabic name *qanât* or the Persian *kariz*.

A qanât is a sloping tunnel that brings water from an underground source in a range of hills down to a dry plain at the foot of these hills. It has an advantage over an open-air aqueduct, in that less water is lost by evaporation on its way from the hill to the plain.

To build a qanât, a line of vertical shafts is dug along the course of the conduit, and the bottoms of these shafts are joined by a continuous tunnel. At various points, other shafts are dug at a slant from the surface to the tunnel, to allow men to go down to maintain the tunnel or to draw water. Finally, when the tunnel reaches its destination, the water is distributed into a system of irrigation channels.

Sargon admired the Armenian qanawât. Although he destroyed them, he brought the secret back to Assyria. In later centuries, qanât irrigation spread over the Near East as far as North Africa and is still used in many places.

Sargon's son Sennacherib[13] proved less aggressive than most Assyrian kings, although he suppressed revolts just as fiercely. The most remarkable thing about Sennacherib was his technical bent.

Sennacherib's empire included not only Assyria proper (northern Iraq) but also Babylonia (central Iraq) and Chaldea (southern Iraq). It included Syria and Phoenicia as well. Palestine was usually tributary, and Assyria for a time ruled Egypt.

The Assyrian kings ruled most of this area through native kings whom they had conquered but left in power on condition that they pay tribute and furnish troops to the King of Kings. The Assyrian kings tried to govern Babylonia through such puppet kings, but the puppets often revolted. In the –690s, Sennacherib deposed one restless Babylonian king and put one of his own sons, Ashur-nadin-shum, in his place. The leaders of the anti-Assyrian faction fled south into the marshes of Chaldea,[14] where they thought they would surely be safe.

However, they reckoned without Sennacherib's engineering genius. He called upon the tributary Phoenician cities, far to the west, for ships. The ships, built in sections, were hauled overland and assembled on the Euphrates. They sailed down the mighty river to Chaldea, where they helped to rout the anti-Assyrians.

But not for long. The Babylonians soon rose again and drove out Ashur-nadin-shum. When the wrathful Sennacherib recaptured Babylon in –689, he massacred the people and sacked and burned the world's greatest city. Not yet satisfied, he dug canals through the city, dammed the Euphrates, and sent the river coursing through these canals. He dumped the temples into the canals and made such a vast muddy morass of the whole area that, for a time, Babylon became a wilderness.

Later, Sennacherib had second thoughts and began to restore the city. His son Esarhaddon continued the work. After the fall of Assyria, Babylon again became the largest city in the world.

The old capital of Assyria had been Ashur on the Tigris, but some early Assyrian kings made Nineveh,[15] farther up the Tigris, an alternate capital. Although Sargon II built a new capital still farther north,[16] Sennacherib abandoned his father's city and adopted Nineveh for his home.

Here a huge wall, faced with stone and pierced by fifteen gates, surrounded the two mounds on which the temples and palaces stood. The rest of the area, an irregular quadrangle about one mile from east to west and three miles from north to south, was occupied not only by the dwellings of the common folk but also by public parks and the orchards of private citizens. For his private garden, where storks clattered and pet lions prowled, Sennacherib designed an improved swape, with a copper bucket and posts of timber instead of dried mud.

To water all his fine plantings, Sennacherib undertook a vast scheme of waterworks. He personally toured the countryside near Nineveh, striding over plains and toiling up mountains, to choose the sites for his constructions.

Ten miles north of Nineveh he dammed the river Tebitu,[17] which descended from the north to flow through the middle of Nineveh and empty into the Tigris. From the reservoir thus created he brought a canal down to the city. Since the grade of the canal was less than that of the river, the canal water arrived at Nineveh high enough to be used for irrigation without hoisting.

To take care of the overflow during the high-water season in spring, Sennacherib installed, northeast of the city, a municipal canebrake, like those of Babylonian cities. He made this marsh into a game preserve, loosing deer, wild boar, and game birds to breed there.

In addition to planting reeds and timber trees, Sennacherib imported from India another novelty: the cotton tree. His inscription proudly announced that: "The mulberry and the cypress, the product of the orchards, and the reeds of the brakes which were in the swamp I cut down and used as desired, in the building of my royal palaces. The wool-bearing trees they sheared and wove the wool into garments."[18]

Sennacherib's first canal sufficed for several years. When the city outgrew it, the king dug another canal to the northwest, where it tapped another stream. He also built over a dozen small canals connected with the Tebitu River and a 12-mile tunnel to bring water to Arbela,[19] a city east of Nineveh.

When these schemes, too, were outgrown, Sennacherib undertook his most ambitious project. He went more than thirty miles from Nineveh, to the watershed of the Atrush or Gomel River. Thence a canal was dug overland to the headwaters of the Tebitu, carrying new water downstream to Nineveh.

Where the canal crossed a tributary of the Atrush-Gomel, near modern Jerwan, Sennacherib built an aqueduct. This was a remarkable piece of construction for its time. It was made of cubes of stone, one cubit (20 inches) on a side. In the actual channel, a layer of concrete or mortar under the uppermost course of stone prevented leakage. The aqueduct crossed the stream on a 90-foot bridge of five pointed corbelled arches, over 30 feet high.

Sennacherib prided himself on completing this canal and aqueduct in a year and a quarter. As work neared completion, he sent two priests to the upper end of the canal to perform the proper religious rites at the opening.

Before the ceremony, however, a minor mishap occurred. The sluice gate at the upper end of the canal gave way, and the water of the Atrush-Gomel poured down the channel without awaiting the king's command.

Sennacherib at once looked into the occult meaning of this event and decided that it was a good omen. The very gods, he believed, were so impatient to see the canal in use that they had caused the breach in the sluice gate.

So King Sennacherib went to the head of the canal, inspected the damage, gave orders for its repair, and sacrificed oxen and sheep to the gods. "Those men," he wrote, "who had dug that canal I clothed with linen and brightly colored woolen garments. Golden rings, daggers of gold I put upon them."[20]

This was no doubt a delightful surprise to the engineers and workmen, who had probably been shaking in their sandals ever since the mishap for fear that the Great King would order all their heads chopped off. He was, indeed, quite capable of it.

In –681, Sennacherib's sons conspired against him. The great engineer-king was lured into a temple in Nineveh and beaten to death with statuettes of the gods. One son, the grimly able Esarhaddon, succeeded Sennacherib and conquered Egypt. He it was who modestly began an inscription:

"I am powerful, I am omnipotent, I am a hero, I am gigantic, I am colossal!"[21]

After Esarhaddon came the melancholy Ashurbanipal, patron of arts and letters, under whom the empire reached its peak. Thereafter it swiftly declined as more and more of the subject peoples revolted and nomadic barbarians, the Cimmerians and Scythians, swept down from the north.

In –612, an alliance of Scythians, Medes, and Babylonians conquered Assyria. In assailing Nineveh, the Medes took advantage of a flood on the Tigris to mount battering rams on rafts. The allies captured the capital and blotted it out. A ferocious yell of triumph went up from the Assyrians' victims:

"Woe to the bloody city! . . . Nineveh is laid waste: who will bemoan her? . . . Thy shepherds slumber, O king of Assyria: thy nobles shall dwell in the dust: thy people is scattered upon the mountains, and no man gathereth them."[22]

Two centuries later, when Xenophon's ten thousand Greek mercenaries passed that way, Nineveh lay in ruins, and the memory of the scourge of Assyria was already fading from the minds of men.

At the time of Assyria's fall, a Chaldean adventurer named Nabopolassar[23] had seized the rule of Babylonia. Under him and his son, the second Nebuchadrezzar,[24] the dynasty ruled much the same area as the former Assyrian Empire. This new empire is sometimes called the Neo-Babylonian Empire, because its capital was at Babylon, and sometimes the Chaldean Empire, because of the nationality of its ruling house.

As Nabopolassar was a usurper, he and Nebuchadrezzar were careful to keep on the good side of the powerful priesthood, who might otherwise have stirred up revolts against them. Hence they lavished vast sums on temples, in Babylon and elsewhere. When Nabopolassar began one of the many rebuildings of Etemenanki, the great ziggurat of Babylon, he made his sons haul bricks like common workmen to show their piety.

Under the energetic rule of Nabopolassar and Nebuchadrezzar, Babylon reached its greatest size and splendor. Nebuchadrezzar came to the throne in –605 and reigned forty-three years. He was the outstanding soldier, statesman, and builder of his time. The indign tales about him in the book of Daniel–how he demanded of his wise men the interpretation of a dream that he himself had forgotten, and how he went mad and ate grass–should not be taken seriously. They are the Judaean scribes' way of getting even with the man who thrice sacked Jerusalem and deported thousands of Jews to Babylonia.

In Nebuchadrezzar's Babylon, the inner city was a slightly distorted rectangle, about a mile and a half from east to west and four-fifths of a mile from north to south. The Euphrates flowed diagonally through the city from north to south, dividing it into the smaller New City on the west and the larger Old City on the east. Moats carried the divided waters of the Euphrates around the inner city, making two triangular islands of it.

Both parts of the city were cut up by a number of broad avenues, crossing almost at right angles and named for the gods–Marduk Street, Shamash Street, and so on. The most important street was the so-called Processional Way, running north and south through the Old City four or five hundred yards from the river. Its true name was "The Street on Which May No Enemy Ever Tread."

Processional Way embodied a marvelous Mesopotamian invention, paving. It was paved with limestone flags and, under these, large flat bricks set in a mortar of lime, sand, and asphalt. Along this street, wagons bearing the images of the gods were wheeled during religious processions. These parades seem to have been the Babylonians' chief public amusement, although they also attended boxing matches.

Such paved processional ways were a regular feature of Near Eastern cities. The Hittite capital of Hattusas[25] had one as early as –1200. At Ashur, the processional way had a pair of grooves in the pavement for the wheels of the sacred wagons, to assure the gods a smooth, safe ride. This was perhaps the world's first railroad. After all, it would not do to have a god's wagon get stuck or his statue be jostled. He would not like it, any more than a modern mortal likes a flat tire; and there is no telling what an angry god might not do.

Because these streets were sacred, special rules governed their use. Sennacherib, the Assyrian engineer-king, placed posts along the processional way in Nineveh, inscribed: ROYAL ROAD. Let NO MAN LESSEN IT. Sennacherib furthermore decreed that any violator should be slain and his body impaled on a stake before his house. Sennacherib's posts sound like the first no-parking signs; but their purpose was probably to stop owners of adjacent lots from extending their houses into the right of way.

For centuries, paving seems to have been confined to processional ways. Little by little it was applied to other important thoroughfares and then to heavily traveled stretches of road outside the main cities. But not until Roman times did any government undertake to pave intercity roads over their entire lengths.

The roads of the earlier empires were for the most part not paved, merely graded. These dirt roads had an advantage over paved roads: they were easier on the hooves of horses than paved roads would have been. When the condition of such roads became bad enough, the officials rounded up corvées of grumbling peasants and compelled them to fill in the holes.

The minor streets of Babylon were probably a tangle of alleys, in which mud-brick houses were jumbled every which way. The Chaldean kings may have entertained advanced ideas of city planning, but they met the same difficulty that has plagued municipal authorities ever since: that it is easy to proclaim a fine city plan but much harder to make people obey this plan when it does not suit them.

An enormous double brick wall surrounded the inner city. The outer wall was 20 to 25 feet thick and probably two or three times as tall as it was thick. The inner wall was thinner but loftier.

These walls were of a type that builders of cities and castles standardized back before –1000 and continued to build down to the coming of cannon, nearly three millennia later. The shape of this standard wall was determined by the availability of stone and brick on the one hand and the size and shape of the wall's human defenders on the other.

The main wall was a solid structure, so thick that attackers could not easily break it down and defenders could move about freely on its top. Along the outer rim of the main wall ran a much smaller wall called a parapet. This was a mere foot or two thick and six feet high. Its upper edge formed a square zigzag pattern called a battlement or crenelation. The crenels, or notches in the battlement, were low and wide enough so that an archer could shoot through them at foes on the ground below; while the merlons, or toothlike projections, were large enough so that the archer could duck behind them when not actually loosing his shaft.

Along the east bank of the Euphrates, several square miles of suburbs formed a huge right triangle surrounding the inner city. A pair of mighty crenelated walls protected this area also. Additional walls lined the river banks and the fortified areas in the old city, which contained the palaces and barracks. Moreover, seventy-five miles northwest of Babylon, the great Median Wall, fifty miles long, stretched from the Euphrates to the Tigris.

Towers rose at intervals from all these walls. Greek historians who wrote of Babylon's walls as a Wonder of the World much exaggerated their height and length. Nevertheless, these fortifications made the city practically impregnable when held by a determined garrison.

Eight fortified gates pierced the inner wall. The center gate on the northern side was the famous Ishtar Gate, the grandest structure of its kind ever built.

This colossal portal comprised a square tower of brick, about 70 feet high and even larger in plan. Cutting through this tower was a lofty vaulted passage, which could be closed off by two pairs of huge wooden doors. On the northward side of the gate proper, flanking the approach, stood two tall, slender towers. North of these, as a first line of defense, rose two smaller towers.

The entire structure was finished with enameled bricks, blue on the towers and green and pink on the connecting walls. On each tower in low brick relief were several vertical rows of brightly colored bulls and dragons, repelling hostile supernatural forces by their frowning glare.

The traveler who strolled into Babylon through the Ishtar Gate found himself on the paved Processional Way between two high brick walls decorated with life-sized lions in bright enameled brick relief. Red-maned yellow lions alternated with yellow-maned white lions. To the right of the stroller rose the walls of a fortified area, extending from the Processional Way to the Euphrates. Here stood palaces, barracks, and administrative offices.

As the traveler emerged from the Ishtar Gate, he might catch a glimpse, over the wall on his right, of greenery. For here were the famous Hanging Gardens, another Wonder of the World. The Hanging Gardens were a splendid roof garden, planted with trees and shrubs atop a princely pleasure house. The roof that upheld the garden was waterproofed by layers of asphalt and sheet lead. The word "hanging," which classical writers used,[27] gives the misleading impression that the gardens were suspended by chains or cables. The words "raised" or "elevated" would give a more correct picture.

Greek and Roman historians attributed the Hanging Gardens to Semiramis, a legendary Assyrian queen supposed to have conquered the whole Near and Middle East and invaded Kush and India. The slender basis for these legends is the fact that for a few years the Assyrian dowager queen Sammuramat (–IX) acted as regent for her son. While Sammuramat may have been an able and forceful ruler, she certainly never did any of the deeds credited to Semiramis.

Diodoros the Sicilian even claimed that Semiramis had built a pedestrian tunnel under the Euphrates at Babylon. It is a pity that archeologists have found no trace of any such tunnel. Think how useful it would be to historical novelists! But, even if Semiramis could have built such a tunnel, she could not, before the invention of pumps, have kept it from filling up with water.

One tale, which may have a foundation of fact, is that Nebuchadrezzar's favorite wife Amytis pined in the Euphratean flatlands for the mountains of her native Media. So the king built the Hanging Gardens as a kind of artificial mountain to please her.

Of Nebuchadrezzar's pleasure dome, only the basement survives. Hence nobody really knows how the Hanging Gardens looked when new.

When Koldewey dug up the basement in the early 1900s, he found the remains of fourteen large stone barrel vaults–true arched vaults. The purpose of these vaults is not known. They may have been used for storage of supplies against a siege, or as summer reception rooms or banquet halls. Being partly underground, they would be cooler than most places in the baking Babylonian summer. Such lavish use of imported stone shows that the building was carefully designed to withstand the effects of time and weather.

In the basement, Koldewey found a peculiar structure:

> . . . a well which differs from all other wells known either in Babylon or elsewhere in the ancient world. It has three shafts placed close to each other, a square one in the centre and oblong ones on each side, an arrangement for which I can see no other explanation than that a mechanical hydraulic machine stood here, which worked on the same principle as our chain pump, where buckets attached to a chain work on a wheel placed over a well. A whim or capstan works the wheel in an endless rotation. This contrivance, which is used to-day in this neighbourhood, and is called a *dolab* (water bucket), would provide a continuous flow of water.[28]

If Koldewey is right, the chain bearing the buckets extended up through the building to the roof garden. Servants, heaving on the windlass bars, kept water flowing up night and day from the well to the king's precious plants.

West of the Hanging Gardens, filling the space between this edifice and the river, stood a building with brick walls having the extraordinary thickness of 70 feet. This was evidently a fortress or keep for a last-ditch stand.

If our traveler continued south on the Processional Way, he would pass on his right the towering, rainbow-hued ziggurat Etemenanki, rising as high as a modern thirty-story building. On his left stood apartment houses up to four stories high.

Turning right on Adad Street and walking towards the river, the stroller would find himself between two great sacred inclosures: Etemenanki on his right and Esagila, the temple of Marduk, on his left. As enlarged and rebuilt by the Chaldean kings, the temple was an L-shaped building that occupied a square about 500 feet on a side.

This temple housed an 18-foot golden statue of Marduk. Although Herodotos says the statue was of solid gold, we may doubt it. Such statues were designed to be carried through the city in carriages during religious festivals, and a solid gold statue of that size would have weighed over fifty tons. Besides being immovable, it would have cost the equivalent of about a hundred million dollars. It is not likely that so much gold was available in ancient times for such purposes. Probably the statue was of wood, covered with gold foil.

From many representations in Babylonian art, we can form an idea of this statue. The bearded, benign-looking king of the gods wore a long robe and a tall crown topped by a circle of feathers, somewhat like those of a Sioux war bonnet. In his left hand he held a scepter and a ring, while his right grasped a scimitar.

At Marduk's feet lay the animal that symbolized him, as the bull symbolized Adad and the lion Ishtar, the Babylonian Venus. Marduk's beast was the dragon or *sirrush*. This was a composite animal, covered with scales, with a snakelike head and neck; the body, forelegs, and tail of a cheetah; and the hindlegs of a colossal bird of prey.

The two great temenoi of Marduk contained many other buildings besides the temple and the tower. There were living quarters for the priests and for pilgrims from out of town, paddocks for the animals used in sacrifices, and so on.

Passing between these two inclosures, the traveler reached another wonder of Chaldean Babylon: the bridge over the Euphrates. Except for Sennacherib's aqueduct at Jerwan, this is the oldest stone bridge of which there is any record. But, whereas Sennacherib merely crossed a brook, Nabopolassar, who built this bridge, spanned one of the world's greatest rivers. Built at a time when the world's few bridges were flimsy affairs of tree trunks, reeds, or inflated goatskins, this bridge was almost as celebrated in the ancient world as the Hanging Gardens. For centuries it was the only structure of its kind in the world.

Since the river has long since shifted its channel, the bases of the piers have been dug up in recent times. The bridge was 380 feet long and rested on seven streamlined piers of baked brick, stone, and timber, 28 by 65 feet in plan. The superstructure was of timber. According to Herodotos it was a drawbridge; one or more of the wooden platforms, which crossed from pier to pier, were taken up at night to keep evil-doers from using the bridge.[29]

The Euphrates bridge had a shortcoming shared by most large ancient bridges. The massive piers took up half the width of the river at that point, seriously impeding the flow of water. At flood time there was danger that the swift current, speeded up still more by this constriction, might scour away the river bottom around the piers until they collapsed, or that débris might gather on the upstream side. Nevertheless, this bridge may be counted the most signal engineering advance of the Chaldean Empire.

After Nebuchadrezzar, the Chaldean Dynasty petered out in faction and incompetence. In –539, Cyrus[30] the Persian wrested Babylonia from the feeble grasp of the aged Nabuna'id,[31] a scholarly king with intense interests in religion and archeology but little in government. Babylon gave up with only token resistance.

Yet the city throve for several centuries more. In fact, under the Persians, Babylonian savants accomplished some of their greatest achievements in astronomy and mathematics.

In –482, Babylon revolted against King Xerxes.[32] The Persian sent his best general, who swiftly retook the city and harshly punished it. Nebuchadrezzar's great walls were partly torn down to make them useless for defense, and Xerxes carried off the eighteen-foot statue of Marduk and melted down the gold for his treasury.

Xerxes is also said to have demolished the temples. Later historians wrote of the temple Esagila and the ziggurat Etemenanki as ruins and blamed Xerxes. Alexander, they said, when he entered Babylon a century and a half later, "commanded the Babylonians to rebuild all the temples which Xerxes had destroyed."[33] He ordered the mountainous piles of brick that marked the ruin hauled away to clear the ground. This was done; but then Alexander died and the work stopped. Since these historians confused the temple and the ziggurat, it is hard to know just what Xerxes is supposed to have done.

Herodotos has another tale to tell. When he visited Babylon thirty-odd years after the revolt against Xerxes, he found both Esagila and Etemenanki standing. A later historian, Diodoros, says merely that the Persians robbed the temples of their golden statues, altars, vessels, and other accessories, and that the ruinous condition of the sacred buildings was due to the action of time.

The best explanation seems to be that Xerxes demolished no temples or ziggurats–to tear down Etemenanki would have been an exceedingly costly project–but that he merely seized all the gold in sight. Later, when grinding Persian taxation caused Babylon to decay, or later still when the city was largely abandoned under the Macedonian Seleucid kings, the priests could no longer afford to keep their buildings in repair. Thereupon they crumbled and dissolved into mud, as do all such structures when left to the elements.

But, to Greek writers, Xerxes–in real life no worse than most despots and better than some–was one of the world's greatest villains because he had attacked Greece. Hence the writers were eager to blame him not only for the misdeeds that he committed but also for many that he did not commit.

Cyrus the Great founded the Persian Empire around –550, and for a little over two centuries the Achaemenid dynasty (named for Achaemenes,[34] an ancestor of Cyrus) ruled the Near and Middle East. This empire followed the usual course of such realms. Periods of peace and prosperity alternated with revolts of the provinces and civil wars between rival claimants to the throne.

The Persian Empire was neither so bad as Greek historians were wont to consider it nor so good as some of its modern apologists say. It was, like all empires, founded on *force majeure*–but it was less oppressive than its predecessors. It was also a despotism–but no other form of large-scale government was then possible. The kings could be fiendishly cruel to their defeated foes–but there were few wholesale massacres and deportations such as the Assyrians had practiced.

Under the Persians, ideas spread hither and thither around the Near and Middle East. The qanât system of irrigation, for instance, spread far and wide from its Iranian homeland. The domestic camel spread from Arabia to Egypt, where it made possible travel and trade across the waterless North African deserts.

The Arabs in their turn learned about irrigation. Under the Sabaean kings they built a huge dam at Ma'rib,[35] near the southwest corner of the Arabian peninsula. This dam, said to have been started by a legendary shaykh,[36] furnished irrigation to its valley for about a thousand years, before it broke down during a great flood.

Although the lordly Persians were not on the whole a technically minded people, they did foster some engineering advances. The Great Kings improved the roads of their predecessors until their whole vast realm was knit by a system of highways. Over these roads the king's messengers galloped in relays, covering as much as a hundred miles a day. These postmen rode "on horseback, and . . . on mules, camels, and young dromedaries," and perhaps drove mail carts as well, depending on the condition of the road.[37]

By far the greater part of the highway system was only graded, not paved. Some of the steepest slopes, however, may have been eased by cuts and fills.

We do hear of a couple of Persian engineers, or at least of Persians placed in charge of an engineering project. When Xerxes led his ill-fated expedition into Greece in –480, he decided to dig a canal through the neck of the Athos peninsula instead of sending his fleet around Mount Athos. A Persian fleet had been wrecked there by a sudden storm eleven years before during Darius' conquest of Thrace; and Xerxes was a cautious, methodical man who had been talked into this rash foray against his better judgment.

Therefore, the king placed in charge of the work two of his nobles,

Bubares and Artachaeës.[38] Both men were related to the royal family; Artachaeës, moreover, was eight feet tall and had the loudest voice in the army. Herodotos explains how the job was done:

> Now the manner in which they dug was the following: a line was drawn across by the city of Sanê; and along this the various nations parceled out among themselves the work to be done. When the trench grew deep, the workmen at the bottom continued to dig, while others handled the earth, as it was dug out, to laborers placed still higher up on ladders, and these taking it, passed it still further, till it came at last to those at the top, who carried it off and emptied it away. All the other nations, therefore, except the Phoenicians, had double labor; for the sides of the trench fell in continually, as could not but happen, since they made the width no greater at the top than it was required to be at the bottom. But the Phoenicians showed in this the skill which they exhibit in all their undertakings. For in the proportion of the work which was allotted to them they began by making the trench at the top twice as wide as the prescribed measure, and then as they dug downwards approached the sides nearer and nearer together, so that when they reached the bottom their part of the work was of the same width as the rest.[39]

Evidently the Phoenicians knew about the angle of repose of the earth of an embankment. But nobody else, including the two princely superintendents, did.

The Persian kings did not rule from any one single capital. Instead, they maintained four capitals and moved about from one to another in the course of a year. The kings and their court probably traveled about, as did medieval European kings, because the transportation of food about the empire was not yet well organized. If the king, his household troops, and his horde of nobles, officials, servants, women, and hangers-on stayed too long in any one place, they would sweep the countryside bare of edibles and cause a local famine.

Three of the four capitals were maintained in the great cities of Babylon, Susa, and Hagmatana.[40] The kings also built a fourth capital, used mainly for ceremonies, at Parsa in the Persian hills of southwestern Iran. The Greeks called this place Persepolis, "Persian City." While each capital had palaces and audience halls, those at Persepolis were the most splendid.

Here Darius, his son Xerxes, and his grandson Artaxerxes[41] labored for decades to make Persepolis a magnificent royal center. The buildings –palaces, audience halls, barracks, treasury, and monumental gateways and staircases–stood on a platform of scarped natural rock and limestone blocks, about 300 by 500 yards in area, which towered 40 feet above the plain and was in turn overshadowed by the Mountain of Mercy behind it.

Of the many buildings on the terrace, the two audience halls were outstanding. Both were square in plan, about 220 feet on a side, with gleaming walls of mud brick covered with gold leaf. The tiles of the roofs, plated with gold and silver, flashed dazzlingly in the clear Iranian air.

Thirty-six columns, 7 feet thick and 65 feet tall, upheld the roof of the older audience hall. These slender columns were of more delicate form than was usual at the time; they made the columns of the Hypostyle Hall of Rameses II look squat and graceless by comparison. The capitals of the Persian columns took a form peculiar to Achaemenid art. Each consisted of a pair of the forequarters of animals–bulls, lions, or composite monsters–kneeling back to back.

The later building, the Hall of a Hundred Columns, was similar, albeit the columns were smaller and more numerous. These kings put up similar buildings elsewhere. At Hagmatana, "not a single plank was left uncovered; beams and fretwork in the ceiling, and columns in the arcade and peristyle, were overlaid with plates of silver and gold, while all the tiles were of silver."[42]

In –331, Alexander the Great burned the buildings at royal Persepolis. Historians disagree as to whether he did this in a drunken rage, as a piece of adolescent vandalism, or as part of a deliberate policy to prove to the world the end of the Persian Empire. Some think it was an accident, but we shall never really know.

Strangely enough, this destruction helped to save some of Persepolis for us, even though Persepolis is a ruin today. The mud-brick walls have dissolved away, and earthquakes have shaken down most of the stonework. Only a few of Xerxes' graceful columns still rise against the bright blue Persian sky.

Nevertheless, much more is left of these palaces than of the royal buildings in the other three Persian capitals. After Alexander's destruction, Persepolis was deserted, since it was at best a small town in a sparsely settled, mountainous land. So the ruins remained much as Alexander left them. The other capitals, however, continued as great cities; and over the centuries the local people carried off the stones of the palaces for their own use, until today almost nothing is left of the royal edifices.

Since the Persians had never produced a class of architects, their kings called upon the subject peoples. Consequently their palaces were built in a mixed style; Scythian gryphons shared the décor with Babylonian winged bulls. It is likely that Darius and Xerxes, with their intense interest in building, kept a firm hand on the over-all designs.

An inscription of Darius at Susa tells how, in building his palace, he used cedar from Mount Lebanon, teakwood from India, stone from Elam, gold from Lydia and Bactria, and turquoise from Chorasmia.[43] It also tells how he used Greek and Lydian masons, Babylonian brick makers, and Median and Egyptian goldsmiths. Each of these kings in turn maintained a studio for the Greek sculptor Telephanes, whose contemporaries ranked him among the leading artists of the time.

The long rows of reliefs on the ornamental stairways at Persepolis furnish priceless information about the costumes of the peoples of the empire. A relief of Darius giving audience, with Crown Prince Xerxes standing behind him, offers an unexpected dividend in technical knowledge. The throne whereon Darius sits is supposed to be of solid gold. But the obvious turnings of the legs and rungs of this chair show two things: first, that the lathe had been invented; and second, that the throne was really of wood with a plating of gold. Nobody ever turned a solid gold chair leg on a lathe!

Another Near Eastern people, who flourished at the time of the Assyrian, Chaldean, and Persian empires, were very active as engineers and technicians. These were the Phoenicians, who dwelt in a chain of city-states–petty kingdoms and republics–along the Lebanese coast at the eastern end of the Mediterranean. They spoke a language very similar to Hebrew, and the present-day speech of Malta is derived from this Punic tongue.

Herodotos knew what he was talking about when he spoke of "the skill which the Phoenicians exhibit in all their undertakings." They were quick to learn from others and ready to transmit their knowledge. Their engineers adopted the Assyrian methods of siege warfare. Their cities were famed for the stoutness of their fortifications. To avoid a shortage of drinking water during sieges, the island city of Arvad[44] took advantage of a fresh-water spring that issued from the sea bottom near the island. The Arvadites set an inverted funnel over the spring and pumped the water to shore through a leather hose.

Phoenician shipwrights also advanced the art of shipbuilding. Egyptian ships of the early dynasties were hardly more than large canoes. When the peoples of the eastern Mediterranean began to sail the sea, they learned that rowing, with the rowers facing aft, was more efficient than paddling.

Ships split into two types: the war galley, long and narrow with many oars and a small sail; and the merchantman, short and tubby with few oars and a large sail. All ships had one single rectangular "square" sail.

The warship needed many oars to dart about in all directions during a battle. The merchantman, on the other hand, needed the space the rowers would otherwise occupy for its cargo and could not afford a large crew of rowers. Most ancient rowers, *Ben-Hur* to the contrary notwithstanding, were free workers, and fairly well-paid ones at that. The use of slaves and prisoners as rowers did not become common until the Renaissance.

There were also ships of an intermediate type, with more oars than a regular merchantman but fewer than a war galley. Sometimes called *myoparones* or "mussel-boats," they were used as naval auxiliaries, as pirate craft, and as merchantmen in pirate-infested waters. An example mentioned by Demosthenes was a twenty-oared trading ship, which plied the Black Sea.

Galleys were used for commerce only under exceptional circumstances. For example, in –VI the Carthaginians claimed a monopoly of all trade in the western Mediterranean and fed interlopers to the fish when they caught them. At this time, certain Greek traders used fifty-oared galleys to run the Carthaginian blockade. They could make a profit in spite of their horde of hungry rowers because the cargo–silver from the mines of Spanish Tartessos–was both compact and precious. Also, such a ship had a fair chance to escape when a Carthaginian galley came crawling like a colossal centipede over the horizon in pursuit.

Carthage had started as a Phoenician colony, founded (according to doubtful traditions) in –814. The Phoenicians had already sailed and rowed their little cockleshell craft the length and breadth of the Mediterranean and set up several other colonies.

During the century following the founding of Carthage, the Phoenicians developed a warship of a new and more formidable type. These may have been the ships that were taken overland in pieces and sent down the Euphrates to help King Sennacherib to put down the Babylonian rebellion. The new warship had a pointed ram at the waterline and was specially braced to withstand the shock of ramming. These ships also had a solid deck over the rowers' heads to carry fighting men.

Whereas earlier warships had but one bank of oars on each side–usually making a total of fifty oars, with one rower to each oar–the new ship had the rowers arranged in pairs on each side. They were seated in some staggered arrangement, each man of the lower bank being inboard of, aft of, or below his seat mate. The exact arrangement has been the subject of endless argument in modern times, though most authorities prefer a vertical stagger. But each rower pulled his own oar.

The new warship came to be known as a diere or bireme.[45] Because it packed more muscle power per foot of length than warships of the old single-banked type, it could go faster.

About −704 a Greek shipbuilder, Ameinokles of Corinth, is said to have carried this principle a step further, by placing his oarsmen in groups of three, also staggered, though nobody knows exactly how. Each man still pulled his own oar. The resulting triere ("three-er") or trireme, driven at speeds up to seven knots[46] by 170 or more rowers, became the standard battleship of the Mediterranean (−VI) and fought the great naval battles among the Greeks, the Persians (whose navy was mainly Phoenician), and the Carthaginians.

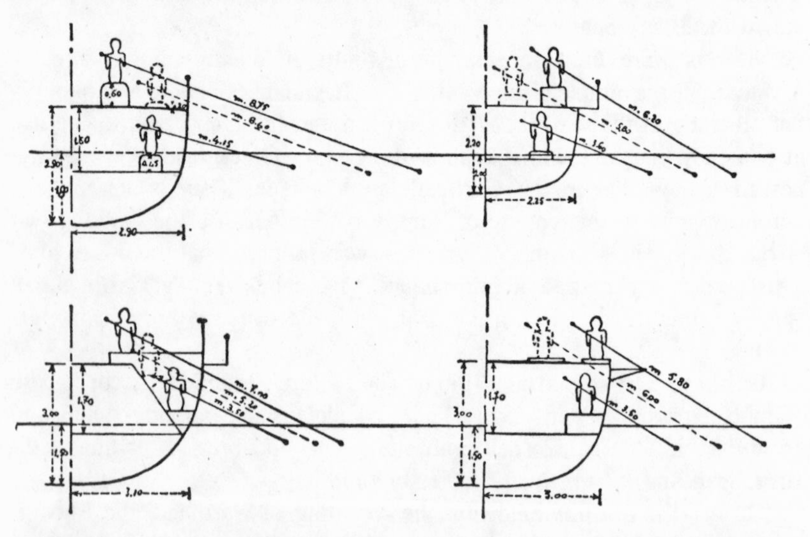

Fig. 2. Four alternative plans suggested for the arrangement of rowers in a classical trireme (*after Ucelli*). The dotted figure is seated farther aft—away from the reader—than his companions. In another possible arrangement, the lowest man is seated farther inboard, so that his oar is the same length as the others.

The Phoenicians also pioneered in large-scale manufacture. Their leading products were woolen goods colored with a deep-red dye obtained from mollusks of the family Muricidae—small sea snails with spiny shells—found along the Phoenician coast. The narrow streets of

Old Tyre reeked with the stench of the dye works. The Phoenicians also turned out metal wares, which were snapped up by eager buyers in distant ports.

From Egypt, where it had long been practiced, the Phoenicians borrowed the art of making glass. The Egyptians and Mesopotamians used glass mainly for glazing beads and tiles and for making small bottles to hold cosmetics. The Phoenicians, using the excellent sand found at the mouth of the Belos River,[47] developed commercial lines of bowls and other vessels. All this glass was cast, built up, or ground; glass blowing came later.

The Phoenicians and their neighbors, the Syrians and the Jews, were the first to make glass cheap enough so that ordinary folk could afford glass drinking vessels instead of cups of pottery or metal. In early Achaemenid times, glass was so precious that only the King of All Kings and his grandees could afford to drink from goblets of glass. But a few centuries later, thanks to the Syro-Phoenician glassmakers, nearly everybody could afford this luxury.

Although the Phoenicians were the leading explorers, seamen, manufacturers, and traders of their time, no people can be good at everything at once. For instance, Phoenician art is of poor quality. The Phoenicians never really tried to develop a distinctive style of their own. Instead, they mass-produced cheap copies of the art works of the Egyptians or the Greeks and peddled these gimcracks to the eager barbarians on the fringes of civilization, from Scythia to Portugal and from Senegal to Britain.

Their literature, likewise, does not seem to have amounted to much, so far as we can judge from the few scraps that have come down. Most of it seems to have perished in the destruction, one after the other, of the Phoenician cities. For Sidon was destroyed by the Persians in –345, Tyre by Alexander in –332, Carthage by the Romans in –146, and Beirut by Tryphon, the Seleucid king, in –140.[48]

Although the Phoenicians were not a particularly warlike people—they were businessmen, not soldiers—they defended their cities with fanatical courage and stubbornness. The Sidonians burned up themselves and their families rather than surrender to Artaxerxes. Naturally, the archives went up in smoke, too.

Other early Mediterranean peoples also accomplished notable feats of engineering before the start of written history. Even though only a few written words have come down from this age in these regions, the silent ruins bear witness to the fact that here, too, men asserted their mastery over matter with all the energy, daring, and craft of a modern engineer.

In the second millennium B.C., in Crete, a line of sea kings, ruling from Knossos, built exquisite unfortified palaces with stone walls and post-and-lintel colonnades. Pillars of tree trunks, installed upside down lest they begin to sprout during the rainy season, supported the lintels. Ceramic drain pipes carried away the water from elegant baths. The pastel murals with which the walls were decorated still show us youths and maidens performing perilous gymnastics on the horns of bulls, while long-haired Cretan men in loin cloths and ladies in off-the-bosom dresses look on.

Smaller palaces rose at Mallia on the north coast and at Phaistos on the southern. Whether these cities were parts of one realm or capitals of rival states we know not. About −1700, an earthquake shook down these palaces. Soon they rose again, more splendid than ever, and stood until the downfall of Minoan civilization about −XV. The likeliest causes for this downfall are the great eruption of the island of Thera, which blanketed the eastern Mediterranean with volcanic ash; and deforestation, which deprived the Minoans of timber for their galleys, on which their sea power depended.

Less gorgeous to look upon than the Cretan palaces but just as important from the engineering point of view was the stone-paved road that stretched away from Knossos and possibly linked all three capitals. This was the first stone-paved road in the European area and perhaps one of the first in the world.

During and after the great days of Crete, kings on the Greek mainland built palaces at Mycenae and Tiryns, fortified with thick cyclopean walls. From a distance these structures look not unlike ruined medieval castles. Up close, you see the crude strength of the massive masonry, of enormous stones with smaller stones plugging the chinks between them, and realize that here are relics of an earlier, simpler, and ruder world.

The Mycenaean kings also built remarkable beehive tombs for departed royalty. These tombs had the form of huge corbelled domes, which, when finished, were completely buried under tons of earth, making artificial hills. Buried with the dead were quantities of jewelry and masks of thin hammered gold.

During the same period, the people of Sardinia built no less than 6,500 thick-walled defensive structures, from single towers to complete castles, using the same cyclopean construction. Interior rooms and passages were corbelled, so that the rooms had a tall conical shape. The prehistoric Maltese, using similar methods, raised nearly a score of huge stone temples.

So, as far as engineering is concerned, the Golden Age of Greece was a natural outgrowth of the technical methods already worked out by various Mediterranean peoples in the days before written history. While mighty monolithic empires rose and fell in the river valleys of the East, the Mediterranean folk–lively, garrulous, enterprising, and often irreverent–grew swiftly in technical skills, until they bade fair to overtake the older cultures of the lands of morning.

THE GREEK ENGINEERS

FOUR

Around –1000, when David and Solomon reigned in Israel, the young Mediterranean civilization of Crete and Mycenae sank under the invasion of the Dorians and other barbarous newcomers. Three to four centuries later, the invaders had mingled with the natives of the rocky isles and rugged shores of the Aegean Sea to form a new folk, the Hellenes or Greeks. The Greeks dimly remembered the cultured era before the invasions in legends like those of the siege of Troy, the wanderings of Odysseus, and the reign of King Minos in Crete. No true, trustworthy history, however, survived from that elder day, because no writings that anybody could read came down.

East of the Greeks, four mighty kingdoms throve: the Lydian, Median, Chaldean, and Egyptian, all soon to be engulfed by the rising Persian Empire. Their influence, brought overland through Lydia and oversea by the Phoenicians, stimulated the Greeks to develop a civilization of their own. Authentic Greek history begins about –700. For the first century, however, this history tells us little save that this seafaring folk founded colonies on many shores of the Inner Seas, from Spain to the Crimea, and maintained a shadowy spiritual unity by athletic meets every four years.

The time from the early –500s to the late –400s is called the Golden Age of Greece, because then the Greeks made extraordinary advances in art, literature, science, philosophy, and democratic government. But the Greeks of the Golden Age did not snatch their ideas out of the thin air of Mount Olympos.

Their land is not rich. It is rugged beyond the conception of anybody who has not seen it; the landscape looks somewhat as the Grand Tetons would look if they were lowered until the sea foamed about their feet. The interior is so cut up by mountain ranges that communication with the gods must have seemed easier than communication with a town across the ridge.

The Greeks borrowed ideas from the Egyptians, the Babylonians, and the Phoenicians, much as these peoples in their time had borrowed ideas from each other. The remarkable thing about the Greeks of the Golden Age is that they made so much of their borrowings so quickly.

In Greece we see the first hint of the connection between engineering and pure science, which has become a commonplace of the modern world. Pure science in Greece, freed for the first time from priestly supervision, developed haltingly. The first surge in science took place in Miletos, on the western shores of Asia Minor. Here in the –500s flourished the astronomer and physicist Thales; his pupil the geographer Anaximandros; and the latter's pupil Anaximenes. The mathematician Pythagoras fled from nearby Samos to escape the tyranny of the local political boss. Later, other Ionian thinkers pondered and wrote on nature, geography, and history.

Hardened by incessant warfare among their tiny city-states, some of the Greeks of European Greece sent help to the Greek cities of Asia Minor when the latter revolted against Persian rule. After crushing this revolt, the Persian king Darius I looked sternly across the Aegean Sea at those he considered troublemakers. He had already, in –512, conquered Thrace and reduced Macedonia to a tributary state with the help of a Samian engineer, Mandrokles, who built a floating bridge across the Bosphorus.

So, in –490, Darius the Great sent an expedition by sea against the Athenians, whom he deemed the worst troublemakers of all. The Persians landed at Marathon, eighteen miles northeast of Athens across the tapering Attic peninsula. The Athenians marched to meet them with their hearts in their mouths. For the Persian force, though small, still outnumbered the Athenians; the Persians were practically unbeaten; and the promised help from Sparta had not yet come.

Here luck and technology came to the aid of the Athenians. The invincible Persians depended on foot archers and cavalry. The archers, shooting from behind a palisade of wicker shields, softened up the foe. Then the horsemen swarmed out and cut the enemy to pieces.

But, as a result of some logistical difficulty like that which sent Teddy

Roosevelt's Rough Riders to Cuba without their mounts, the Persians' horses had failed to reach the scene. So the world's most dashing cavalry were condemned to stumble about on foot as best they could.

Moreover, Greek bronzesmiths had already developed a new suit of armor for the Greek heavy infantryman or *hoplites*. A bronzen helm with a towering horsehair crest and projections to guard his nose, cheeks, and neck protected his head. A bronze cuirass, molded to fit his manly form, inclosed his torso. A kilt of leather straps studded with bronze buttons warded his loins, while bronzen greaves protected his legs. His shield was a circular structure of wood and leather, a yard across, with a facing of thin bronze. The Greek soldier's main offensive weapon was a short stabbing spear. He used his short broad chopping sword only when his spear was lost or broken.

The Persians had no such panoply. Later, in Xerxes' time, they began to fit heavy cavalry with shirts of iron scale mail. But it is unlikely that any of Darius' soldiers were so equipped. Most of them, probably, went into battle simply in their uniform hats, coats, and trousers, with a spear and a buckler.

In addition the Greeks, like the Assyrians before them, had developed close-order drill. Instead of rushing forward in a disorderly mob, with each captain leading his own little knot of fighters, the Greeks had learned to march in ranks and files to the tune of flutes and to dress their lines, so that a company presented a bristling, impenetrable hedge of spears and shields. They may even have marched in step.

When the Athenian line advanced, the Persian archers loosed their withering blast. The Athenians failed to wither; they plodded ahead, arrows bouncing from their bronze defenses. Then they charged. Once at close quarters, they had their unarmored foes at a grievous disadvantage.

In the center the Persians drove back the lobster-shelled Hellenes by weight of numbers and fierce fighting. But the Athenian wings closed in; the fight became a massacre; and the Persians fled to their ships, leaving over six thousand on the field.

To the Persians, this was a minor border skirmish. To the Greeks, on the other hand, it was an event of great moment, for it gave them the courage to face the much greater host that Xerxes led into Greece twelve years later.

In –480, Xerxes decided to end the squabble with these belligerent Westerners by extending his sway over European Greece. So he ordered a bridge built over the Hellespont. The bridge was built but destroyed by a storm. Xerxes had the engineers beheaded and appointed others, headed by the astronomer Harpales.[1]

Warned by their predecessors' fate, the new crew built a bigger bridge with larger safety factors. The engineers anchored 674 galleys in a double line. They connected each line by two cables of flax, weighing 50 pounds a foot, and four cables of papyrus. Long planks were laid at right angles to the cables, brush was piled on the planks, and earth on the brush. Over this bridge Xerxes' vast army–perhaps more than 150,000 soldiers, with several times that number of noncombatants–passed in safety, most of them to leave their bones in the stony soil of Hellas.[2] After his navy was crushed at Salamis, Xerxes took one of his three army corps back to Asia Minor; another was smashed by the Greeks at Plataia; and, of the remainder, many died of starvation during the retreat.

Athens emerged from the struggle as the leading state of Greece. To wage the war against Persia, which dragged on for another thirty years, the Athenians and their allies formed the Delian League. Members sent money for ships and arms to the League's treasury at Delos.

Soon, however, the Athenians moved the treasury of the League to Athens and transformed the League into an Athenian empire. Thus began the Golden Age. Perikles, one of history's greatest statesmen, rose to power in Athens. In his charm, dauntless courage, many-sidedness, high-mindedness tempered by low political cunning, and sagacity flawed by extravagance and a weakness for the grandiose, he reminds one of Franklin D. Roosevelt.

Now, Perikles was a ruthless imperialist, though that was not then considered wrong. Determined to make his beloved Athens the most beautiful city on earth, as well as the world capital of the arts and sciences, he spent the Delian treasury on a huge program of building. When the other members of the League protested, Perikles told them that so long as Athens protected them it was none of their business what was done with the money.

In –480, Xerxes' soldiers had burned the old temples on the Athenian Akropolis. During the –440s and –430s, Perikles retained the leading artists and architects of the time to cover the Akropolis with temples, shrines, and statues, the very ruins of which today provide one of the world's most splendid sights.

The Akropolis, a great ship-shaped flat-topped rock, towers over the city. Worshipers climbed the steep western slope along a zigzag path lined with statues. At the top, this path led through a monumental gateway, the Propylaia, remarkable for the use of wrought-iron bars to reinforce the marble ceiling beams. This is the first known use of metal structural members in building.

The aperture of the Propylaia framed the Parthenon,[3] or temple of Athena, a hundred yards away and a little to the right. The Parthenon, built by the architects Iktinos and Kallikrates, covers an area 101 by 228 feet. In building it, the architects deliberately used clever optical illusions. Many of the lines that one would expect to be straight are not. The columns have a slight bulge and lean slightly inward. The steps surrounding the temple bulge slightly upward in the middle. The columns at the corners are a little thicker than the others lest, having only the sky behind them, they appear thinner. The architects artfully exaggerated the natural perspective to make the temple look even larger and grander than it really is.

Similar refinements were used in later temples. Impressive as the surviving Greek temples are, however, the emotions to which they give rise are probably due less to the artistic skill of Iktinos and his colleagues than to the associations that they conjure up. To an educated modern man, a Greek temple is not just a mass of masonry, of good workmanship if somewhat primitive design. Instead it is Zeus and Aphrodite, Jason and Achilles, nymphs and centaurs, Marathon and Salamis, Homer and Pheidias, Perikles and Plato, and all the other glowing images that the concept "ancient Greece" evokes, as if it were one of Keats's

> Charm'd magic casements, opening on the foam
> Of perilous seas, in faery lands forlorn.[4]

For the Parthenon's statues, Perikles hired Pheidias, the leading sculptor of the age. Pheidias had already made a thirty-foot bronze colossus of Athena Polias,[5] which stood about a hundred feet beyond the Propylaia, to the left of the Parthenon.

For the main hall of the Parthenon, Pheidias created a huge gold-and-ivory statue of the goddess. He also executed, for the pediments at the ends of the roof, two groups of statues representing scenes from the myths of Athena. And he supplied the many matchless reliefs for the frieze around the main wall and for the metopes around the entablature (the structure just above the columns and below the roof).

Later, in Elis, Pheidias made a gold-and-ivory statue of Zeus for the temple at Olympia. This statue was listed among the ancient Wonders of the World.

There are various tales of Pheidias' fate. According to Plutarch, whose account seems the most plausible, Pheidias returned to Athens but was there attacked by Perikles' political enemies. These accused him of stealing some of the gold for the statue of Athena and of impiously carving his own face on one of the warriors in the reliefs. The sculptor was accordingly put in prison, where a sickness carried him off.

For almost a thousand years, the Parthenon shed its glory on Athens. Then, like many other temples, it was converted into a Christian church. This involved removing statues, cutting doorways through the walls, and bricking up the spaces between the columns to make new walls. The gold-and-ivory Athena disappeared, so today we can only guess at its appearance from what we know of the conventions of Greek religious art.

Later, under the Turks, the Parthenon became a mosque; but it was still in fair repair in 1687. At that time the Turks, at war with the Republic of Venice, stored gunpowder in the temple. A German gunner with a besieging Venetian army dropped a mortar shell through the roof, igniting the stored powder and blowing out the whole central part of the temple.

After the city fell, the art-loving Venetian general, Francesco Morosini, tried to remove the chariot of Athena from the West Pediment. But the workmen, unused to such a task, dropped the sculpture and smashed it.

When the Earl of Elgin was British ambassador to Turkey in 1801, he found that the Athenians of that less artistic age were feeding the remaining sculptures into limekilns. Getting permission from the Turkish government, which cared nothing about such things, Elgin removed most of the remaining sculptures and shipped them to England. When the ship transporting the marbles ran on a rock off Kythera and sank, Elgin hired divers to bring up the sculptures and finally got them to their destination. There he sold them to the British Museum for £36,000—little more than half of what he had spent in collecting them.

Thirty years later, when Greece became independent, the Greeks demanded the sculptures back, but to this day the Elgin Marbles continue to dwell in London.

In +XX the Greeks, with the help of American donations, began to put the ruins of the Parthenon back together. Reassembly and restoration have continued slowly on the Akropolis ever since. Framed in the doorway of the Propylaia, the Parthenon, even in ruins, is still a breathtaking sight.

However, the traveler who wants to see Greek temples more or less intact does better to study the temple of Hephaistos (the so-called

"Theseum") in Athens, or the temples at Paestum in southern Italy, and at Segeste and Agrigento in Sicily. But, while these temples are structurally better preserved than the Parthenon, those in Italy and Sicily lack its finish. For the western Greeks, not having the excellent marbles of Aegean quarries, coated their temples with stucco. This has now worn off, leaving a rough and pitted surface.

Of the other Greek temples, some, like those at Selinunte in Sicily, have been shaken down by earthquakes. Some, like the temple of Zeus at Olympia, were pulled down on orders of the Roman emperor Theodosius II to suppress competitors of the new state cult of Christianity (+V). And many were demolished by Byzantines, Crusaders, or Turks who wanted the stone for building and fortification.

To recapture the full beauty of the Parthenon when new, you can find faithful small models in museums, such as the Metropolitan Museum of Art in New York. There is even a life-sized replica of this temple, in concrete, in Nashville, Tennessee, the interior of which is used as an art museum. The Nashville Parthenon, alas, is defective in two regards. First, the concrete is not very good and is already crumbling. Second, whereas on the side walls of the original Parthenon no two reliefs were alike, the builders of the Nashville replica saved money at the cost of authenticity by making the friezes on the two sides duplicates.

Greek temples followed a design that spread all over the Mediterranean world, lasted for centuries, was revived in Renaissance Europe, and is still used in modified form in some modern art museums, banks, churches, and memorials.

The earliest Greek temples were small boxlike buildings of brick or stone, housing a rough-hewn image of a god. Over the temple rose a low-gabled roof of wood, except that in a few temples the center was left open to the sky.

The Greek temple was strictly a god-house; only the priests were allowed inside. To shelter worshipers from the rain, the side walls and roof extended forward to form a small porch. A pair of posts cut from tree trunks held up the roof over this porch. The Greeks never invented a true roof truss, wherein a series of beams are joined together to form a rigid structure made up of triangles. Instead, they relied upon an elaborate system of posts and lintels.

As time went on, the Greeks built more splendid temples. Marble columns took the place of wooden posts. Stone replaced brick in the walls. Stone architraves—the long horizontal members resting on top of the columns and holding up the roof—took the place of wooden lintels.

Statuary filled the pediments—the triangles formed by the gables at the ends of the building. Roof tiles were substituted for wooden shingles. The roof, now upheld by rows of columns running clear around the building, extended out in all directions.

Temples grew larger and larger. The Didymeion near Miletos covered an area 160 by 360 feet and was surrounded by a double row of 60-foot columns. Of about the same size was the Artemision of Ephesos, better known from the Bible (Acts xix) as the temple of Diana of the Ephesians.

This temple began as a small shrine (–VIII) and grew by successive rebuildings. For the rebuilding of –600, the architect Chersiphron devised an ingenious scheme for moving column drums to the site. Fearing that if he loaded them on carts, the carts would get stuck, he fitted a wooden frame around each column drum with a pivot on each end. Then he had each drum pulled by a team of oxen, rolling along on its side like a lawn roller.

A still larger version of the Artemision endured from –540 to –356, when a youth named Herostratos, craving eternal notoriety, set fire to it. The final version, with one hundred 60-foot columns, remained intact until the Goths destroyed it in +262.

In such major structures, Greek architects avoided the use of mortar. Instead, they trimmed their stones to an extremely accurate fit and bonded the marble blocks together with I-shaped iron cramps. After chiseling slots in the adjacent parts of the blocks, they inserted the cramps and poured molten lead into the space between the iron and the stone to make all secure. Ancient buildings from classical times are often pockmarked where greedy men of later ages chiseled out these cramps to sell the metal.

Like the Egyptians, the Greeks persisted in using the architectural forms that they were accustomed to from the days of wood. All the details of the entablature, with its cornices, friezes, and so forth, were copies in stone of wooden structural elements. Greek builders even imitated the pegs that held the ancestral wooden structure together by adding little stone knobs called "drops."[6]

In spite of the emotional effect that Greek temples have on the modern beholder, from the strictly engineering point of view Greek temple design remained comparatively static and unprogressive. The architects not only failed to invent the roof truss, but they did not, until late Roman times, try out the arch, long known in the East.

The earlier Greek architects developed two styles of temple, distinguished mainly by the shape of the capitals at the tops of the columns.

The Doric capital, much used on the Greek mainland, had a simple, bulging, cushion-shaped surface.[7] The supreme example of the Doric temple was the Parthenon.

The Ionic column, first used in Ionia, had a more ornate capital, with a pair of spiral ornaments on each side. These spirals, called volutes, are probably derived from some of the leafy forms that the Egyptians gave the tops of their columns. Each style of temple had elaborate rules of proportion and detail. Doric columns, for instance, were supposed to be about eight diameters tall, while Ionic columns were nine times as tall as their diameter.

At last the architect Kallimachos (–V) got bored with endless minor variations of the same two orders of Greek temple and invented a new order, the Corinthian. Vitruvius tells a charming story about this invention:

A girl, a native of Corinth, already of age to be married, was attacked by disease and died. After her funeral, the goblets which delighted her when living, were put together in a basket by her nurse, carried to the monument, and placed on the top. That they might remain longer, exposed to the weather, she covered the basket with a tile. As it happened the basket was placed upon the root of an acanthus. Meanwhile about spring time, the root of the acanthus, being pressed down in the middle by the weight, put forth leaves and shoots. The shoots grew up the sides of the basket, and, being pressed down at the angles by the force of the weight of the tile, were compelled to form the curves of volutes at the extreme parts. Then Callimachus, who for the elegance and refinement of his marble carving was nick-named *catatechnos* ["artificial"] by the Athenians, was passing the monument, perceived the basket and the young leaves growing up. Pleased with the style and novelty of the grouping, he made columns for the Corinthians on this model and fixed the proportions.[8]

In actual fact, the Ionic capital evolved into the Corinthian by gradual stages, by the addition of more and more plant elements. In time the ornate leafy splendor of the Corinthian diffused about the Greek world. Later, the Romans took it up and spread it from Spain to Lebanon; and men of the Renaissance, loving its showy intricacy, brought it down to the world of today.

Meanwhile, inside the temples, the crude wooden eidolon of the earliest temples became a beautiful statue of marble, bronze, or a combination of materials. Marble statues were painted to look lifelike; leaving a marble statue in its natural corpselike pallor is a modern idea. Starting with crude imitations of stiff Egyptian statuary, the Greeks of the Golden Age produced the greatest sculptors of all time.

Never has any people gone in for statuary with such enthusiasm. At first their sculptors made statues of gods and heroes, then statues of everybody of importance, until each small town was decorated with hundreds of statues and every city with thousands. Roman generals plundered Greece of thousands of statues, but there were still thousands left.

Only a handful of these statues, however, have come down intact. What we have, with few exceptions, are fragments and late Roman copies. Those Greek originals that have survived whole usually did so by being buried—accidentally or on purpose—or being sunk to the bottom of the sea. What Greece still has in abundance are the marble bases of statues. In the upper surface of such blocks appear a pathetic pair of footprints, where the statue's feet once rested.

Most of the statues disappeared during the Dark and Middle Ages, when Christianity taught people to despise pagan art. Rulers melted the bronzen statues for money, while peasants broke up the marble ones with sledge hammers and cooked the pieces in limekilns to make mortar.

Another characteristic Greek public structure was the theater. The Cretans of the Minoan Age seem to have devised the first places where people sat on tiers of benches to watch a public performance; in this case bull-grappling. All that we know of this perilous procedure is what we can infer from Cretan paintings. We do not know whether the Minoan bullfight was mainly a religious ceremony, a sport, or a means of getting rid of prisoners, criminals, and other unwanted persons.

Theaters in Greece evolved out of simple open spaces for the performance of religious rites. Greek drama developed from these rites. A site in the shape of a half-bowl was soon found to be the most convenient. At first, terraces dug in the hillside served as seats. Then wooden seats were provided, and finally stone seats. In democratic states, the theater also served as a place for the citizens to meet and be stirred to wise or foolish action by their orators.

The first theaters had two parts: the *theatron* or "seeing place," the sloping semicircular part where the spectators sat; and the *orchêstra* or "dancing place," the flat section in the middle where the action took place.

Later was added the "booth" or *skênê,* whence our word "scene." This was a building, at first of wood, facing the audience across the orchestra. There the actors kept their costumes and props and thence they entered the stage.

In the Golden Age, the enlarged skênê was equipped with advanced stage machinery for producing special effects. There was a hoist for lowering to the stage an actor dressed as a deity, when the play called for a god to step in at the climax and set things to rights, much as the Lone Ranger once did on television. Hence the term *deus ex machina.*

The skênê might also have wings on which were mounted triangular wooden prisms, the *periaktoi,* whereon were painted scenes to correspond to the acts of the play. The prisms were turned to show the appropriate scene. A similar scheme is used in advertising displays today.

The Greeks knew the wave nature of sound and developed a theory of acoustics for use in planning their theaters. The acoustics of the theater at Epidauros are so good that today tourist guides demonstrate them by striking a match while standing in the orchestra; the scratch is easily heard by tourists on the upper benches. Architects also placed vases of bronze or pottery about a theater, with their open ends pointing towards the orchestra, to act as resonators.[9]

In addition to all his building and his fostering of art, Perikles was probably responsible for the Long Walls, a noted fortification at Athens. These walls extended four miles from the city to the Saronic Gulf. They are thought to have been as high as those of Peiraieus, the port of Athens, which were 60 feet tall. Originally two walls were built, one from Athens to the harbor just west of the peninsula of Peiraieus, while the other inclosed the small seaport of Phaleron. Then a third wall was built between the other two, close to the Peiraic Wall, and the Phaleric Wall was allowed to decay.

These Long Walls enabled Athens to hold off the Spartans during the Peloponnesian War. The superior Athenian navy kept the Spartans from interfering with Athens' foreign trade, while the Long Walls prevented the powerful Spartan army, which lacked siege machinery, from cutting off Athens from its seaport. When the Spartans destroyed the Athenian fleet at Aigospotamoi, they made the defeated Athenians tear down the Long Walls, but the walls were soon rebuilt. Corinth built a pair of similar walls to its seaport.

Next, Perikles decided to develop the Peiraic peninsula, with its three excellent natural harbors, as the port of Athens. For this purpose he imported an architect, Hippodamos of Miletos, to lay out a new port city.

Hippodamos planned Peiraieus on the grid plan, long known in Mesopotamia and India, with straight wide streets crossing at right angles. Whether Hippodamos got his ideas by travels in the East we know not. He was an individualist who dressed eccentrically and wrote one of the first utopian tracts, setting forth his constitution for an ideal city. Among other things he proposed "to honor those who discover anything which is useful to the state."[10] This might be considered the first proposal for a patent law. In later centuries, grid plans like Hippodamos' became usual for new Greek cities.

The Periklean Age, so swiftly progressive in other respects, was a time of anti-scientific reaction in Athens. The initial surge of scientific effort across the Aegean in Ionia had spent itself. The Ionians' doctrines had at best been highly speculative; with few facts to go on and little idea of experiment, they made so many guesses about the problems of nature that they were bound to be right sometimes. But at least they sought rational answers to mundane problems instead of resorting to dreams, oracles, and other supernatural sources.

Not that Perikles himself was anti-scientific. He was a friend and patron of all Athenian intellectuals, scientists included. One of his friends was the elderly Ionian philosopher, Anaxagoras of Klazomenai, who had settled in Athens. Here Anaxagoras taught that the sun was "a mass of red-hot metal . . . larger than the Peloponnesos," and "that there were dwellings on the moon, and moreover hills and ravines."[11] Because of these teachings, Perikles' enemies brought a legal action against Anaxagoras on grounds of impiety. Perikles saved his life, but Anaxagoras had to pay a heavy fine and leave Athens.

Periklean Athens harbored plenty of philosophers, but most of these were not scientifically inclined. Even the great Sokrates was anti-scientific, priding himself on "refusing to take any interest in such matters and maintaining that the problems of natural phenomena were either too difficult for the human understanding to fathom or else were of no importance whatever to human life."[12]

Sokrates' pupil Plato,[13] although an enthusiast for mathematics, was otherwise even more anti-scientific than Sokrates. He sneered at experimental science:

> . . . that knowledge only which is of being and of the unseen can make the soul look upwards, and whether a man gapes at the heavens or blinks at the ground, seeking to learn some particular of sense, I would deny that he can learn, for nothing of that sort is science . . . in astronomy, as in geometry, we should employ problems, and let the heavens alone . . .[14]

In other words, the "true" astronomer should dredge correct knowledge of the universe out of his inner consciousness, without bothering about stars and planets.

There were, however, some real scientists in the Golden Age besides Anaxagoras, and some of them were engineers as well. One, for instance, was Archytas of Taras (Roman Tarentum, modern Taranto), a Greek city in southern Italy. Archytas was seven times elected president of his city, yet he found time to solve a number of problems in mathematics and mechanics and to make several inventions, including a child's rattle and a mechanical bird that flew by compressed air. Some historians of technology think he invented the screw.

Archytas was a friend of Plato despite the latter's bias against science, especially science of practical utility. Between −367 and −357, Plato twice visited Syracuse, then under the *tyrannos* Dionysios the Younger. (The Greek word meant "boss" or "dictator," whether or not his rule was oppressive or tyrannical in the modern sense.) Here Plato tried to put into effect his theories of making kings into philosophers and philosophers into kings by preaching to Dionysios (*if* Plato's letters that the story of the visit is based upon are genuine). At last the tyrannos lost patience, put Plato under house arrest, and would probably have had him killed had not Archytas sent a galley to rescue him.

While the Greeks took over foreign engineering ideas and improved upon them, Greek engineering was, on the whole, not so remarkable as other Greek achievements. Although the professions of architect (*architektôn*), engineer (*technitês*), and machinemaker (*mêchanopoios*) were recognized as respectable, Greek technical achievements of the Golden Age were rather modest. One reason, no doubt, was that no Greek city-state commanded the wealth or the manpower to execute grandiose public works, like those of the watershed empires.

Most Greek roads, for example, were mere tracks. Some of these tracks followed the beds of Greece's many dry rivers, streams that ran with water after winter rains but were dry the rest of the year.

Elsewhere, a road was often a pair of ruts—dug, chiseled, or simply worn—in the stony soil. Sometimes the ruts were lined with cut stone and provided with switches and sidings like those of a railway. Each rut was about 8 inches wide and 3 to 6 inches deep. The gage—the distance between the centers of the ruts—varied in different places from 4′ 6″ to 4′ 11″. Hence a Syracusan who brought his chariot to Athens found that his wheels were too far apart to fit the ruts.

The streets of most Greek cities were muddy, filthy alleys, although a few were paved with slabs of fieldstone, with a dressing of mortar. There was no attempt at drainage. The stroller harkened to the cry of *exitô!* ("Coming out!") which meant that a load of refuse was about to be thrown into the street.

The private houses of even the larger cities were small and undistinguished. They followed the Near Eastern pattern with a blank wall of mud brick toward the street and all the rooms opening on an interior courtyard.

Some public works were undertaken to improve the land. The steep, rocky slopes and short, swift, intermittent rivers of Hellas did not lend themselves to irrigation. But the Greek engineers undertook to drain marshes, not only because they needed more land for farming but also because they had a vague feeling that bad air or "harmful spirits"[15] rising from the marshes caused disease. They were not entirely wrong, except that the disease-carrying agent was not air or spirits but the malaria-bearing mosquito.

At the time of the Persian Wars (–V) the leading statesman of Akragas[16] in Sicily was the magnate and philosopher Empedokles. This brilliant if eccentric thinker devised the theory that all matter is made of four elements: earth, air, fire, and water. Empedokles also improved the health of the people of Selinous[17] by draining the local marshes.

About the same time, the architect Phaiax provided Akragas with an elaborate system of water channels and tanks to furnish the citizens with water from nearby sources. Phaiax's water mains can still be explored, albeit with some labor and risk.

In Greece proper, marshy Lake Kopaïs in Boiotia presented a difficult engineering problem. This lake had several inlets but no outlet on the surface. The water escaped through underground channels in the surrounding limestone mountains. From time to time, these channels became blocked by silt, or an earthquake cut them off by faulting. Then the lake rose and flooded the towns on the marge.

From prehistoric times, efforts were made to drain the lake by driving two artificial tunnels under it. Later, Alexander sent an engineer to clear the channels, and other engineers continued these efforts into late Roman times. Then, with the general decay of public works, the lake was allowed to fill up. In the 1890s the channels were once more opened, so that today the lake bed is solid farmland.

Another pick-and-shovel project, at which several Greek engineers had a try without success, was digging a canal across the Isthmus of Corinth, which joins the Peloponnesos to the rest of Greece. Several ancient rulers, beginning with Periandros, tyrannos of Corinth about −600, attempted this task, but all found the job too big for them. Instead, small ships were pulled across the isthmus on the *diolkos,* a kind of tramway, by means of rollers.

Herodotos told of what he considered the three greatest engineering works of the Greeks in his day. All were on the island of Samos. One was a temple, and as for the other two, built under the tyrannos Polykrates:

> One is a tunnel, under a hill 900 feet high, carried entirely through the base of the hill, with a mouth at either end. The length of the cutting is almost a mile—the height and width are each eight feet. Along the whole course there is a second cutting, thirty feet deep and three feet broad, whereby water is brought, through pipes, from an abundant source to the city. The architect of this tunnel was Eupalinos, son of Naustrophos, a Megarian. Such is the first of their great works: the second is a mole in the sea, which goes all round the harbor, nearly 120 feet deep, and in length over 400 yards.[18]

This tunnel, explored in the 1870s and 80s, turned out to be smaller than Herodotos had said. It was 3,300 feet long and about 5.5 feet in width and depth, with a trench for the clay pipe 3 to 25 feet deep. The tunnel was cut from both ends. The crews missed making a perfect join in the middle by 20 feet horizontally and 3 feet vertically. Hence the tunnel has a kink in the middle. But, to make even that close a meeting, Eupalinos must have used surveying instruments, though just what kind is not known.

The Golden Age of Greece was brought to a close by Sparta's defeat of Athens in the Peloponnesian War (−431 to −404). Sparta was a small town in a broad flat valley, which lay between towering mountain ranges in the southern Peloponnesos.

The original Spartans had conquered various neighboring peoples, made serfs of them, and kept them cowed by terrorism. To make their rule secure, the later Spartans completely militarized their society. After their victory over Athens, they ruled all Greece for a time.

We think of the Spartans as brave to excess, grimly dutiful, curt, somber, brutal, and stupid. Yet their smiths are said to have made the first locks and keys of the modern type. Earlier keys were simple curved rods of wood, bone, or metal to be thrust through a hole in the door in order to tease the bolt back.

Moreover, a recent theory holds that the Spartans owed their success in war to having invented steel, or at least to having been the first to equip all their soldiers with steel weapons (–VII), at a time when everybody else was still using weapons of soft wrought iron or even of bronze.

Now, steel is an alloy of iron and carbon. Wrought iron–soft, malleable, and hard to melt–contains very little carbon, usually less than 0.1 per cent. On the other hand, cast iron–hard, brittle, and easy to melt–contains 2.0 to 4.0 per cent of carbon. Steel, with an intermediate carbon content, more or less combines the hardness of cast iron with the workability of wrought iron, while it is stronger and springier than either.

Nearly all the iron smelted by ancient smiths would today be classed as wrought iron. Their method was to dig a pit in a hillside, line it with stone, fill it with iron ore and wood or charcoal, and ignite the fuel. When the fuel had all burned up, a porous, stony, glowing mass would be found among the ashes. This was fished out and hammered, amid a shower of sparks, to compact the iron and squeeze out the impurities. The finished lump, called a "bloom," was about the size and shape of a large sweet potato.

In time men learned to make the fire hotter by blowing on it with a bellows and to build a permanent furnace of brick instead of merely digging a hole in the ground. Steel was made either by smelting iron ore with a large excess of charcoal, or by packing a piece of wrought iron with charcoal and cooking it for days until the iron absorbed enough carbon, or by melting low-carbon wrought iron with high-carbon cast iron so that the resulting mixture should have the medium carbon content desired.

Cast iron was invented in China (–IV). In the West, cast iron, though known in classical times, never became easy to make until improved furnaces, either derived from China or invented independently, appeared in medieval Europe, and it never became really cheap and common until +XIX.

Although the Peloponnesian War devastated Greece, led to fierce civil conflicts, and brought down the Athenian Empire, scientific progress speeded up after this war. Pure science took another spurt. This advance was mainly identified with Plato's pupil Aristotle (–IV) and the school he founded at Athens. Engineering also advanced and, in –III, science and engineering flowered as never before under the Ptolemies, the Macedonian kings of Egypt.

The Greek armies of the Golden Age did not use elaborate engines of war. Hence walled cities were almost never captured save by treachery or by a long siege to starve out the defenders.

An armed man atop a wall had a great advantage over another man who tried to climb up. The man above could shoot an arrow, cast a javelin, thrust with a pike, or drop stones, hot pitch, melted lead, or red-hot sand. If none of these stopped the climber, the defender could push the ladder over; defenders kept forked poles handy for this purpose. Therefore, to attack a wall defended by even a meager number of able soldiers, the attacker needed siege engines.

Towards the end of the Golden Age we begin to hear about such engines in Greece. Perikles is said to have attacked the walls of Samos with battering rams in movable sheds, built for him by Artemon of Klazomenai (–441). In the Peloponnesian War the Boiotians, fighting on the side of Sparta, attacked the wooden stockade of Delion with a homemade flame thrower:

> They sawed in two and hollowed out a great beam, which they joined together very exactly, like a flute, and suspended a vessel by chains at the end of the beam; the iron mouth of a bellows directed downwards into the vessel was attached to the beam, of which a great part was itself overlaid with iron. This machine they brought up from a distance on carts to various points of the rampart where vine stems and wood had been most extensively used, and when it was quite near the wall they applied a large bellows to their own end of the beam, and blew through it. The blast, prevented from escaping, passed into the vessel which contained burning coals and sulphur and pitch; these made a huge flame, and set fire to the rampart, so that no one could remain upon it.[19]

Siege engines were already old at the start of the Golden Age of Greece; the Assyrians had developed the art to a high degree. There is an old theory that Homer's "Trojan horse" was nothing but a battering ram, set up at the siege of Troy by the Achaeans' oriental allies.

The Phoenicians, having learned about siege machinery from the Assyrians, spread this knowledge about the Mediterranean. Classical writers tell of Phoenician engineers who introduced battering rams suspended by chains from the roofs of wheeled sheds. This combination they called a "ram tortoise." To withstand the impact of heavy stones, the roofs of these engines were braced by timbers as thick as a foot square.

The Carthaginians soon began using such engines against the Greeks of Sicily. Both peoples had colonized this island, and each strove for centuries to evict the other. This use of engines of war stimulated the Siceliot Greeks to one of the most remarkable engineering efforts of ancient times.

In –399, Syracuse–the New York of the Greek world–was ruled by the shrewd, ruthless, nearsighted Dionysios the Elder. This tyrannos (father of the Dionysios who persecuted Plato) anticipated later dictators by wearing a steel vest. Dionysios planned to attack the Carthaginian colonies in Sicily. To make sure that he had the advantage,

> . . . he gathered skilled workmen, commandeering them from the cities under his control and attracting them by high wages from Italy and Greece as well as Carthaginian territory . . . he divided them into groups in accordance with their skills, and appointed over them the most conspicuous citizens, offering great bounties to any who created a supply of arms.[20]

Thus was the first ordnance department launched; and so also was the now notorious "military-industrial complex."

One task of Dionysios' research teams was to develop warships larger than the standard triere. The dictator's "purpose was to make weapons in great numbers and every kind of missile, and also tetreres [ships with four banks of oars] and penteres [ships with five banks], . . . being the first to think of the construction of such ships."[21]

We have no details of the progress of these projects, because ancient shipbuilding seems to have been carried on entirely by rule of thumb. There is very little solid information in all classical literature about shipbuilding and seamanship. Probably, down to early modern times, shipbuilding was in the hands of families or guilds who kept their lore secret and passed it down by word of mouth. Hence the earliest known book on shipbuilding dates from 1536.

Nevertheless, it is tempting to try to reconstruct details of these superwarships on the basis of today's knowledge. For instance, the most obvious first step would be to build ships that were simply enlarged trieres, with the oars in four or five banks instead of three. But difficulties would arise.

In a triere the oars, though worked from different levels, were (by an ingenious arrangement of the rowers) nearly all of the same length. With four or five banks, however, the oars of the uppermost banks must have been longer than the others. Such oars would have had a different natural period and would have been hard to keep in time with the rest. In fact, it probably proved difficult to row at all with such a bristling mass of oars, no matter how arranged.

Perhaps as a result, the ships of these new types spread but slowly. The records of the Athenian dockyards for the year –330 show 393 trieres of the standard type against 18 tetreres and no penteres. But five years later, the figures are 360 trieres, 50 tetreres, and 7 penteres.

What probably caused the sudden increase in the use of these new ships was that some forgotten genius suggested putting more than one man on an oar. We can imagine that various combinations of men and oars were tried, because similar combinations were actually tested by the Mediterranean powers in medieval and Renaissance times. These later mariners found that in the smaller galleys, with rowers in groups of two or three, it was more efficient to have each man pull his own oar.

In such medieval galleys, however, the rowers were staggered horizontally instead of vertically. The two or three men sat on one bench at an angle to the length of the ship, so that the inboard man was well aft of the outboard one. Thus his oar passed in front of his seat mates and their oars–as long, that is, as the rowers kept strict time.

In the larger medieval galleys, it was found most effective to range the oars in a single bank and have four or five men pull each oar. So that the rowers could handle them, such large oars or sweeps were fitted with cleats for the rowers to grasp.

Intermediate arrangements, such as four or five men on one bench, pulling two oars, did not work well. Besides being easier to manage, the use of multi-rower oars or sweeps was cheaper. On a sweep, only one man had to be a trained, well-paid professional oarsman. The rest could be labor of the cheapest kind.

At any rate, we may assume that in the −340s or −330s, some shipwright–probably a Phoenician, since Phoenicia and Phoenician Cyprus were active in marine invention at this time–began to build ships of a wholly new type. These were ships with one or two banks of oars and two to five men per oar. On each side was a frame or outrigger called an *apostis,* in which the oars were pivoted. These new ships soon took the place of Dionysios' marine hedgehogs. Thereafter names like "tetrere" ("fourer") or "quadrireme" and "pentere" ("fiver") or "quinquireme" no longer referred to the number of banks of oars but to the number of files of rowers on each side. By the end of −IV, shipwrights were building ships of every rate up to tenners and eleveners.

Another of Dionysios' teams invented the catapult. This first catapult was essentially a large crossbow mounted on a pedestal. It shot a dart like a massive arrow, up to six feet long.

Two years later, Dionysios took a battery of his new weapons to the siege of Motya, a Carthaginian colony at the western end of Sicily. When a Carthaginian commander, Himilkon, brought a fleet from

Africa to help the Motyans, he burst into the harbor and found the Syracusan ships drawn up on the beach.

> Himilkon attacked the first ships, but was held back by the multitude of missiles; for Dionysios had manned the ships with a great number of archers and slingers, and the Syracusans slew many of the enemy by using from the land the catapults which shot sharp-pointed missiles. Indeed this weapon created great dismay, because it was a new invention at the time. As a result, Himilkon was unable to achieve his design and sailed away to Libya . . .[22]

So began the story of artillery. Using his catapults to clear the walls of Motya and attacking with battering rams and movable belfries of the Assyro-Phoenician type, Dionysios forced an entrance to the city. Although the Motyans fought with bitter heroism, at last the Greeks beat down their resistance and began to massacre them. Dionysios stopped the massacre, not from feelings of pity but from motives of thrift, for he wanted the Motyans alive to sell as slaves.

It took several decades for knowledge of catapults to spread around the Mediterranean. Aineias the Tactician, who about –365 wrote *On the Defense of Fortified Positions,* barely mentions catapults. A few years later somebody took a catapult dart to Sparta as a curio. When King Archidamos III saw it, he cried:

"O Herakles! The valor of man is extinguished!"[23]

Thus Archidamos uttered the first recorded protest against the mechanization of war. Such protests have gone on unheeded ever since. By –350 the arsenal of the city of Athens contained a pair of dart throwers, and in the following decade King Philip of Macedon made liberal use of them.

The early catapults had a sophisticated mechanism for cocking and releasing. On the pedestal was mounted a slanting wooden beam with a large groove in its upper surface. This was called the *syrinx* or trough. The bow was fastened by brackets to the forward or upper end of the trough.

Sliding back and forth in this groove of the trough was a smaller beam, with a smaller groove in its upper surface. This smaller beam was called the *diostra,* "projector" or "slide." At the after end of the slide was a crosshead of metal. This crosshead bore a trigger mechanism, with a hook or finger to hold the bowstring. The hook could be raised to release the string by twisting a handle, striking a knob, or pulling a lanyard.

At the after end of the trough, a windlass provided means for pulling back the slide against the resistance of the bowstring. On the sides of the crosshead of the slide were two pawls or dogs. As the slide was drawn back to the cocked position, these pawls clicked over the teeth of two narrow bronze racks, one on each side of the trough.

To shoot, one started with the slide in the forward position. The bowstring was engaged with the hook. A crew of men then pulled the slide back by means of the windlass, bending the bow. The exact distance the slide was drawn back depended upon the range desired. When the commander decided that the slide was back far enough, the windlass was slacked off. The pawls of the crosshead engaged the teeth of the rack, so that the tension of the bow could not pull the slide forward again.

The dart was placed in the small groove on the upper surface of the slide, either before or after cocking, and the engine was discharged by releasing the bowstring from its engagement with the hook. The string snapped forward, leaving the slide behind and sending the missile on its way. Then the slide had to be pulled forward by hand or by windlass in order to shoot again.

During the following century, a number of improvements were made in catapults. Some were built in large sizes for shooting balls of stone or brick, weighing from 10 to 180 pounds. Brick had the advantage that it usually broke on impact, so that the foe could not readily pick up the balls and shoot them back. Today, near the cave called the Ear of Dionysios in Syracuse, Sicily, you can see a pile of stone catapult balls a foot in diameter, weighing about one talent (60 to 70 pounds) apiece.

Stone-throwing catapults, first mentioned in connection with Alexander's siege of Tyre (–332), may have been invented in Phoenicia. They were equipped with a pouch or strap in the middle of the bowstring to hold the missile. Dart throwers (*oxybeloi*) were used against men; stone throwers (*petroboloi* or *lithoboloi*) against structures like ships and siege towers.

The effective range of heavy, one-talent stone throwers was less than 200 yards, but dart throwers could shoot much farther. We hear of dart throwers casting their missiles 600 to 800 yards, or twice the effective range of the stoutest longbow. Catapults built in modern times on the ancient models have failed to exceed 500 yards, but this failure may be due to lack of experience on the part of the builders.

Another development was a small, portable, one-man dart thrower or crossbow. The earliest crossbow was called a *gastraphetês* or "belly weapon" because of the curved crosspiece at the butt end, which the arbalester braced against his chest. It was also called a hand catapult and a scorpion. Although the crossbow was well known from –IV on,

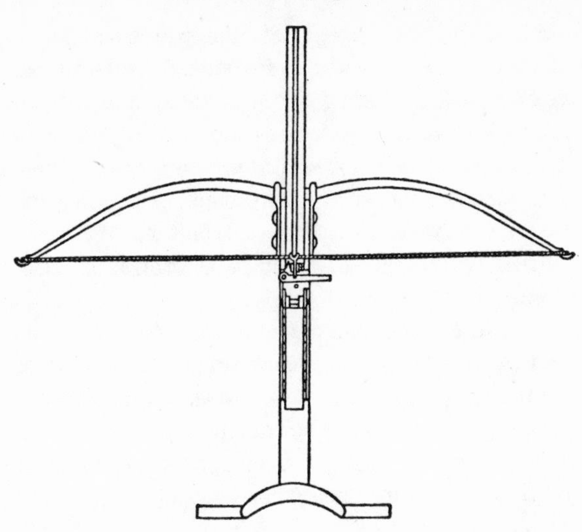

Fig. 3. Heron's *gastraphetês* or crossbow (*after Diels & Schramm:* Herons Belopoiika).

it never attained the popularity in classical times that it achieved in medieval Europe. The only detailed description of an ancient crossbow that we have is by Heron of Alexandria (+I), who describes a rather cumbersome device with a bow of horn and the elaborate cocking mechanism of larger catapults.

Almost nothing is known about the inventors of these devices, save that one early catapult engineer was a Pythagorean philosopher of Taras, Zopyros, who may have flourished in early –IV.

Further improvements in catapults appeared in late –IV. Demetrios the Besieger, one of the generals who fought over the pieces of Alexander's empire, mounted light dart throwers, shooting 27-inch darts, on ships.

Another improvement was a change in the motive power of the catapult. Instead of using a solid bow of wood or other springy material, the catapult was built with a massive frame surrounding the trough. Through each side of this frame was threaded (with a huge iron needle) a braided rope of hair, up and down until two stout skeins had been built up. A pair of rigid throwing arms, thrust through the skeins, took the place of the arms of the solid bow.

The skeins could be made either of horsehair or of human hair. Poor women sold their hair for skeins, and during sieges the women of the besieged city often donated their hair to the defense.

The torsion catapult, with skeins of hair, proved more efficient than the flexion catapult with a solid bow. Because the torsion catapult could throw a heavier missile, or could shoot farther, than a flexion catapult of the same size, it largely replaced the latter.

The torsion catapult, however, was more complicated than the flexion catapult and harder to manage. For instance, its skeins would go slack with long use or wet weather, so the catapult had to have means for tightening them. Therefore flexion catapults continued in use, as later allusions to them show.

To describe these engines, the ancients used the terms *katapeltes* (or *catapulta*) and *bal(l)ista*. These meant simply "thrower" or "shooter" —that is, any missile weapon. These words were applied indiscriminately to all catapults. Although some ancient and modern writers have tried to limit one word or the other to some particular class of catapults, they have not done so in any consistent way.

The art of poliorcetics or siegecraft advanced swiftly during –IV. Cities that had been protected by walls of mixed stone, brick, and wood now built walls of solid stone to withstand the new siege engines and dug deep ditches in front of these walls to keep the engines at a distance. Philip II of Macedon took a corps of siege engineers, headed by Polydos the Thessalian, on his forays.

Later, Polydos' pupils Diades and Charias performed the same office for Philip's son Alexander the Great. Diades wrote the first known book on siege engines and invented the "crow" (*korax*). This was a flying bridge attached to the top of a belfry so that, when the tower rolled up to the enemy's wall, the bridge could be lowered across the gap.

The most spectacular display of siegecraft in the ancient world took place in –305, eighteen years after the death of Alexander the Great. Alexander left, to squabble over his empire, two widows and a pack of rapacious Macedonian generals. One widow murdered the other; then the generals murdered all of Alexander's surviving relatives.

These generals, called the Diodochoi or Successors, made themselves kings over various parts of the empire. For over two centuries they and their descendants fought one another, none being strong enough to possess himself of the whole but none willing to content himself with less.

In –305 one of these generals, Demetrios called Poliorketes, "the Besieger," attacked the island of Rhodes because the Rhodians refused to join him and his father, the fierce and crafty old Antigonos the One-eyed, in an attack on Egypt. At that time, another general named Ptolemaios or Ptolemy ruled Egypt.

Demetrios landed on the island of Rhodes with forty thousand soldiers and a force of pirates who had joined him in hopes of loot. He set up a fortified camp and attacked the city of Rhodes, at the north end of the island, from the seaward side. For this attack he prepared four floating siege engines, each consisting of a pair of galley hulls lashed together. Two of these engines bore four-story towers, while the other two carried catapults. He also armed small galleys with catapults in penthouses.

By a surprise attack at night, Demetrios seized the main mole guarding the harbor. Here his master gunner Apollonios mounted huge catapults throwing 3-talent (180-pound) catapult balls. Demetrios tried to force the defenses of the harbor with troops from landing craft, covered by a heavy bombardment.

Though greatly outnumbered, the Rhodians fought fiercely and countered every move by the attacker. Their own ships broke through the spiked log boom with which Demetrios protected his engines and sank two of them. Demetrios prepared a super-tower mounted on the hulls of six galleys, but a storm overturned this engine as it was being towed into position, and the Rhodians took advantage of the storm to recapture the mole.

Then Demetrios attacked from the land side. He built eight tortoises or wheeled sheds to be pushed up to the wall to protect the engineers while they filled up the ditch and undermined the wall. He built two enormous ram tortoises, each housing a ram 180 feet long and worked by a thousand men.

Demetrios' largest war machine was a colossal belfry designed for him by the engineer Epimachos. Different accounts give it different sizes, but it seems to have been 100 to 150 feet high and 50 to 75 feet square on the base. It had nine stories, each loopholed for catapults to shoot through. The loopholes were protected by shutters in the form of big leather cushions stuffed with wool, which could be raised from within. Inside were two sets of ladders, one for traffic up, the other down. On each level stood a water tank with buckets for putting out fires. The whole contraption was pushed on eight huge iron-tired, castor-mounted wheels by 3,400 of Demetrios' strongest soldiers.

Demetrios moved his engines forward and attacked the wall, but the Rhodians beat back the efforts of his men to swarm through the gaps. Then the Rhodians moved all their catapults to one section of the wall and, by showering the belfry with incendiary missiles in a sudden night bombardment, set it afire.

By the time Demetrios had pulled his engines out of range, repaired them, and prepared to attack again, the Rhodians had another trick up their sleeves.

Before the war began, the municipal architect of Rhodes had been Diognetos. A Phoenician, Kallias of Arados, came to Rhodes and lectured on his wonderful new machine for defending cities. This was a revolving crane to seize hostile siege engines, hoist them into the air, and drop them down again inside the city. Impressed, the Rhodians fired Diognetos and gave Kallias his job.

So, when Demetrios' 180-ton monster neared the city, the Rhodians told Kallias to go ahead with his revolving crane. But Kallias had to admit he was baffled. Then the Rhodians fired Kallias and begged Diognetos to take his old job back. After holding out for a while, Diognetos agreed to save the city on condition that he should have the belfry if he could capture it.

One account says that he mobilized the Rhodians to go out at night and pour liquids—water, mud, or sewage—into the ditch in front of the section of wall at which the belfry was aimed. The adjacent field thus became a bog, in which the advancing tower stuck fast. Another story says he tunneled under the field. When the wheels of the belfry passed over these tunnels, the wheels sank into the ground. In any case, the belfry was stopped. Demetrios launched more attacks, but without artillery support these were beaten off.

At last, after a siege of more than a year, Demetrios signed a treaty with the Rhodians and sailed away to other battles. At Thebes he built a belfry so heavy that it could be moved only a quarter of a mile in two months.

Diognetos brought the captured belfry into the city and set it up in a public place with an inscription:

DIOGNETOS DEDICATED THIS TO THE PEOPLE
FROM THE SPOILS OF WAR

Afterwards, the Rhodians sold the timber, bronze, and iron from Demetrios' war engines and used the money to build the Colossus of Rhodes, a 100-foot statue of the sun god, to whom they had prayed for deliverance.

Centuries later, a stone thrower of a new type appeared. This had a single arm which, impelled by a torsion skein, flew up in a vertical plane against a padded stop. At the end of the arm was a sling or spoon to hold a catapult ball. This engine is called an *onager* (Latin for the

Asiatic wild ass) because of that animal's mythical habit of kicking stones back at its pursuers. All that is known about the origin of the man-made onager is that it came into use some time in the first three centuries of the Christian era. In any case, it soon replaced the heavy two-armed stone thrower.

Under the Roman Empire, armies also used light catapults on wheels as field artillery. They were not very effective, because they were too heavy and bulky in proportion to their fire power. Such mobile catapults had been tried out as far back as –207 by the Spartan general Machanidas. But at the battle of Mantinea, the Spartans lost to the Achaean League and Machanidas was slain before the catapults had a chance to shoot.

Ancient armies did not usually carry complete large catapults with them, because such bulky objects would have slowed them down too much. Instead, the gunners brought along the skeins, slings, metal fittings, and other parts that could not be improvised. Then, when a siege began, they cut down trees and built their engines on the spot.

With the fall of the West Roman Empire in +V, the two-armed torsion catapult drops out of sight. Perhaps the decline of European engineering at this time made the building of so complex an engine impractical. However, the flexion dart thrower, with a solid bow, continued in use through the Dark and Middle Ages. It was used at the siege of Rome by the Goths in +537 and at that of Paris by the Northmen in +886.

After printed treatises on ordnance began to appear during the Renaissance, and after cannon had already made catapults more or less obsolete, engineers like Leonardo da Vinci and Agostino Ramelli (+XVI) still showed catapults of the crossbow type in their books. One of Ramelli's designs is that of a compound siege catapult with six bows. Biringuccio, writing in the 1530s, notes that explosive bombs "can also be thrown from ballistas as the ancients used to do or, if desired, with guns as the moderns do." Biringuccio's "ancients" are the technicians who lived more than a century before his time.[24]

While the two-armed torsion catapult disappeared in the West, the crossbow survived there. A simple hunting crossbow is shown on two monuments of Roman Imperial times in Gaul, and William of Normandy took a company of crossbowmen to the battle of Hastings. On the other hand, the crossbow died out in the East. Although a tenth-century Byzantine writer mentions it, by the time the Crusaders brought it to Constantinople, in +XI, the Byzantines looked upon it as a new weapon.

The onager also survived into the feudal era. At that time it was

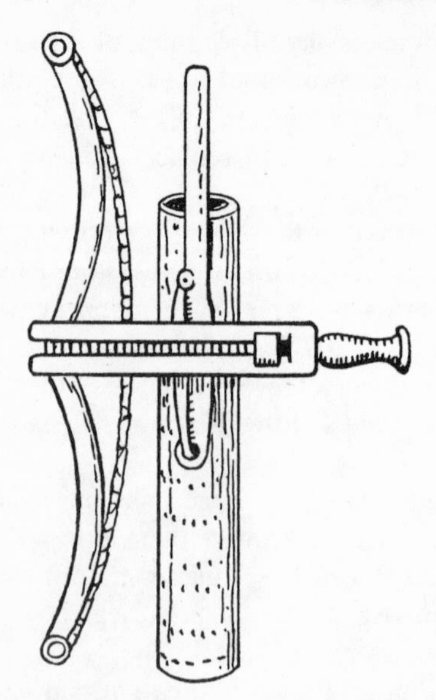

Fig. 4. Crossbow with quiver, from a Gaulish funerary monument of Imperial Roman Times at Puy, France (*from Diels*).

called the mangon or mangonel, from the Greek *manganon,* "device" or "contrivance." It decided the outcome of one battle at least. During the Albigensian Crusade, in 1218, Simon de Montfort besieged Toulouse. While he was riding around the city walls one day, a crew of women, manning a mangonel, let fly at him. The stone smashed his head like an egg and ended the siege.

The onager in turn gave way to the trebuchet or counterweight catapult, first mentioned in Spain in early +XII. The trebuchet had a pivoted throwing arm with a sling on the long end and a heavy counterweight on the short. This catapult had the advantage over skein-powered catapults that wet weather did not affect its performance. Sometimes the range could be adjusted by varying the size of the counterweight or by shifting the counterweight towards or away from the fulcrum. Sometimes the counterweight was assisted by a crew of soldiers, pulling on ropes attached to the short end of the throwing arm. In a catapult of a very simple type, men, pulling on the short end of the arm, provided all the energy.

Knowledge of catapults also reached the Far East. They were known in China by 1004 and in India by 1300, and there were literary allusions to the catapult or *pau* in China several centuries before +1000. When the Polo brothers first visited the court of Kublai Khan in the 1260s, the Italians charmed the Mongol emperor by building much larger catapults than the Mongols were used to, throwing 300-pound stones, for use against the city of Hsyang-yang. One missile smashed a house to kindling and caused the city's surrender. In Chinese chess one piece is still called "the catapult." The Cambodian kings who built Angkor mounted small catapults on elephants.

On the other hand, China knew the crossbow even before it was invented in the West. Writing in early –V, the Chinese general Sun Wu[25] mentions the weapon. And when in –35 the impetuous Chinese general Chên Tang defeated and slew the troublesome Hunnish king Jïjï,[26] the Chinese army used crossbows. This is undoubtedly a case of independent invention.

Soon, however, the whole art of catapult artillery was swept away by the discovery that "This villainous salt-petre" could "be digg'd / Out of the bowels of the harmless earth," and that "these vile guns,"[27] even in their crude fourteenth- and fifteenth-century forms, multiplied the power of the bombardier many times over.

So ends the story of Greek engineering, from the beginnings of classical history down to the time of Alexander. All in all, to the Greeks of that period, we owe more in the fields of art, literature, philosophy, logic, politics, and pure science than in the field of engineering. But neither was their engineering negligible. Moreover, Dionysios' brilliant and sinister idea of hiring men to invent machines of war was to persist down the ages and to culminate in the vast and secret military research and development projects of today.

In the period after Alexander, the comparative backwardness of Greek engineering, compared with the Greeks' other attainments, came to an end. Soon the Greeks–though usually living outside of Greece–led the world in this respect, as we shall presently see.

THE HELLENISTIC ENGINEERS

FIVE

In –IV, Alexander son of Philip, king of Macedonia, subdued all of Greece. Then he led an army of Greeks and Macedonians to the conquest of the mighty Persian Empire. From the rocky shores of Ionia, his columns pounded past the bustling Phoenician seaports to the shimmering sands of Egypt, and from Egypt to the ancient ziggurats of Mesopotamia, the tiger-haunted jungles of Hyrkania, and the lonely steppes of Central Asia.

The Persians, ruled by a well-meaning but hesitant mediocrity of a king, were smashed in three thunderous battles and many sieges and skirmishes. Having taken the throne of the King of Kings, Alexander led his army of Macedonians, Greeks, and Persians over the Afghan crags and into the Indus Valley to vanquish the glittering rajas and their lumbering elephants.

By −323, Alexander had conquered a realm as great as that of the first Darius. He had encouraged the intermarriage of Macedonians with Persians and had laid grandiose plans for further conquests, explorations, and public works. Then, not yet thirty-three, he suddenly died of (probably) malaria, in Babylon.

Alexander's generals soon liquidated the conqueror's kinsmen and carved up his empire. The leading kingdoms of the Successors were Egypt, under the Ptolemies; Macedonia, ruled by the descendants of Demetrios Poliorketes; and the Seleucid kingdom–Syria, Mesopotamia, and Iran–under the line of the general Seleukos. Several kingdoms waxed and waned in Asia Minor, of which the most important was Pergamon.

Although most of Alexander's Macedonian officers soon discarded the Persian wives he had found for them, his hoped-for mixture of Greeks with orientals soon took place anyway. Alexander and the Successors founded scores of new cities in the conquered lands. They encouraged thousands of Greeks and Macedonians to settle in these cities side by side with Persians, Syrians, Egyptians, and other native peoples. Hellenes swarmed out of barren Greece to serve in the armies and bureaucracies of the Successors, forming a ruling class in the new kingdoms.

The interloping Greeks soon mingled with the native upper classes. Greek culture influenced the orientals, while oriental ideas affected the Greeks. The brilliant Graeco-oriental civilization that resulted is called the Hellenistic.

The Hellenistic Age was in many ways like our own century. It was a time of immense intellectual ferment, of travel and tourism, of scholarship and research, of popular outlines and lectures, of clubs and societies, of invention and promotion. It was a time on one hand of a scramble for the wealth created by the advance of technology and the spread of commerce, and on the other of communistic revolutionary movements for the division of this wealth. Some of the fine arts, such as playwriting, declined from the high standards of the Golden Age; but science and engineering flourished as never before.

Another "modern" feature of the Hellenistic Age was a love of the grandiose. Hellenistic kings armed their soldiers with longer spears and massed them in bigger phalanxes than ever before; they built more sumptuous temples and palaces; they erected taller buildings and statues; they organized more splendid parades and committed more dastardly crimes.

The Successors and their descendants fought many wars, but these wars were less ferocious and destructive than many wars have been. The kings fought the other Hellenistic kings—usually their brothers-in-law—in a somewhat sporting, gentlemanly spirit. They tried many novel military expedients: huge phalanxes on the Macedonian model, with the soldiers wielding 21-foot pikes; elephants, which often defeated their own side by stampeding back through the ranks; even Arabs swinging 6-foot swords from the backs of camels.

As metalworking techniques advanced, iron began to take the place of bronze for armor. Alexander the Great is the first man known to have worn an iron helmet. Dionysios the Elder of Syracuse, who invented the ordnance department, had already pioneered with the iron corselet, which he wore under his tunic to foil assassins. By the end of –IV, the iron cuirass had become common. To save weight, however, most classical cavalrymen continued to prefer a corselet made of several layers of linen canvas glued together and molded on a form; the result was much like a modern laminated plastic.

Meanwhile, in the little-known West, Rome grew from city-state to nation and from nation to empire, until it engulfed the entire Hellenistic world. In –30, Rome conquered the last of the Successors' kingdoms, Egypt, and brought the Hellenistic Age to a close. But, for three centuries, the lands of the eastern Mediterranean were the scene of some of the liveliest and most interesting developments in the entire history of ancient science and technology.

In addition to Alexander, another man had an equal effect on the flowering of Hellenistic science and engineering. This was Alexander's old tutor, Aristoteles of Stagyra (–384 to –322), whom we call Aristotle. By his researches, writings, and teachings, Aristotle gave all the sciences a push so vigorous that it kept them spinning for centuries.

Few men have affected the thought of the world more than Aristotle. He was the first encyclopedist and also the founder of the scientific method. His method was neither pure theorizing, like that of Plato, nor the mere gathering of data, like that of Herodotos. Instead, he creatively combined both. Although Aristotle often went wrong in applying the scientific method, he made the necessary beginning.

At seventeen, Aristotle arrived in Athens from a Greek provincial town at the northwest corner of the Aegean Sea. He joined Plato's classes and for twenty years listened to Plato's discourses. He may even have become Plato's assistant. There are rumors that Aristotle once quarreled with Plato and tried without success to set up his own school; but their differences were soon patched up.

After Plato died, Aristotle and his fellow-pupil Xenokrates crossed the Aegean to settle at Assos in Asia Minor. Here Aristotle began lecturing. One of his hearers was a local magnate, Hermias the eunuch, who had become tyrannos of the town of Atarneus. Aristotle married Hermias' niece and took his bride to Lesbos for two years of honeymooning while studying marine biology. It would be interesting to know what the princess thought of a husband who spent his days wading in tidal pools and his nights cutting up sea-things on the kitchen table.

In –342, Aristotle heard that Philip II of Macedon was looking for a tutor for his son Alexander. Aristotle got the job, either because his father had been physician to Philip's father or because he played a part in a plot between Philip and Hermias against the latter's Persian overlords.

In any case, the Persians discovered the plot and killed Hermias. For seven years, in a small Macedonian town, Aristotle tutored Alexander and the latter's young friends. We do not know just what Aristotle taught Alexander, or how effective his teaching was; but it is not likely that he found the headstrong and violent young prince a docile or studious pupil.

When Philip was murdered and Alexander became king, Aristotle went back to Athens. Since Xenokrates was now running Plato's school, called the Academy from the name of the park where it met, Aristotle set up his own school in another park, the Lyceum.[1] This school was also known as the Peripatetic because, like Plato, Aristotle liked to walk about as he lectured.

Here Aristotle taught and wrote for thirteen years: a lean, dandyfied man with a lisp and a tart sense of humor. Once, when a chatterbox, after flooding Aristotle with talk, asked:

"Have I bored you to death with my gabble?"

Aristotle replied: "No, by Zeus, for I wasn't listening to you!"[2]

When Alexander conquered the Persian Empire and invaded India, he may have subsidized Aristotle's researches. There are tales that Alexander sent Aristotle an elephant and other specimens from the East. But their relationship was soured by the fate of Aristotle's nephew Kallisthenes, who had gone off to the East as a member of Alexander's staff. The headstrong and tactless Kallisthenes irked Alexander by refusing to worship him and by making a public scene over the matter. Alexander accused Kallisthenes of treason and had him thrown into prison, where he soon died.

During his Athenian period, Aristotle wrote nearly all of his works that have come down to us. He was one of the world's most prolific writers, composing the equivalent of 50 to 100 modern books. But, of this huge output, we have only a fraction. Most of what we have consists of huge treatises–actually, a series of extended lecture notes–on science, politics, history, morals, and literary criticism.

Although Aristotle could write well enough when he chose, these treatises, not being meant for publication, make no concessions to the reader. For long stretches they are wordy, dry, and maddeningly dull, devoted to the tedious elaboration of the obvious.

As a scientist, Aristotle was strongest in biology, sociology, psychology, and logic. His greatest contributions of all were in biology, thanks to those years of wading the Lesbian lagoons. Yet he passed along many old-wives' tales, which he failed to check by observation. Although he was twice married, he said that men have more teeth than women; it never occurred to him to ask either of his wives to open her mouth for a count.

In the physical sciences, Aristotle was much less successful. He did marshal arguments for the roundness of the earth in a way that settled the question. Otherwise, he managed to be wrong in nearly everything he did in physics, meteorology, and astronomy. He argued that the earth is at the center of the universe; that the heavenly bodies never change; that earthquakes are caused by winds trapped inside the earth; and that atoms do not exist.

After Aristotle died, his treatises passed into the hands of a follower who handed them down to his descendants, treasured but unread, until they were brought out and published about –80. Other copies besides these probably circulated, but we cannot trace them.

Some of these treatises were revised by later writers who added paragraphs or whole books to them. Furthermore, like many modern professors, Aristotle probably permitted his pupils to do much of his routine research and to write up their results under his by-line. Moreover, many later writers put Aristotle's name on their own writings for reasons of prestige. Hence there are never-ending disputes as to whether certain writings are truly Aristotle's.

One of these doubtful works furnishes the reason for talking so much about Aristotle. This is a short article called *Mêchanika,* or *Mechanics*, included in collections of Aristotle's "minor works." Most scholars who write about Aristotle either ignore the *Mechanics* or deny that Aristotle wrote it. If it was by any known author, they say, it was probably by Straton of Lampsakos. Some of their arguments seem to be circular. They say: Aristotle could not have written the *Mechanics* because he was not interested in the subject, and we know that he was not interested in the subject because he never wrote about it.

True, Aristotle had the well-to-do Greek gentleman's snobbish disinclination to experiment or invent, because experimentation and invention involved manual work, and manual work was fit only for slaves and "base mechanics." These inferior persons should never be admitted to citizenship, said Aristotle, because "no man can practise virtue who is living the life of a mechanic or laborer."[3]

As for the authorship of the *Mechanics,* the best guess is probably that of the late George Sarton: "The *Mêchanika* attributed to Aristotle is probably of Straton's time or even of later date, yet a part of it may be

Aristotelian."[4] If this is an unsatisfactory way to leave the world's first known engineering treatise, that cannot be helped.

Straton was a pupil of Aristotle who eventually headed Aristotle's school. It came about in this way: When Alexander died (–323) certain Athenians attacked Aristotle because of his Macedonian connections. Aristotle prudently retired to Euboia, where he soon died.

Theophrastos the botanist, a longtime pupil and friend, took over the Lyceum. Although Theophrastos was an even more prolific writer than Aristotle, time has been less kind to him. Of his enormous volume of writings on many subjects, we have only about a dozen articles or essays, most of them fragments of larger works, and two long treatises on plants.

When Theophrastos died about –287, the school came under the headship of the physicist Straton, who had been living in Alexandria as tutor to the Ptolemaic crown prince. Although a worthy successor to Aristotle and Theophrastos, Straton fared even less well at the hands of time. We know nothing of his personality save that he was a thin, sickly man; his writings have entirely disappeared except for a few fragments quoted by later writers.

Straton did correct some of Aristotle's many blunders in physics. For instance, he realized that falling bodies move at an ever-increasing speed (that is, they are *accelerated*) and not, as Aristotle thought, at a uniform speed proportional to their weights.

Furthermore, Aristotle had rejected Demokritos' theory of atoms, saying that there were no such things. He also asserted that a vacuum could not exist, either, because a falling body fell at a speed that varied inversely as the resistance of the fluid (air or liquid) it fell through, and if there were no fluid the body would move at infinite speed.

Straton, accepting atoms, explained that vacuum was simply the empty space between atoms. In air, this space decreased when the air was compressed, forcing the atoms closer together, and increased when the air was rarefied. Hence Straton discovered the "spring of air," which Boyle and Mariotte, nearly 2,000 years later, scientifically studied and measured.

Whether written by Aristotle, or Straton, or both, the *Mechanics* is the oldest-known engineering textbook. It begins:

> Our wonder is excited, firstly, by phenomena which occur in accordance with nature but of which we do not know the cause, and secondly by those which are produced by art despite nature for the benefit of mankind. Nature

often operates contrary to human expediency; for she always follows the same course without deviation, whereas human expediency is always changing. When, therefore, we have to do something contrary to nature, the difficulty of it causes us perplexity and art has to be called to our aid.[5]

To explain the law of the lever, our author goes off into several paragraphs of wordy theorizing about the mystical properties of the circle. Aristotle had caught this habit from Plato, and Straton may have caught it from Aristotle.

Our author does, however, make an interesting point. He is excited by the fact that, in a train of gears, each gear wheel turns in a direction opposite to those with which it meshes. He says: ". . . some people contrive so that as the result of a single movement a number of circles move simultaneously in contrary directions, like the wheels of bronze and iron which they make and dedicate in the temples."[6] A diagram shows a train of three gear wheels, represented by circles.

This is the first mention of gear wheels. Our author does not say whether these are toothed cogwheels or smooth wheels in which rotation is transmitted by friction only. Presumably the friction gear came before the toothed gear, but we have no dates.

Probably our author's wheels were smooth. Later, somebody learned that roughening the rims reduced slippage and, from this roughening, gear teeth evolved. This step may have taken quite some time, for cutting and filing a pair of toothed gear wheels, having the right numbers of teeth and transmitting rotation without jamming, is not an easy task. Because it is the easiest type of gearing to make, the first gearing to be used successfully may have been a pair of gears with shafts at right angles, one of the gears being a crown gear.

As for the "wheels of bronze and iron," this is explained by statements in the works of two later engineers, Philon of Byzantium and Heron of Alexandria. The first remarked that: "The ancients used many [wheels] of this sort; when they wanted to enter the temple, they sprinkled their clothes with water squirted from this wheel; then they moved it by hand, because they thought that in touching copper they purified themselves." And Heron added: "In the porticoes of Egyptian temples, revolving wheels of bronze are placed for those who enter to turn round, from an opinion that bronze purifies."[7]

Here is a minor piece of priestly magic, comparable to the prayer wheels of Tibet. The worshiper performs a simple mechanical action, namely turning a wheel, and draws out so much holiness per revolution.[8]

Mention of these wheels in the *Mechanics* favors Straton's authorship, at least of this passage. For, Heron identifies these wheels with Egyptian temples, and I know of no reference to them in Greece. And, whereas Straton spent years in Egypt, Aristotle never went there.

The statement in the *Mechanics* also suggests that somebody had figured a way to get more salvation per revolution out of the prayer wheels by mounting two or more so as to form a train of friction gears. It is a sobering thought that all later clockwork and other gearing may be descended from this petty piece of priestly hocus-pocus.

The author of the *Mechanics* then goes back to the lever and discusses the geometry of the beam balance. He notes that dishonest merchants had discovered how to rig such a balance or scale to cheat their customers:

And thus dealers in purple [dye], in weighing it, use contrivances with intent to deceive, putting the cord out of center and pouring lead into one arm of the balance, or using the wood towards the root of a tree for the end towards which they want it to incline, or a knot, if there be one in the wood; for the part of the wood, where the root is, is heavier, and a knot is a kind of root.[9]

He applies the law of the lever to the oars of galleys. He makes a wrong guess as to why a ship sails faster if the sail is hoisted higher; he thinks it has to do with the leverage of the mast against the hull, whereas the real reason is that winds usually blow faster higher up. This however brings him to another pregnant observation:

Why is it that, when sailors wish to keep their course in an unfavorable wind, they draw in the part of the sail which is nearer to the steersman, and, working the sheet, let out the part towards the bows? Is it because the rudder cannot counteract the wind when it is strong, but can do so when there is only a little wind, and so they draw in the sail? The wind then bears the ship along, while the rudder turns the wind into a favoring breeze, counteracting it and serving as a lever against the sea. The sailors also at the same time contend with the wind by leaning their weight in the opposite direction.[10]

The author is groping towards an explanation of how a ship can sail into the wind. If his explanation seems confused, blame him not; no satisfactory explanation was possible until Stevin discovered the triangle of forces in the 1580s.

The important thing about this paragraph, however, is that it gives the earliest indication that men had learned to sail close-hauled, at an angle against the wind. Almost any ship can sail at a right angle to the wind. But, to sail at *less* than a right angle to the wind—that is, to head up-wind—you must be able to pull the sail taut and to clew it around so that it lies almost parallel to the keel.

Sailors measure the angle between the keel of a ship sailing close-hauled and the direction of the wind in points. A point is one-eighth of a right angle, or 11¼ degrees. For thousands of years, all sailing ships had the simplest possible rig: one square sail. With such a sail one can sometimes sail into the wind, but less than a point (11¼° less than a right angle) and then only if the ship is handy, with a deep keel (to keep it from sliding sideways) and not too much upperworks. Also, the sail must not be too baggy. The best that modern sailing ships can do is two points (22½° less than a right angle) into the wind with square rig and three points (33¾° less than a right angle) with fore-and-aft rig.

Therefore most ancient sailing was done before the wind. The skipper sat in port until the wind blew the right way. Ships were built with high sterns because most waves struck them from behind. If caught at sea by a calm or a change of wind, the sailors struggled on under oar power.

As late as +X, sailing effectively against the wind in northern European waters was so unusual that when a Viking chieftain named Raud the Strong made use of this method to escape from Olaf Trygvasson, the bloodthirsty and fanatical Christianizing king of Norway, the king was sure that Raud must be using witchcraft. This so enraged the pious king that when he finally caught Raud, he had him killed by the unusual method of stuffing a viper down his throat.

In the Mediterranean of –300, however, seamen had found that, by clewing the sail around until the yard was almost parallel to the keel, they could sail into the wind by tacking; that is, sailing close-hauled into the wind on a zigzag course. With the ships of the time, however, it took a weary lot of sailing to gain a comparatively small distance up-wind. To sail a full point or more into the wind, one needs either a fore-and-aft sail, or masts at both ends of the ship to control its direction.

Some ancient merchantmen carried a foresail called the *artemôn,* slung beneath a slanting bowsprit like the water sail of the large ships of +XVII and +XVIII. But this sail was good only for holding the bow down-wind in sailing before the wind. A few large ancient merchantmen were real two-masters—one is shown in an Etruscan tomb painting of –VI—but this design never became general until medieval times.

A central rudder, instead of the steering oars or quarter rudders used in classical times, also helps to keep a ship headed up-wind. What happened when a ship could not gain distance to windward is shown in a novel of Roman times, Achilles Tatius' *Kleitophon and Leukippe* (+III). The narrator, young Kleitophon of Tyre, tells of being wrecked by a storm on a voyage to Alexandria:

> . . . a wind blew upwards from the sea full in the ship's face, and the helmsman bade the sailyard be slewed round. The sailors hastened to effect this, bunching up half the sail upon the yard by main force, but the increasing violence of the gusts obstructed their efforts; for the rest, they kept enough of the full spread to make the wind help them to tack.[11] As a result of this, the ship lay over on her side, one bulwark raised upward into the air and the deck a steep slope, so that most of us thought that she must heel over when the next gale struck us. We transferred ourselves therefore to that part of the boat which was highest out of the water, in order to lighten that part which was down in the sea, and so if possible, by our own added weight depressing the former, to bring the whole again to a level; but all was of no avail . . . For some time we thus ineffectually struggled to bring to an equilibrium the vessel thus balanced on the waves: but the wind suddenly shifted to the other side so that the ship was almost sent under water, and instantly that part of the boat which had been down in the waves was now violently thrown up, and the part formerly raised on high was crushed down into the waters . . . the same thing happening a third and a fourth, nay, many times, we thus imitated the motion of the ship.[12]

This passage is plainly a landlubber's account of an unsuccessful effort by the crew to keep from being blown ashore by tacking against the wind.

Aristotle-Straton's treatise also describes a swape with a leaden counterweight. The author goes on to discuss the roller, rolling friction, the wedge, the pulley, the sling, the capstan, and the windlass. The fact that he gets confused when he tries to discuss friction is not surprising, because the laws of friction were not worked out until recent centuries.

The author also asks: "How is it that dentists extract teeth more easily by applying the additional weight of a tooth-extractor than with the bare hand only?"[13] His explanation shows that, in his day, dentists pulled teeth with forceps not very different from those of our own time.

Having discussed all the simple mechanical-advantage devices except the screw, our author also asks how one can raise a weight greater than the force of one's pull by means of multiple pulleys. So we know that builders had learned to use more than one pulley at a time on a given load, though they may not yet have had the compound pulley block holding two or more pulleys in the same frame.

Finally, he asks many shrewd questions about the breaking stress in pieces of wood of different shapes, and about the mysteries of motion. He cannot answer these questions; for the answers, the world had to wait for Galileo (+XVII) and his successors.

After Straton died, the Peripatetic school declined in importance. It continued to lead a harmless but barren existence until in +529 the Christian emperor Justinian, as part of his persecution of pagans and heretics, closed all the philosophical schools of Athens.

Greece prospered briefly, following the conquests of Alexander, when returning soldiers brought home the loot of Asia and when those who settled in the new Greek cities of the Macedonian Empire sent to Greece for the commodities they were used to.

But when these new communities learned to grow and make their own commodities, Greece entered a long period of economic decline, combined with a shrinkage of population. This decrease seems to have been due partly to emigration, as the new and more easterly centers of the Graeco-oriental world beckoned ambitious youths, and partly to the Greek custom of tossing unwanted babies, especially girls, on a rubbish-heap to perish. During the Hellenistic Age this custom became as destructive as a plague.

Likewise the scientific center of the classical world shifted, in early –III, from Athens to Alexandria, the booming capital of Ptolemaic Egypt. While material prosperity is not a sufficient condition for intellectual advance, it helps.

Over the kingdom of Egypt reigned the amiable but shrewd and far-sighted Macedonian king, Ptolemaios son of Lagos, who had grown up with Alexander. Like most of the Ptolemies he was stocky and stout, with a bull neck, deep-set eyes under beetling brows, and a high-bridged beak of a nose. He ruled a land that even then harbored several millions: the teeming peasantry of the Nile Valley and Delta, wild tribes of sand-dwelling nomads, and settlements of Greeks, Macedonians, and Jews.

The government was an extreme bureaucratic absolutism, which held monopolies on oil, textiles, banking, and other forms of commerce. Crushing taxation, which fell heaviest on the peasantry, supported the glittering court and the gleaming army. The peasants often revolted but never effectively enough to drive out their Macedonian overlords.

When Alexander visited Egypt in –331, he had with him a Rhodian or Macedonian architect named Deinokrates. This Deinokrates had come to Alexander's headquarters with letters of recommendation.

When the king's officials put him off, he captured Alexander's attention by appearing at a public audience dressed like Herakles, with club and lion's skin. Then he explained that he wanted to carve Mount Athos into a statue, presumably of Alexander, holding a city in its left hand and in its right a bowl into which all the streams of the mountain should drain.

This grandiose plan, anticipating Mount Rushmore, delighted the king. But on second thought Alexander asked if the neighborhood had enough good wheat fields to furnish such a city with food. When told it did not, he vetoed the plan but kept the resourceful Deinokrates with him.

In Egypt, Alexander commanded Deinokrates to lay out the city of Alexandria on the site of a sleepy Egyptian fishing village, Rhakotis, which stood on a spit of land, a mile and a half wide, separating the Mediterranean from swampy Lake Mareotis. The latter is one of a chain of lakes, which extend across the northern part of the Nile Delta and which then swarmed with brigands and hippopotami. Here was built one of the most famous cities of the ancient world, and one of the most important in the history of science and engineering.

Creation of this city entailed vast engineering works. Twenty miles eastward, a small branch of the Nile emptied into the sea. A canal was dug to connect this branch with Alexandria, so that river craft could sail directly from Alexandria to Memphis and Upper Egypt.

To seaward of the new city lay a chain of rocky islets. Pharos, the largest of these, was a mile and a half long. By joining some of these islands to a peninsula and to each other, Alexander's successor Ptolemaios created a splendid harbor. Then he divided the harbor in half by a mile-long breakwater and causeway, the Heptastadion or "Seven-Furlonger," stretching from Pharos to the shore. The eastern or Great Harbor contained the naval dockyards and an inner harbor for royal yachts; the western harbor was for fishing vessels and other small craft.

At the eastern tip of Pharos, Ptolemaios' architect Sostratos of Knidos (–III) erected a sky-scraping tower as a landmark for ships. This was the famous Lighthouse of Alexandria, one of the original Seven Wonders of the World.

Although most ancient architects and engineers had to be content with remaining anonymous, some of them managed to put their names on their structures when nobody was looking. A story tells us that, when Sostratos built the Lighthouse, the king as usual wanted his name alone to appear on the work. Sostratos craftily inscribed on the stone:

SOSTRATOS SON OF DEXIPHANES OF KNIDOS
ON BEHALF OF ALL MARINERS
TO THE SAVIOR GODS

Then he covered this inscription with a layer of plaster, on which was chiseled the customary royal inscription. In time the plaster peeled off, removing the name of the king and exposing that of the architect.

Finished in the reign of the second Ptolemaios, the Lighthouse stood between 380 and 440 feet tall, compared with 480 feet for Khufu's pyramid and 555.5 for the Washington Monument. It was built in three sections: the lowest square, the intermediate octagonal, and the highest cylindrical. Helical stairways led to the top, and the lowest section contained fifty rooms. The tower itself came to be called the Pharos from the island on which it stood.

At a later date the Pharos was certainly used as a lighthouse; that is, as a tower on whose summit a fire was kept at night to help navigation. There is some question, however, as to whether such a light was kept burning in early Ptolemaic times, or whether the tower guided ships to harbor by daylight only. It would indeed have helped Mediterranean mariners in this way, as it was visible long before the low flat coast of the Delta rose out of the turquoise sea. From an account by an obscure classical geographer of another genuine lighthouse, which may have existed before the Pharos, I think it probable that the Pharos carried a beacon fire from the start.[14]

The tower stood for fifteen hundred years, guiding ships to port. Although stories that it cast its beams afar by subtle combinations of lenses may be dismissed as legends, a simple flat metal mirror behind the flame, to strengthen its light, is not impossible.

Our best descriptions of the Pharos come from a pair of medieval Muslim travelers. The Spanish Moor Idrisi,[15] who was in the eastern Mediterranean about 1115, wrote:

We notice the famous lighthouse, which has not its like in the world for harmony of construction or for solidity; since, to say nothing of the fact that it is built of the excellent stone of the kind called *al-kadhdhân,* the courses of these stones are united by molten lead, and the joints are so adherent that the whole is indissoluble, though the surge of the sea from the north incessantly beats against the structure. The distance between the lighthouse and the city is one mile by sea and three miles by land. Its height is 300 cubits of the *rashâshi* standard, each equal to three spans, making a height of 100 fathoms, whereof 96 are to the lantern and four for the height of the lantern. From the ground to the middle gallery measures exactly 70 fathoms, and from the gallery to the top of the lighthouse, 26.

One climbs to the summit by a broad staircase built into the interior, which is as broad as those ordinarily erected in minarets. The first staircase ends about halfway up the lighthouse, and thence, on all four sides, the building narrows. Inside and under the staircase, chambers have been built. Starting from the gallery, the lighthouse rises, ever narrowing, to its top, until at last one cannot always turn as one climbs.

From this same gallery one begins to climb again, to reach the top, by a flight of steps narrower than the lower staircase. All parts of the lighthouse are pierced by windows to give light to persons ascending and to provide them with firm footing as they climb.

This building is singularly remarkable, as much on account of its height as its solidity; it is very useful, in that it is kept lit night and day as a beacon for navigators throughout the whole sailing season; mariners know the fire and direct their course accordingly, for it is visible a day's sail [100 miles] away. By night it looks like a brilliant star; by day one can perceive its smoke.[16]

When Yûsuf ibn-ash-Shaykh, another Spanish Moor, visited the tower in 1165, he found it no longer used as a lighthouse. Instead, a small mosque had been installed on top in place of the beacon. Faith had triumphed over utility. Being an experienced builder and architect, ibn-ash-Shaykh carefully measured the tower, and it is his figures that we rely on today. We cannot, however, translate them confidently into modern feet, because he gave the measurements in cubits and we do not know which of several possible cubits he meant.

In the century following ibn-ash-Shaykh's visit, an earthquake brought the Pharos crashing down. The ruins were still to be seen in +XV, but now they have all disappeared. All one can see today are some sea-washed rocks along the shore of the island, which is now firmly joined to the mainland by the piling up of sand along Ptolemaios' seven-furlong causeway.

As with many ancient buildings and monuments, people made up legends about this structure after the original had disappeared. One story about the Pharos is that the Byzantine Emperor, wishing to harm the Caliph's trade, sent word to the latter that treasure was hidden in the tower. Eyes agleam with avarice, the Caliph set his men to demolishing the Pharos, and not until it was half destroyed did he perceive the trick that had been played upon him.

Alexandria itself was laid out on the typical Hellenistic gridiron plan, with its long axis east and west. Two main avenues, 46 feet wide and bordered by colonnades, crossed at right angles and divided the city into quarters. The longer of these avenues, Kanobic Street, ran from the Gate of the Sun at the eastern end of the walled city to the Gate of the Moon at the western end.

The royal parks and palaces occupied Point Lochias and the area around the base of this promontory. Not far away–though experts disagree as to which way–stood the Sema (a tremendous tomb that Ptolemaios built for the body of Alexander), the Museum, the Library, and other public buildings.

Because Alexandria has been continuously occupied from Alexander's time to the present day, few relics of Hellenistic times survive. And, because it is now a large, heavily built-up city, with over a million people, it cannot be thoroughly dug up by archeologists to settle doubtful points. Many maps have been published, showing the city as the cartographer thought it was in ancient times. But the wide differences among these maps show them to be largely based upon simple guesswork.

The Greeks called the city "Alexandria near Egypt,"[17] for the land on which it stood was not considered part of Egypt proper. The population was mixed, and the various nationalities lived in their own sections: the Egyptians in Rhakotis to the southwest, the Jews in the Delta Quarter to the northeast, and the Greeks in the Broucheion in between.

The polyglot population got the reputation of being lively, quick-witted, humorous, and irreverent, but lacking the more solid virtues. Though of little worth as soldiers, they were all too ready to riot. The different communities rioted against each other, and all together rioted against any king or emperor whom they disliked. Sometimes one of the more ferocious tyrants, a Ptolemy VII or a Caracalla, lost patience with this fickle and turbulent folk and ordered a massacre. But the Alexandrines never learned.

Ptolemaic Alexandria was not only a teeming political and commercial center but also the world's scientific capital. The city gained this eminence as a result of a genetic accident, which made the first three Macedonian kings not only able statesmen but also genuine intellectuals. Ptolemaios I Soter was a historian; Ptolemaios II Philadelphos dabbled in zoölogy; and Ptolemaios III Evergetes was a mathematician.

Philadelphos was perhaps the most brilliant of the entire line: sickly of body and luxurious in tastes, but immensely shrewd and versatile. He it was who remarked that the trouble with being a king was that you had to kill so many people who had not done anything really wrong, but whose existence was harmful to the state. He staged magnificent parades to amuse the effervescent Alexandrines. One of these displayed the animals from his private zoo, including a polar bear (however such a beast got to Egypt!), and a gilded phallus 180 feet long.

After the third Ptolemy, the line began to run down. Thereafter, incompetent Ptolemies did more harm than the occasional able ones could repair. The last of the line, the famous Cleopatra VII, was considered a wonder because she was the first of the dynasty to learn Egyptian.

The man who enabled the Ptolemies to achieve their most important accomplishment was an exiled Athenian politician, Demetrios of Phaleron. This Demetrios Phalereus had studied in the Peripatetic school in Athens, entered politics, rose to prominence by his oratory, and became rich. Later, Kassandros, a Successor who ruled Greece and Macedonia, made Demetrios governor of Athens.

For ten years, Demetrios ruled Athens as Kassandros' puppet, giving the Athenians sound, moderate rule. He boasted that "in his city all things were abundant and cheap, and every one had plenty to live upon."[18] He also indulged his taste for a luxurious and licentious private life, although he issued puritanical decrees forbidding other Athenians to do likewise.

During a war between two Successors, Demetrios Poliorketes (who later besieged Rhodes) landed in Greece and seized the port of Peiraieus (−307). Demetrios Phalereus fled to Egypt. The fickle Athenians, who had filled Athens with statues of Demetrios Phalereus, now destroyed the statues, condemned Demetrios Phalereus to death *in absentia,* and turned to the other Demetrios with such fawning worship that even that vain young man was disgusted.

Arriving at the court of Ptolemaios Soter, Demetrios Phalereus soon worked himself into the position of the king's literary adviser. In this capacity he suggested that the king set up a universal library, to hold copies of all the books in the world.

A book in those days was a papyrus roll, handwritten or dictated by the author and copied by scribes. The roll was a long strip of squares of papyrus glued together, edge to edge and wrapped around a wooden dowel rod at one end. Plinius the Elder tells how this writing material was made:

> The process of making paper from papyrus is to split it with a needle into very thin strips made as broad as possible, the best quality being in the center of the plant. . . . Paper of all kinds is "woven" on a board moistened with water from the Nile, muddy liquid supplying the effect of glue. First, an upright layer is smeared on to the table, using the full length of papyrus avail-

able after the trimmings have been cut off at both ends, and afterwards cross strips complete the lattice work. The next step is to press it in presses, and the sheets are dried in the sun and then joined together . . .[19]

The squares varied in size, most of them being 9 inches to a foot on a side. Although some rolls were as much as 150 feet long, containing the entire *Iliad* or *Odyssey,* such rolls were awkward to handle and rarely used. In Plinius' time, the standard roll was of twenty sheets, which gave a strip 15 to 20 feet long. Long works were divided into "books" according to the amount of text that could be conveniently written on rolls of standard size. One such "book" formed a cylinder about 6 inches in diameter and contained the equivalent, approximately, of ten to twenty thousand words of modern English text. The text was written in columns, about as wide as those of modern books.

Papyrus paper was a glossy, crackly, golden-brown substance, brittle and fragile. Because the material was so perishable, because only a small number of copies of most books were ever made, and because the roll was awkward to handle and easy to drop and damage, only a small fraction of the writings of classical times has come down to us.

However, King Ptolemaios' buyers scoured the Mediterranean for valued books. Travelers arriving in Egypt were compelled to give up any books they had. These books were copied, the originals placed in the Library, and the copies given the travelers.

Later, Ptolemaios III persuaded the Athenians to lend him the original, autograph copies of the plays of Aischylos, Sophokles, and Euripides for copying. Once he had them, the king kept them and sent back the copies, cheerfully forfeiting the fortune of fifteen talents, which he had deposited as bond for the return of the originals.

Founded by the first Ptolemaios, the Library reached its definitive form under the second, who appointed the first of a long line of chief librarians. All the rulers of this dynasty (save perhaps the seventh Ptolemaios, who favored the native Egyptians against the Greek ruling class) fostered the Library and added to it.

At its height, the Library held nearly three-quarters of a million rolls. Many of these, however, must have been duplicates, because there were not enough authors in the ancient world to produce so many separate titles.

While it endured, the Library made Alexandria the unquestioned intellectual capital of the world. In building the Library, the Ptolemies made a far greater contribution to civilization than all their palaces and parades. Many rulers have sought eternal fame: some by conquest and massacre, some by building grandiose temples and tombs, some by forcibly converting multitudes to their particular creed, and some by imposing a host of strangling rules and restrictions on their subjects. But few rulers have ever succeeded in doing so much good with so little suffering as the Ptolemies did in building up the Library of Alexandria.

As for Demetrios Phalereus, who started the whole thing, he did not fare so well. Ptolemaios I had two sons, both named Ptolemaios. When the question of succession arose, Demetrios Phalereus favored the elder son, a gloomy and violent youth surnamed Keraunos, or "Thunderbolt." But Ptolemaios I named the younger son as his successor. Prince Thunderbolt left the court, murdered his way to the Macedonian throne, and soon perished in battle with Celtic barbarians. When Ptolemaios II became king he banished Demetrios Phalereus to Upper Egypt, where the Athenian died of snakebite.

A series of fires and depredations during the Roman period gradually destroyed the Library. As the books were stored in two or more buildings, no single fire consumed them all. When Julius Caesar occupied Alexandria in –48, Cleopatra urged him to help himself to the books, and he took away hundreds or thousands to be shipped to Rome.

Then Alexandria revolted against Caesar and Cleopatra. In the fighting, either the books that Caesar had taken or those in one of the Library buildings, or both, were burned. When Antonius formed his connection with Cleopatra, he stole and gave her the 200,000-roll library of Pergamon to replace the losses.

The Library probably suffered further damage when Aurelianus suppressed a revolt in Alexandria in +272; when Diocletian put down another revolt in +295; and again in +391 when Bishop Theophilus, another bloodthirsty fanatic of the Hitlerian type, led a Christian mob to the destruction of the temple of Serapis, where some of the books were kept. The remaining rolls were finished off by the Arabs of the Muslim general 'Amr ibn-al-'Âṣ when he captured the city in +646.

A story relates that 'Amr wrote his Khalîfah asking what to do with these books of the infidels. He received the reply that if they agreed with the holy Qur'ân they were superfluous, whereas if they disagreed with it they were pernicious, so it were well in any case to destroy them.

Modern apologists for the Arabs have denied this story and put all the onus of the destruction on the Christians. Christian apologists, on the other hand, have striven to exculpate the godly Theophilus and put the blame back on the Muslims.

In fact, we shall never know just how many books were destroyed at each devastation. Nor shall we know to what extent the destruction was due simply to the agents of time and neglect—mice and mold, thieves and termites—which were suffered to work their will unchecked when, with the rise of Christianity, governments lost interest in the preservation of mundane writings. All we can say for sure is that monotheism proved as deadly a foe of learning as war and barbarism.

With the rise of Christianity and Islam, the ancient custom of burning the books of one's foes, to torment them or simply to enjoy the bonfire, was aggravated by the fanatical animus of dogmatic theology. For example, the Christian Roman emperor Valens commanded a general burning of non-Christian books (+373). The Muslim Arabs destroyed the books of the Zoroastrian Persians when they conquered Iran (+637). The Crusaders burned the books of Muslim learning, to the number of over 100,000, when they captured Tripoli (1109). The Spaniards did likewise when they reconquered Andalusia from the Moors (+XV); Cardinal Jiménez, a successor to Torquemada as Grand Inquisitor, had a haul of 24,000 books burned at Granada. And Diego de Landa, Bishop of Yucatán, topped off the record in the 1560s by burning the entire native literature of the Mayan Indians, on the ground that "they contained nothing in which there were not to be seen superstition and lies of the devil."[20] *Tantum religio potuit suadere malorum.*[21]

Perhaps the biggest single loss to classical literature occurred when the Crusaders treacherously seized Constantinople in 1204 in order to carve up the Byzantine Empire into feudal domains, incidentally opening the way for the Turkish conquest of southeastern Europe. Hundreds of classical works, which had survived till then, went up in flames at last. Small though the extant fraction of ancient literature is, the wonder is that any survived at all.

Closely connected with the Library was the Museum. The word "museum"[22] means "shrine of the Muses." The Peripatetic school in Athens had centered around such a shrine, which was also used by the school as a library, because in Athens a school had to have a religious basis in order to gain the protection of Athenian law.

The Museum of Alexandria was the nearest thing to a modern university that the ancient world experienced. There was at least one building where specimens could be displayed, experiments performed, and lectures heard. In this building a number of scholars, paid from the royal treasury (and later by Roman governors) studied, wrote, and taught. A priest of the Muses headed the college.

More exact we cannot be, in the absence of hard information. Sometimes the terms "Library" and "Museum" are used interchangeably, as if they were one and the same institution; but we do not know just how this center of learning was administered.

The scholars seem to have had a good deal of freedom, like those in the present Institute of Advanced Studies at Princeton. However, they sometimes presumed too far on their freedom of speech. When Sotades of Maroneia wrote a ribald verse about the marriage of Ptolemaios Philadelphos to his sister, the poet was jailed. He escaped, but Philadelphos' admiral recaptured him, put him into a leaden jar, and dropped him into the sea.

Much brilliant scientific work was accomplished in the Museum during its first three centuries. Eratosthenes calculated the size of the earth with amazing accuracy. Hipparchos compiled his great star catalogue and invented latitude and longitude. Aristarchos of Samos, a pupil of Straton, had the boldness to put the sun instead of the earth at the center of the solar system; Copernicus only borrowed the idea 1,800 years later. Herophilos and Eristratos launched the sciences of anatomy and physiology by dissecting corpses until unrest among the Egyptians, whose religious feelings were outraged by this practice, led the king to forbid dissection.

While the savants of the Museum were advancing the pure sciences, other men were making similar progress in engineering. For raising heavy weights, for instance, builders no longer had to depend upon long sloping ramps. They lifted stones directly by cranes, consisting of one or more poles fastened together at the top and a block and tackle with pulleys. By the end of the Hellenistic Age, hoists contained as many as five pulleys, giving a five-to-one mechanical advantage.

The rope was pulled by means of a capstan or by a treadwheel. This was a large drum-shaped wheel in which, like squirrels in a cage, men walked up the curving inner side to make the wheel turn. Although the treadwheel had long been used in Mesopotamia for raising water, its use as a hoist for building materials was probably a Hellenistic invention. Two Hellenistic engineering treatises, by Philon the Byzantine and Biton, tell about it. Philon proposed to use it to work a bucket chain for raising water, which is plausible enough.

Biton, however, wanted to move a belfry by means of this mechanism. Simple calculations show that this must have been an armchair invention that would not work in practice, because the siege tower would be far too heavy to be moved by the number of men that could be crammed into the apparatus. Still, this is the first proposal for a self-propelled vehicle, wherein the source of power is wholly contained within the vehicle. However impractical, Biton's design is the remote ancestor of the automobile.

We know what a classical treadwheel looked like, because a clear picture of a treadwheel working a crane appears in a first-century relief on a Roman funerary monument and elsewhere in Roman art. The relief on the monument, commemorating the building of the tomb, shows the treadwheel with four men inside it and a fifth climbing in, while two others lend muscle by pulling ropes attached to the outer ends of the spokes. Two other workmen have affixed palm fronds to the top of the crane, presumably to celebrate the end of the job.

Other engineering problems arose in building the enormous statues admired in Hellenistic times. Hellenistic ideas of sculpture differed from those of earlier centuries. Statues showed particular individuals with all their physical peculiarities. Faces took on realistic expressions. The Olympian calm of the busts and statues of the Golden Age gave way to realistic smiles, scowls, and grimaces of pain—such as the tortured expression on Skopas' famous statue of Marsyas, a mortal who lost an argument with a god and was about to be flayed as a forfeit.

Hellenistic sculpture is often said to represent a decline from the "idealism" of the earlier period. I suspect that the earlier sculptors failed to give their works expression and individuality, not because of idealism, but because they did not know how. The techniques had not yet been worked out. Hence these earlier sculptors tended to make all their men look alike, just as people in pictures illustrating magazine stories today often show a monotonous sameness.

A leader of the realistic school was Lysippos of Sikyon, a town on the northern coast of the Peloponnesos. Lysippos and his brother Lysistratos are credited with several advances in the technique of sculpture, such as the life mask and the lost-wax method of casting. In addition, Lysippos is said to have turned out the huge total of 1,500 statues. But the only one whereof we have even a copy is his Apoxymenos, or Man Scraping Oil from Himself.

Lysippos also made a couple of colossi for the city of Taras in Italy. The first was a Herakles; the second and larger one, a 60-foot Zeus. To keep the prevailing wind from blowing down the larger statue, Lysippos put up a column on the windward side of the statue to break its force.

Of Lysippos' pupils, Eutychides executed the famous Fortune of

Antioch and perhaps the renowned Winged Victory of Samothrace, which now stands in the Louvre in Paris. Another pupil was Chares of Lindos, a minor city on the island of Rhodes. Chares created the most celebrated statue of antiquity, the Colossus of Rhodes.

As you recall, after Demetrios Poliorketes gave up his siege of Rhodes and sailed away, the Rhodians sold the materials of Demetrios' engines of war and used the proceeds to build a colossal statue of the sun god, Helios-Apollo. It took Chares twelve years to complete the work, which cost 300 talents, or about 20,000 pounds of silver. That was a lot of money in those days.

The Colossus is said to have been 70 cubits (90 to 120 feet, depending on which cubit is assumed) high. It may be compared with the Statue of Liberty, which stands 151 feet from base to torch or 111 feet from heel to crown. While the exact location of the Colossus is not known, some scholars believe that it stood near the site of the existing Mosque of Murad Reis, at the north end of the old city of Rhodes; others, that it stood near the site of the Castle of the Knights of Rhodes farther south.

A battered bas-relief found on Rhodes probably shows what the Colossus looked like. A sun god, nude but for a cloak draped over his left arm, stood on a pedestal, gazing eastward over the sea. A spiky crown of solar rays encircled his head, and his right hand was raised to shade his eyes against the beams of the rising sun. He was probably, like Alexander, clean-shaven but with his hair rather long.

A work on the Seven Wonders of the World exists under the name of Philon of Byzantium, a noted Hellenistic engineer, albeit some scholars doubt if Philon wrote it. This little treatise gives a good idea of Chares' methods.

First, the skeleton of the statue consisted of two or three stone columns extending up through the legs and the drapery to the trunk and head. Stone architraves joined these columns at the top. From the columns a spiny armature of iron rods extended out to the surface of the statue. The outer surface was made of plates of bronze, hammered into shape and riveted to each other and to the iron braces.

To get these plates up to the site of the work, Chares and his crew simply piled a great mound of earth around the statue and walked up a spiral path to the top. As the work rose, the mound was enlarged until the last rivets were in place. Then all the earth of that 100-foot man-made hill was shoveled up and borne away in baskets.

Chares' Colossus stood for fifty-six years, arousing the admiration of all for its beauty as well as its size. Then, in –224, an earthquake overthrew it, and the "colossal wreck" lay on the ground until the Saracen conquest. In +656, an Arab general, Mu'awiyyah, scrapped it and shipped the bronze to Syria. There a Jewish merchant of Edessa bought it and carried it off on 900 (or 980) camels, presumably to be turned into trays and lamps.

Philon says that the statue contained 500 talents (15 tons) of bronze and 300 talents (9 tons) of iron, and that building the Colossus caused a temporary scarcity of bronze. His figure for the bronze is probably wrong, because the bronze in that case would be only one-sixteenth of an inch thick, which seems too flimsy to withstand the wind. Moreover, the number of camels that carried away the bronze could easily bear over 200 tons. Probably the bronze was about an inch thick and weighed over 200 tons. Compare the Statue of Liberty, of the same general size, which weighs 225 tons, including 100 tons for the copper sheeting. One would expect Bartholdi's Liberty to weigh somewhat less than Chares' Helios, because Bartholdi had steel girders to work with while Chares did not.

Some well-known stories about the Colossus, which appeared long after the statue was built, can be safely rejected. One is the statement of Sextus Empiricus (+II), who said that at first the statue was planned to be half its eventual height. When the city decided to double the height, Chares asked for only twice the original fee, forgetting that the material would be increased eightfold. This error drove him to bankruptcy and suicide.

It seems, however, incredible that a man with the engineering skill that Chares must have had should not have known the square-cube law. This law states that, if you increase the dimensions of an object while keeping its shape the same, the area increases as the square of the dimensions while the mass and volume increase as the cube. That is why no flying animal has ever exceeded about 30 pounds in weight, and Sindbad's roc, which bore off elephants in its talons, would be quite impossible.

Other tales about the Colossus appeared in the Middle Ages, more than a thousand years after Chares' time and several centuries after the remains of the statue had been junked. The best-known of these says that the Colossus bestrode the harbor,

> . . . the brazen giant of Greek fame,
> With conquering limbs astride from land to land.[23]

Its feet rested on the ends of two moles, so that ships passed between its legs. This legend, perhaps suggested by the remains of fortifications on the moles, is impossible for engineering reasons. For one thing, the wide-spread legs would rise at a slant, and the interior stone columns, which braced the whole statue, could not have stood at such an angle. For another, Chares' method of construction, if applied to such a site, would have required filling the harbor with an earthen mound, which would have made the harbor useless for twelve years or ruined it for good. Furthermore, despite the statue's size, only very small ships could have entered the harbor between the legs. And finally, no ancient authority says anything about such a dramatic pose.

Other medieval tales averred that the Colossus was 900 feet tall (technically impossible) and that it had a beacon in its head (unlikely because of the difficulty of getting fuel up to the beacon). Recently the Governor of the Dodecanese Islands announced plans for building a replica of the Colossus in modern Rhodes, using bronze-colored aluminum for the skin.

In the reign of Ptolemaios II Philadelphos (–285 to –247) there dwelt in Alexandria a man named Ktesibios. The son of a barber, he grew up in his father's trade but showed an early bent for gadgeteering.

One day Ktesibios wished to mount a mirror in his father's shop so that it could be pulled up and down, like a window sash, without the mechanism's showing. Therefore he installed a wooden channel under a ceiling beam with a pulley at each end. The cord from the mirror ran up, over one pulley, along the channel, over the other pulley, and down. At the other end of the cord was a leaden weight, sliding up and down in a tube.

All went well—except that when the mirror was pulled up, the counterweight trapped air below as it descended. And the air, escaping from the tube, emitted a musical sound. This gave Ktesibios the idea of building musical instruments worked by pneumatic machinery.

From this beginning, Ktesibios became the Edison of Ptolemaic Alexandria. Although nothing more is known of his personal life, and the book he wrote is lost, several later writers described his inventions. It is a guess, though not unlikely, that he joined the Museum as one of the Ptolemies' subsidized savants. He did construct a singing cornucopia for the funerary monument that Philadelphos erected to his sister-wife Arsinoê about –270.

Ktesibios' main inventions were the force pump, the hydraulic pipe organ, the musical keyboard, the metal spring, and the water clock. Vitruvius describes the pump:

> It follows now to describe the machine of Ctesibius, which raises water to a height. It would be of bronze. At the bottom would be twin cylinders, a small distance apart, having pipes converging in the shape of a fork, meeting in a vessel in the middle. In this vessel would be valves, accurately fitted over the upper openings of the pipes, which stop up the openings of the pipes and do not allow that which the air has forced into the vessel to escape . . . Pistons, smoothly turned and treated with oil, are inserted into the cylinders from above; and thus confined, they are worked with rods and levers. As the valves close the openings, the air and water in the cylinders will be driven onwards . . .[24]

The beam that worked the pistons was pivoted between them, so that when one piston rose the other fell. Without repeating all of Vitruvius' exposition, we can say that he describes a perfectly practical pump, which could be built to his specification. Remains of several pumps of this kind, dating from the classical period, have been found. In one from Metz, the cylinders were of lead encased in a block of wood.

Ktesibios' organ consisted of the following main parts: 1. A two-cylinder air pump, like the water pump previously described. 2. An air vessel shaped like an inverted funnel or bowl, into which the pump forced air, and into which water was admitted under pressure from the bottom, to keep the air pressure and hence the flow of air through the organ pipes constant. 3. A series of tubes leading up to the organ pipes. 4. A set of pipes. 5. A set of valves separating the tubes from the pipes, each valve consisting of a disk with a hole, so that in one position the flow of air is cut off by the solid part of the disk and in the other the hole in the disk registers with the tube and the pipe so that the air flows freely through. 6. Finally, a keyboard for operating these valves. The invention included not only the main idea of the organ, but also the hydraulic means for keeping the air pressure constant and the keyboard for selecting the pipes to sound.

Moreover, Ktesibios kept his valves in place by means of iron springs —the first known metal springs. People had of course long known the springy qualities of wood and horn and made use of these virtues in bows. But this was the first use of metal for the purpose. However, iron springs did not prove satisfactory for a long time to come, because an iron spring requires special heat treatment, and only the most skillful smiths could make such a spring.

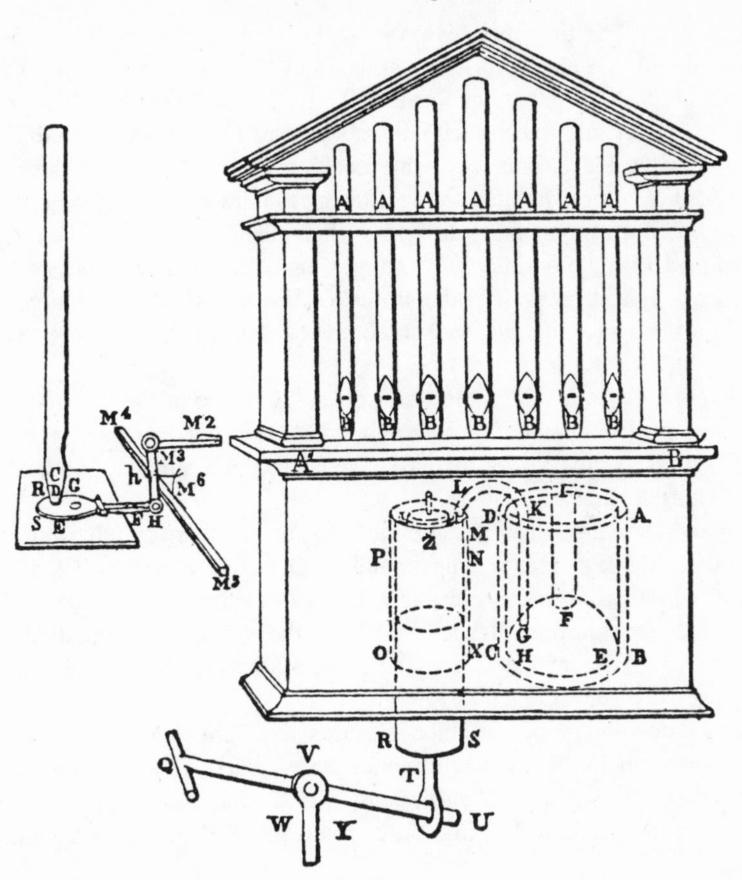

Fig. 5. The pipe organ of Heron of Alexandria, built after the design of Ktesibios (*from Woodcroft's edition of Heron's* Pneumatics).

In addition, Ktesibios tried bronze springs in a catapult that he designed. In this machine, instead of the throwing arms being thrust through skeins of hair, these arms were pivoted. When the string was drawn back, the short ends of the arms were forced against bronze springs.

Ktesibios also used his new-found knowledge of pneumatics to design another catapult, worked by compressed air. The arms were pivoted as in the previous design. But, when the string was drawn back, instead of bending springs, the short ends of the arms forced pistons into air-tight cylinders.

No catapults like those of Ktesibios are known to have ever been used in warfare. The probable reason is that, although the inventor's ideas were sound, the metal-working standards of the time were not up to the demands that he put upon them. Nobody, for instance, could make a piston and cylinder accurately enough so that the cylinder would hold its air while the engine was being cocked. Even 2,000 years later, in +XVIII, Matthew Boulton thought his mechanics were doing well when they bored steam-engine cylinders without an error greater than "the thickness of an old shilling."[25] At this time England was the world's foremost industrial nation, the leader of the Industrial Revolution, so you can imagine what classical standards of shop work must have been.

Ktesibios also invented the water clock. A glance at the sun may tell a peasant all he needs to know about the time, but increasingly busy, organized city life calls for increasingly accurate and minutely subdivided reckoning of time.

A rod stuck upright in the ground or projecting horizontally from the top of a wall serves as a simple sundial. Or rather, it would if the sun followed the same course every day, instead of shifting up and down the sky with the seasons. Hence the tip of the shadow follows different courses on different dates, and the place where the shadow falls must be marked off accordingly.

A contemporary of Ktesibios, Berosos the Babylonian, invented an improved sundial. Berosos represented a new class of men: learned and intelligent non-Greeks or "barbarians" who adopted a veneer of Greek culture and were absorbed into the mainstream of Hellenistic intellectual life.

Berosos started as a priest of Marduk in Babylon. But mighty Babylon was fast decaying, especially after Antigonos One-eye sacked it in –310. So Berosos sought his fortune in the West, when so many Greeks and Macedonians were seeking theirs in the opposite direction. In Athens and Kôs he taught astronomy and astrology, for the science and the superstition were still one in those days. Later, at the court of Seleukos' son Antiochos, Berosos wrote in Greek a history of Babylonia.

Berosos' hemicyclic sundial was a hemispherical bowl carved in a block of marble. From the inner rim of the bowl, a pointer or gnomon extended out horizontally to the center of the circle. Lines inscribed on the inner surface of the bowl showed the course followed by the tip of the shadow in different months. In an improved form of this sundial, possibly developed by Aristarchos of Samos, most of the bowl was cut away, leaving only the section over which the shadow of the gnomon actually played. The sundial of the modern type, with a slanting gnomon pointing towards the pole star, is a medieval Muslim invention.

The best sundial, however, worked only in the daytime and then only on clear days. Hence the clepsydra, also called a night clock or a winter clock, appeared in ancient Egypt. In its original form it was a jar with a hole in the bottom, into which a measured amount of water was poured. When the water had all run out, time was up.

The Athenians used the clepsydra to measure the time allowed orators. If a man was being tried for his life, the jar was filled; but, in trivial cases, only a little water was poured in. If a speech was interrupted, the outflow hole was stopped up until the speaker could resume.

Despite his contempt for practical applications of science, Plato is said to have rigged up an alarm clepsydra to signal the start of his classes at dawn: "It is said that Plato imparted a hint of the water clock's construction by having made a time-piece for use at night which resembled a water-organ, being a very large water-clock."[26]

A German scholar figured out that Plato's clock was probably a jar with a siphon. Water ran into the jar until it reached the curve of the siphon. Then the siphon emptied the jar all at once into another vessel, and air escaping from the other vessel blew a whistle.

The clepsydra had a shortcoming, unimportant in such a crude timepiece but significant when efforts were made to construct a more accurate water clock. The rate of flow of water through a hole in the bottom of a jar is not always the same. Water flows faster if the head of water–that is, the depth of the water at the orifice–is greater. Hence if two dippers of water are poured into the jar, they do not take twice as long to flow out as one dipper, but a little less, because the first dipperful flows out under higher pressure than the second. Also, naturally, the clogging of the hole by dirt would slow down the flow.

Ktesibios solved these problems. His orifices were made of gold, or of gemstones bored through. Therefore the hole could not be stopped by rust or dirt, and it could easily be cleaned without being worn away to a larger size.

To keep the rate of flow constant, Ktesibios changed the system of using a single vessel. Instead, he set up three vessels, of which the first emptied into the second and the second into the third. The first vessel was kept full. The second vessel had an outlet in the bottom and an overflow outlet partway up the side, like the overflow outlet in a modern bathtub. Since water rose in the second vessel only up to the overflow hole, the water in this vessel stood at a constant depth, and the rate of flow was constant.

But, since neither the first nor the second vessel was allowed to run dry, how did Ktesibios measure time? In the third vessel, a drum-shaped float of cork floated on the water. As the third vessel filled, the float rose. By means of a rack and pinion gear, the rising float turned a shaft. By gear wheels driven by this shaft, "figures are moved, pillars are turned, stones or eggs let fall, trumpets sound, and other side-shows."[27]

Although Vitruvius says that Ktesibios used gear wheels (*tympani,* literally "drums") the making of such wheels was still in its infancy. Because of the practical difficulty of making good toothed gears, it is likely that most early clock motions were transmitted instead by pulleys and strings, as was the case with the water clocks built a thousand years later under the Caliphate.

A favorite time-marking mechanism with the early clock makers was a bird that moved and sang on the hour, exactly as does the bird in a modern cuckoo clock. The earliest clocks were designed to mark the hours by noisy alarums, not to subdivide the times between. When

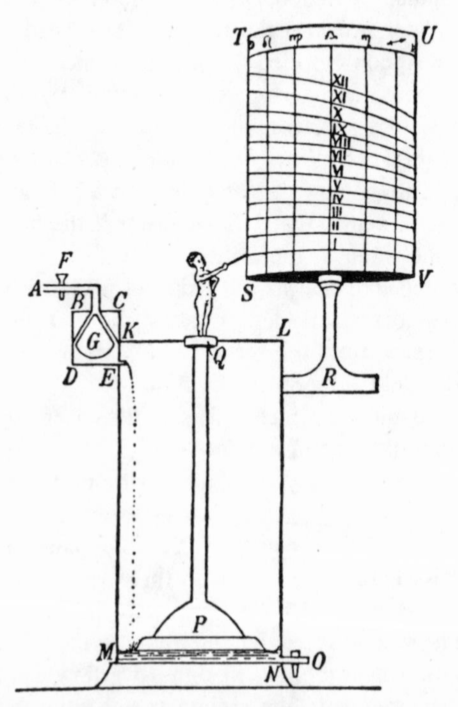

Fig. 6. The parastatic water clock of Ktesibios (*from Diels*). The pillar at the upper right can be turned to allow for variations in the length of hours in different seasons.

Shakespeare's Cassius in *Julius Caesar* says: "The clock hath stricken three" (Act 2, sc. 1) he is not anachronistic after all.

Ktesibios did, however, also invent a clock that showed the time more exactly, just as a sundial does in good weather. This "parastatic" clock had the same system of vessels as the first clock. A staff, rising from the center of the float, bore on its top a figure holding a pointer. This pointer indicated hours, marked on a pillar.

Ktesibios had to overcome a complication that would not trouble him today. In his time, an hour was defined as one-twelfth of the day from sunrise to sunset. As the day is longer in summer than in winter, the daytime hours were also longer in summer than in winter.

One method of adjusting the clock to hours of different lengths was to control the flow of water by an adjustable cone-shaped valve. Another was to drill the outlet hole of the intermediate vessel in a bronze disk that could be turned in its mounting, so that the hole could be raised or lowered and the head, and hence the rate of flow, varied.

Still another method was to mount the pillar bearing the scale of hours so that it could be turned. Instead of a single scale, a series of scales or a graph was inscribed on the pillar. By turning the pillar, the pointer could be made to indicate longer or shorter hours as desired.

Vitruvius describes another water clock, though without stating whether Ktesibios or one of his successors built it. One improvement in the mechanism was to connect the float with the revolving shaft by a cord wound around a drum and counterweighted, instead of by a rack and pinion.

In this "anaphoric" clock, the main shaft turned a bronze disk on which was engraved a chart of the heavens. The ecliptic–the sun's apparent path among the stars–was represented by a circle, off-center from the disk, with 365 small holes in it. Each day a little metal sun was moved from one hole to the next. Hours were represented by a grid of bronze wires mounted in front of the disk. The owner told time by seeing where the metal sun stood with relation to the wires representing the hours. In late +XIX, a fragment of such an anaphoric clock (of about +II) was found in eastern France, and another piece was turned up near Salzburg.

This device, by the way, shows why "clockwise" is the direction it is. If you face the sun at noon in the northern hemisphere, you face south. If you face south, the sun appears to rise on your left and set on your right. Hence the inventor of the anaphoric clock, when he set up his little model sun to imitate the real one, naturally had it rise on the left and set on the right, too. The anaphoric clock evolved into the astrolabe, a portable medieval instrument that could be used as a watch, compass, and astronomical calculating device.

A later colleague of Ktesibios, Andronikos of Kyrrha, built an elaborate timekeeping structure at Athens. This is the Horologium, or Tower of the Winds, which still stands a few blocks north of the Akropolis. It is a small eight-sided building with a sculpture on each side representing one of the winds. Originally a gnomon projected from each side, with a set of markings beneath it for telling time. There was a water clock inside, and on top stood a weathervane in the form of a bronze triton.

Shortly after Ktesibios, Philon of Byzantium flourished. He may have been Ktesibios' pupil. At least, he visited Alexandria and Rhodes and knew the leading engineers in both cities. A sentence in one of his writings has been taken to imply that he also knew Ktesibios.

Philon wrote several works, but of these we have only two books, and parts of two others, of his treatise *Mêchanikê Syntaxis,* or *The Ele-*

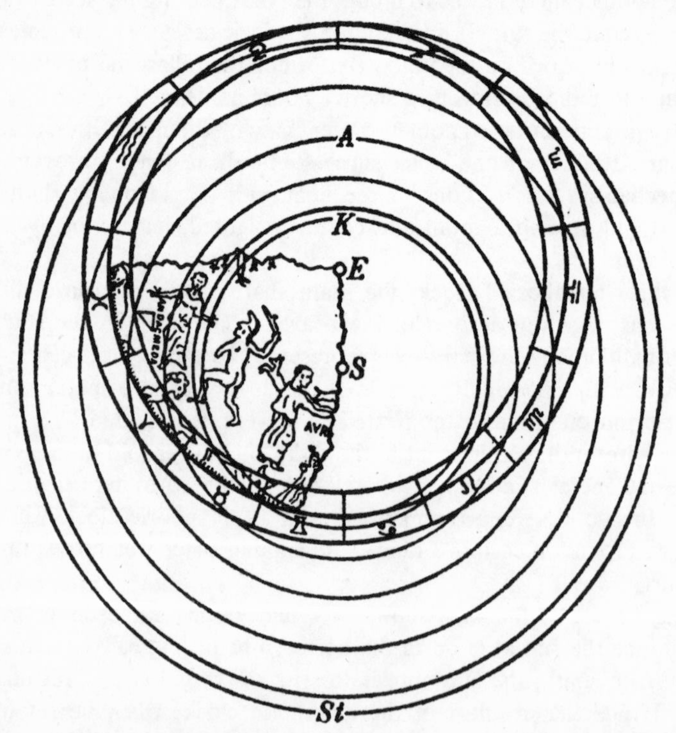

Fig. 7. The Salzburg fragment of the dial of an anaphoric clock, like that described by Vitruvius (*from Diels*).

ments of Mechanics. These fragments tell us about pneumatic devices, catapults, military preparations, and siegecraft.

Philon describes several catapult designs of his own, all fairly conventional except that one has a pair of wedges built into the structure. By means of these wedges, the skeins, which were liable to go slack, could be tightened by a few blows of a mallet without taking the whole contraption apart.

Philon also describes Ktesibios' metal-spring and compressed-air catapults and tells about the rapid-fire repeating catapult which Dionysios of Alexandria built at Rhodes. In this last, the darts were contained in a hopper over the groove in the slide and fell into it one by one, as in a nineteenth-century Gatling gun. The string was pulled back and released over and over by turning the windlass at the after end. This windlass was connected with the cocking mechanism by a pair of linked chains, one on each side of the trough, passing over two pairs of pentagonal nuts. Although mechanically impractical, this is the first known attempt at a chain-and-sprocket drive, like that of a bicycle.

Philon, like other Hellenistic engineers, loved to invent marvelous mechanical toys, and his *Pneumatics* describes some of these devices. For one thing, Philon was fascinated by siphons. By means of siphons he made pitchers, basins, and other vessels automatically empty and refill themselves. Bronze figures of men and animals drink, pour wine, and perform other acts. Trick pitchers pour wine, water, or a mixture of the two.

Several chapters describe mechanical wash basins, worked by a system of counterweights, strings, and pulleys. In the most elaborate, a bronze hand holds out a piece of pumice stone–the Greek equivalent of soap–for the user. When the latter takes the stone, the bronze hand disappears, and enough water flows out the spout into the basin to wash the man's hands. Then the flow of water ceases and the hand comes out with another stone. In another design, a copper animal, such as a horse, stands in the basin with its head down. When the man has finished washing, the animal drinks the basin dry.

Other devices include cups from which the wine poured into them suddenly runs out into a false bottom, to the astonishment of the drinker; a constant-level oil lamp; a waterproof lantern; an inkwell hung in gimbals so that it stays upright at all times.

Many of these entertaining machines were meant neither to do useful work nor to advance scientific knowledge, but merely to enable the rich Alexandrines who bought them to amaze and amuse their guests by displays of parlor magic. Ktesibios began this fad with his elaborate time-sounding devices, and most of the great Heron's machines had a no more practical aim.

However light-minded these machines now seem, it does not follow that they were really useless. On the contrary, some of the important machines of later times probably evolved from them. It is ironic that war on the one hand and frivolous ostentation on the other have given rise to many of the most vital devices and techniques of our civilized life.

Philon also describes a portentous invention with great practical possibilities: the water wheel. In some of his designs, the wheel merely forms part of one of his magical displays of twittering birds. One little overshot wheel powers a copper holy wheel, like that mentioned by Aristotle-Straton. Only, instead of the worshiper's moving the wheel, the flow of water turns the wheel. The worshiper grabs the wheel and stops it, while holy water squirts out of a hole in the axle.

In addition, Philon shows a practical application of the idea: a chain of buckets driven by an undershot water wheel with a series of spoon-shaped spokes arranged in a circle around the hub. In introducing this

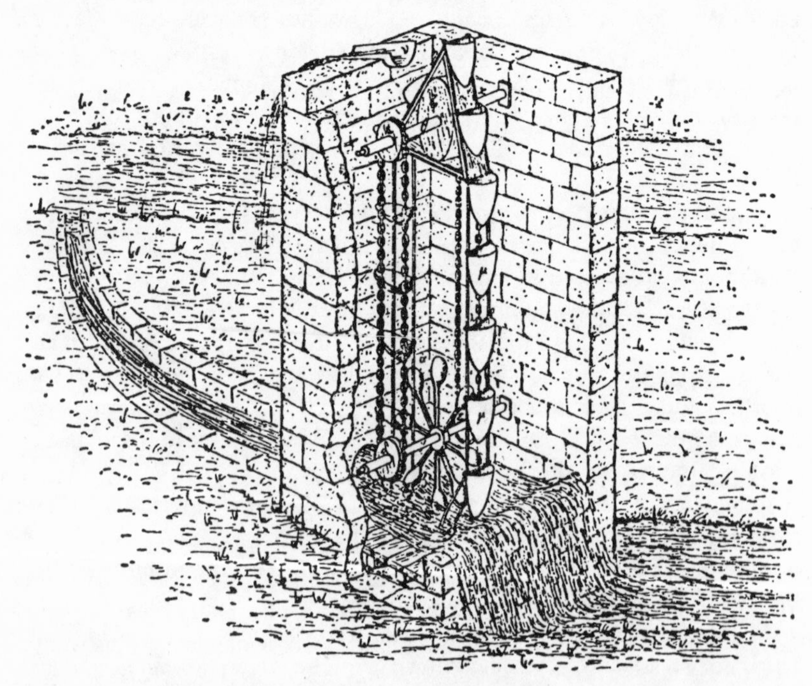

Fig. 8. The bucket-chain water hoist powered by an undershot water wheel described by Philon of Byzantium (*after Carra de Vaux*).

water-hoisting apparatus, Philon remarks that the wheel "can be applied to many other uses." And, in describing one of his toy whistling water wheels, he says it shall have, around its rim, "openings like the openings of water wheels without paddles."[28] This implies that water wheels *with* paddles were known. The two passages suggest that Philon knew the noria—the undershot water wheel combined with a drum-shaped water-raising wheel, later described in detail by Vitruvius. And we know that, soon after Philon's time, the water wheel was applied to the grinding of grain.

If these water-wheel designs really go back to Philon, which is not certain, then this water hoist is the first recorded case of putting the energy of running water to practical use. It was man's second step (the sail was his first) in the art of getting useful work directly from the forces of nature, instead of from the muscles of men and animals. It was a step of vast importance, because it showed how power could be concentrated.

Previously, one could increase the power used on a task only by increasing the number of men or animals, whose muscles furnished this power. But men and animals are bulky. It is hard to apply the power of large numbers of them at once in a small space, and the difficulty of getting them to pull in unison increases with the numbers employed. Moreover, they all require care and feeding.

When men learned to use the power of water and wind, it became possible to concentrate much more power in less space than had been the case before and thus easily to perform tasks that had been difficult or impossible. The art of getting useful work from the forces of nature has evolved steadily from Philon's time through the invention of the windmill, the steam engine, the turbine, and so on to the nuclear power plants of our own day.

The skill of Philon and his colleagues, the "mechanical wizards" of Alexandria, became famous in the Hellenistic world. Hence, just as the people of that time exaggerated the powers of poisons, they exaggerated those of mechanical automatons.

For instance, it was rumored that Nabis, the communist dictator of Sparta (about –200), kept a robot, which looked exactly like his wife Apega and was covered with spikes hidden under its clothes. When a rich Spartan refused to give Nabis the money he demanded, Nabis would present the man to the robot, which seized the victim in a spiky embrace and hugged him until he either gave in to Nabis' demands or expired. Despite the story's absurdity, as sober a historian as Polybios believed it.

Another belief was that one could, by placing magnets in the ceiling of a building, cause an iron statue to float in mid-air. Plinius says:

> The architect Timochares had begun to use lodestone for constructing the vaulting in the Temple of Arsinoê at Alexandria, so that the iron statue contained in it might have the appearance of being suspended in mid-air; but the project was interrupted by his own death and that of King Ptolemaeus who had ordered the work to be done in honor of his sister.[29]

Later writers asserted that such floating statues actually existed in various temples. Whether the writers were merely improving on Plinius, or whether some such statue did "float" by means of a fine wire, we cannot now tell. The Muslim tradition of Muḥammad's coffin, floating poised betwixt heaven and earth, probably stemmed from this same Hellenistic source.

The one thing we can be sure of is that, for mechanical reasons, no such statue could be so poised by magnets of the kind the ancients had. The story belongs with a group of legends about the mysterious powers of magnetism, along with the tale of the Magnetic Mountain. The latter started with the Egyptian geographer Claudius Ptolemaeus (+II). Ptolemaeus stated that in the Far East, near the Isle of Satyrs, lay an island mountain of lodestone, which destroyed all ships passing near it by drawing out the nails that held them together. The story was repeated many times: in the medieval legends of Virgil the Magician, in the tale of Duke Ernst of Swabia, and in the geography of the Moor Idrisi.

The yarn may have been invented to explain the fact that so many ships and boats in the Indian Ocean are built without nails, by sewing planks together or fastening them with wooden pegs. The real reason for such shipbuilding methods is simply that the vessels are built in countries so poor in mineral wealth that the shipwrights cannot afford to use iron.

The greatest Hellenistic engineer and one of the greatest intellects of all time was Philon's contemporary, Archimedes of Syracuse (–287 to –212). Archimedes studied in Alexandria, where he came to know several of the leading scientists of that teeming hive, such as Aristarchos the astronomer.

Eventually Archimedes returned to Syracuse, where he spent the rest of his life. Although he was not rich, the fact that he was a kinsman and friend of the tyrannos Hieron II assured him leisure in which to think and experiment. A mathematician and engineer, he is the one man of classical times who in sheer brain power stands on the same level as Aristotle.

In the field of mathematics, Archimedes discovered the ratio of the surface of a right circular cylinder to that of a sphere inscribed in it. He proposed a new system of numerals to handle large numbers. He calculated the ratio of the circumference to the diameter of a circle (π) to be between $3\frac{1}{7}$ and $3\frac{10}{71}$, the most accurate calculation up to that time. He also established the laws for finding the centers of gravity of plane figures and made many other mathematical discoveries. In addition, he did some work in astronomy.

Archimedes' discoveries in engineering were equally momentous, although Plutarch would have us believe that the Syracusan looked upon such practical applications of his knowledge with gentlemanly disdain. According to Plutarch, Archimedes:

> . . . possessed so high a spirit, so profound a soul, and such treasures of scientific knowledge, that though these inventions had now obtained him the renown of more than human sagacity, he yet would not deign to leave behind him any commentary or writing on such subjects; but, repudiating as sordid and ignoble the whole trade of engineering, and every sort of art that lends itself to mere use and profit, he placed his whole affection and ambition in those purer speculations where there can be no reference to the vulgar needs of life; studies, the superiority of which to all others is unquestioned, and in which the only doubt can be whether the beauty and grandeur of the subjects examined, and of the precision and cogency of the methods and means of proof, most deserve our admiration.[30]

This is typical Platonic snobbery. Many ancient writers, such as Cicero, express the idea that the essence of a true gentleman is his refusal to have anything to do with things of practical utility.

Did Archimedes share this attitude? We cannot tell. His works on hydrostatics and on "sphere making" certainly touched on engineering, Plutarch to the contrary notwithstanding. And even if Archimedes paid lip service to the Platonic ideal of gentlemanly uselessness, it is hard to believe that he could have been so good at engineering and have done so much of it if he had not really liked it.

In engineering, Archimedes founded the science of hydrostatics and discovered "Archimedes' law," that a body partly or wholly immersed in a fluid loses weight equal to the weight of the fluid displaced. Weighing a king's crown led to this discovery and to the first measurement of specific gravity:

Hiero [the tyrannos Hieron II] was greatly exalted in the regal power at Syracuse, and after his victories he determined to set up in a certain temple a crown vowed to the immortal gods. He let out the execution as far as the craftsmen's wages were concerned, and weighed the gold out to the contractor to an exact amount. At the appointed time the man presented the work finely wrought for the king's acceptance, and appeared to have furnished the weight of the crown to scale. However, information was laid that gold had been withdrawn, and that the same amount of silver had been added in the making of the crown. Hiero was indignant that he had been made light of, and failing to find a method by which he might detect the theft, asked Archimedes to undertake the investigation. While Archimedes was considering the matter, he happened to go to the baths. When he went down into the bathing pool he observed that the amount of water which flowed outside the pool was equal to the amount of his body that was immersed. Since this fact indicated the method of explaining the case, he did not linger, but moved with delight he leapt out of the pool, and going home naked, cried aloud that he had found exactly what he was seeking. For as he ran he shouted in Greek: "*Heurêka, heurêka!*"[31]

Then, following his discovery, he is said to have taken two masses of the same weight as the crown, one of gold and the other of silver. When he had done this, he filled a large vessel to the brim with water, into which he dropped the mass of silver. The amount of this when let down into the water corresponded to the overflow of water. So he removed the metal and filled in by measure the amount by which the water was diminished, so that it was level with the brim as before. In this way he discovered what weight of silver corresponded to a given measure of water.

After this experiment he then dropped a mass of gold in like manner into the full vessel and removed it. Again he added water by measure, and discovered that there was not so much water; and this corresponded to the lessened quantity of the same weight of gold compared with the same weight of silver. He then let down the crown itself into the vase after filling the vase with water, and found that more water flowed into the space left by the crown than into the space left by a mass of gold of the same weight. And so from the fact that there was more water in the case of the crown than in the mass of gold, he calculated and detected the mixture of the silver with the gold, and the fraud of the contractor.[32]

Archimedes also worked out the law of the lever, proved it mathematically, and developed the theory of mechanical advantage. In telling his cousin Hieron of this discovery, he boasted: "Give me a place to stand on and with a lever I will move the whole world!"[33]

The tyrannos challenged him to prove his assertion. Archimedes did so by single-handedly launching one of the largest ships in the world. He turned a windlass connected to the ship by a series of compound pulleys;

perhaps he even invented the compound pulley block with two or more pulleys in it.

The ship launched by Archimedes was but one of several ships of unprecedented size, which the Hellenistic kings built in –III. At this time there took place in the eastern Mediterranean a naval race like the race in battleships of the first half of +XX, before the developments of the Hitlerian war rendered the battleship obsolete.

This race began when Demetrios Poliorketes, the Successor who had besieged Rhodes and who had already constructed an elevener before that campaign, directed his Phoenician shipwrights to build a thirteener, a fifteener, and a sixteener. The other Successors followed hasty suit, because one of these monsters outmatched several smaller ships. When Demetrios was overthrown (–285), the first two Ptolemies, then jointly ruling Egypt, got the fifteener, while Lysimachos, the Successor who ruled Thrace and western Asia Minor, got the sixteener. When Demetrios' descendants regained the throne of Macedonia, they obtained the sixteener and kept it down to the end of the kingdom (–168). Then the victorious Romans took it to Rome as a trophy.

Meanwhile, in Egypt, Ptolemaios II Philadelphos built a twentier and a pair of thirtiers. These held the record until his grandson, Ptolemaios IV Philopator, who reigned from –221 to –205, built a fortier. This was the largest war galley of all time.

Philopator, an odious character in most respects, had a passion for spectacular ships. He built a 300-foot pleasure barge: a flat-bottomed floating palace 45 feet wide. Next he built the fortier, 420 feet long and 57 feet wide. The ship had a double bow, a double stern, and seven rams. Four thousand men rowed it, those of the upper bank pulling 57-foot oars counterweighted with lead. It never fought, but its sheer unwieldiness inspired another capital invention, the dry dock:

At the beginning [the fortier] was launched from a kind of cradle which, they say, was put together from the timbers of fifty five-bank ships, and it was pulled into the water by a crowd, to the accompaniment of shouts and trumpets. Later, however, a Phoenician conceived the method of launching by digging a trench under the ship near the harbor, equal in length to the ship. He constructed for this trench foundations of solid stone seven and a half feet in depth, and from one end of these foundations to the other he fixed in a row skids, which ran transversely to the stones across the width of the trench, having a space below them six feet deep. And having a sluice from the sea, he let the sea into all the excavated space, filling it full; into this space he easily brought the vessel, with the help of unskilled men; . . . when they had barred the entrance which had been opened at the beginning, they again pumped out the sea-water with engines. And when this had been done, the ship rested securely on the skids aforementioned.[34]

We can be sure that this pioneer dry dock did not have lock gates of the modern type but was closed and opened by the laborious method of dumping earth into its entrance and digging the earth out again.

The water was probably pumped out of the dry dock by means of the Archimedean screw, which Archimedes is said to have invented during his stay in Egypt. This pump, still commonly used for irrigation in Egypt, is a cylinder divided inside by a helical or screw-shaped partition running its whole length. The screw is mounted slantwise so that its lower end dips into the water to be pumped. When the cylinder is turned, water is trapped in each turn of the helix, raised to the upper end, and spilled out.

In Syracuse, meanwhile, the tyrannos Hieron undertook a program of shipbuilding with more practical ends in view. In Hieron's day, while Sicily was still independent, it derived its prosperity from the export of wheat. So Hieron built a fleet of wheat transports.

These ships were so successful that Hieron determined to build a ship that should combine the virtues of all types in one hull. It should be a combination of war galley, royal yacht, and freighter.

Under Archimedes' supervision, a Corinthian architect named Archias built the great *Syrakosia,* "Lady of Syracuse." This was a twentier with three masts—perhaps the first three-masted ship in the world. The hull was sheathed in lead to discourage marine growths. Armament included turrets for archers and a catapult throwing three-talent balls. There were stalls for horses, a tank for fish, a hold for wheat and other cargo, a set of luxurious cabins for the dictator and his friends, and everything else that the fertile minds of the builders could think of.

After Archimedes had singlehandedly launched the *Syrakosia* by means of his windlass and pulleys, faults of the design transpired. For one thing, she was too big for most of the harbors of Sicily. It is also a good guess that Hieron found that a ship designed to act as a warship, yacht, and merchantman all at the same time would not be very good at any of these tasks. It would be too slow for a warship, too costly for a merchantman, and too crowded for a yacht. In the end, Hieron sent the *Syrakosia* as a gift to Ptolemaios Philopator, who liked to collect such nautical freaks.

Many typewriter ribbons have been worn thin in argument over the arrangement of the oars on these gigantic galleys, which were not exceeded in size by any other ships until +XVIII. No surviving ancient writings describe the arrangement; no ancient works of art or diagrams show this arrangement; and no ancient ships have come down in such a shape as to settle the question.

Consider Philopator's fortier. Forty banks of oars, one above the other, is ridiculous. Such a ship would be as high as it was long, and the oars of the upper banks would be quite unmanageable.

But forty men on one oar would be quite as impractical. As each rower needs at least a yard of oar for himself and his elbows, the length of the loom grasped by the rowers would be at least 120 feet, which means an oar with a total length of over 500 feet! And the upper oars of Philopator's fortier were only 57 feet long.

The probable answer to this puzzle is that of Dr. Lionel Casson, that the fortier was a catamaran–a ship with two hulls joined by a common deck. If each of the twin hulls had three banks of oars on each side manned by, from top to bottom, eight, seven, and five rowers respectively, that would give forty files of rowers. It is possible that the other super-galleys, such as the twentiers and thirtiers, were also catamarans.

Actually, the greatest beam that any ancient ship is known to have had was found on one of two enormous barges, which the Roman emperors kept on Lake Nemi. The larger ship was 240 feet long and 78 in beam; the other only slightly smaller.

For many centuries, these barges lay on the bottom of the lake, resisting all efforts to raise them. About 1446, the versatile Renaissance architect, Leon Battista Alberti, sent divers down to explore the wrecks and tried without success to haul them up.

In 1535, another attempt was made by Francesco De Marchi, who examined the ships in person by means of a primitive diving suit. This device, invented by Guillaume of Lorraine, was a wooden bell reinforced by metal hoops, suspended from a boat on the surface and covering the upper half of the diver.

The intrepid De Marchi had a difficult time. He suffered from a nosebleed, caused by the change of pressure, and got entangled in his harness. He could see practically nothing through the small window of the bell because the water was so murky. Most maddening of all, a host of little fishes, drawn by the remains of the lunch of bread and cheese that De Marchi had brought down with him, swarmed around and nibbled his exposed flesh in spite of his efforts to shoo them away.

The attempts of Alberti, De Marchi, and several later explorers to

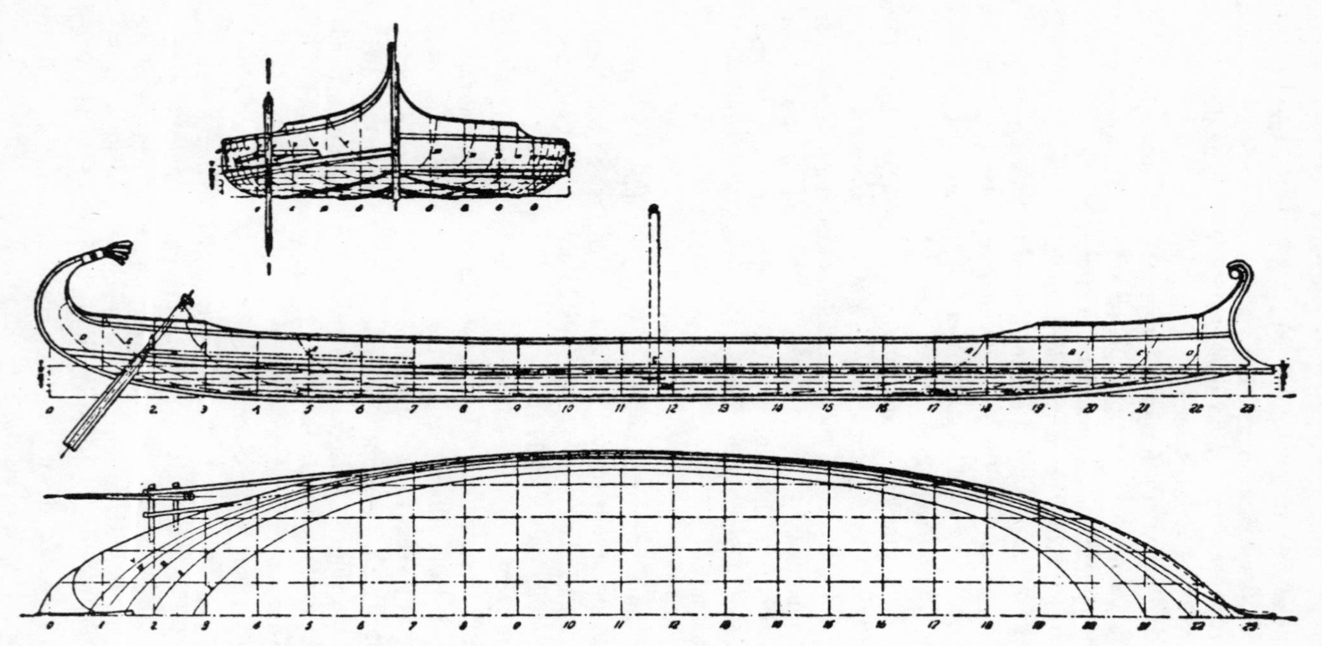

Fig. 9. The hull of the smaller of the two ships of Lake Nemi, as restored by the Italian Ministry of Marine.

raise the ships merely damaged them further. Many parts and other relics were brought up–timbers (which the peasants promptly stole for firewood), nails, bronze ornaments and statues, and pieces of tiling and mosaic–most of which disappeared into private collections. Not until the present century were the entire ships recovered.

Between 1927 and 1932, Italian archeologists, using an ancient Roman drainage tunnel, drained Lake Nemi. The project suffered many delays and misadventures; a gale sank the barge with the pumps, and once the work was stopped for seven months while the lake filled up again. But at length the ships were exposed to the light of day. By special four-track railroads laid in the mud of the lake bed, the archeologists hauled the two hulls out on shore and built a beautiful museum around them.

These ships are something of a mystery. No surviving ancient writer mentions them, and they were not rediscovered until 1444. For a long time the most popular theory was that the mad emperor Caligula built them as pleasure barges on which to stage orgies.

When the ships were recovered, the archeologists marveled at the graceful form that the hulls possessed despite their broad beam, and at the care and skill with which they had been built. The pine of the hulls was protected by vermilion paint, tarred wool, and a triple layer of leaden sheathing. The ships were steered by 37-foot quarter rudders or steering oars, two on the smaller ship and four on the larger. The two extra rudders of the larger ship are thought to have been mounted at the bow, to steer the ship when the rowers were backing it. Each ship had two auxiliary keels on each side of the main keel, and most metal parts were gilded. Parts of bronze and iron had resisted corrosion amazingly well, as a consequence of the purity of the metal.

Although the hulls were sound, the superstructures had collapsed. This collapse was the result partly of the great weight of marble, bronze, and other ornamental substances that they bore, and partly of earlier attempts to raise the ships.

When the lake was drained, many interesting objects, which had formed parts of these ships, were recovered. These findings did not, however, settle the question of how the ships were propelled. In large ancient galleys the oars were pivoted, not in the ship's side, but in a projecting frame or outrigger called the *apostis,* extending out over the water and giving the rowers nearest to the ship's side enough mechanical advantage to make their strength effective.

The larger of the two Nemi ships had an apostis along each side, so that it may have had at least one bank of oars. By measuring the space available inside the hull, one expert concluded that there were four or five men to each oar. Another, however, thought that the oars had but one man apiece and that the rowers sat out on the apostis. The smaller ship had no apostes and was probably towed about the lake.

The most extraordinary find was a pair of turntables revolving on small wheels arranged in a circle. Under one platform the wheels took the form of eight bronze balls, each having a pair of trunnions, which fitted into journals in the turntable. Under the other, the wheels comprised eight tapered or conical wooden rollers similarly journaled. The purpose of these turntables is not known, but they may have been statue bases.

These wheels are not quite true ball and roller bearings. True ball and roller bearings roll freely in a circular groove called a race, without any trunnions or journal bearings. Still, nothing more like this is known until +XVI, when Francesco Martini and Leonardo da Vinci suggested similar anti-friction bearings and millwrights and clockmakers put them to use.

Another technical triumph of the Nemi ships was a bucket-chain hoist, designed as a bilge pump. At the top, this pump had a horizontal shaft on which was mounted a drum, over which the chain of buckets passed. At each end of this shaft was a wooden flywheel with a crank handle. At least, such a flywheel was found, although not all scholars agree that it formed part of the pump.

The meaning of this design is that this is the first known example of a crank mechanism, outside of hand mills or querns. Cranks were used more and more from +I on. The Emperor Julian's physician Oreibasios (+IV) mentioned cranks in his account of machines for resetting dislocated limbs. It is also a good guess that in Roman and medieval times, cranks were employed in racks and other torture machines–a grim field in which to study the evolution of machinery!–because they were useful in applying carefully graded degrees of force. A sketch in the Utrecht Psalter (about +850) shows a man sharpening a sword on a grindstone, which another man turns by means of a crank. So by +IX the crank had evidently become a common device.

There was also an elaborate hydraulic installation, with a force pump, lead piping, and a finely made bronze valve. Two anchors were found. One of the common classical type was made of oak, with iron-tipped flukes and a stock (the crosspiece that keeps the flukes vertical so that one shall dig in) of lead. A wooden anchor must obviously be weighted somehow to make it sink. The other anchor was of iron with a wooden

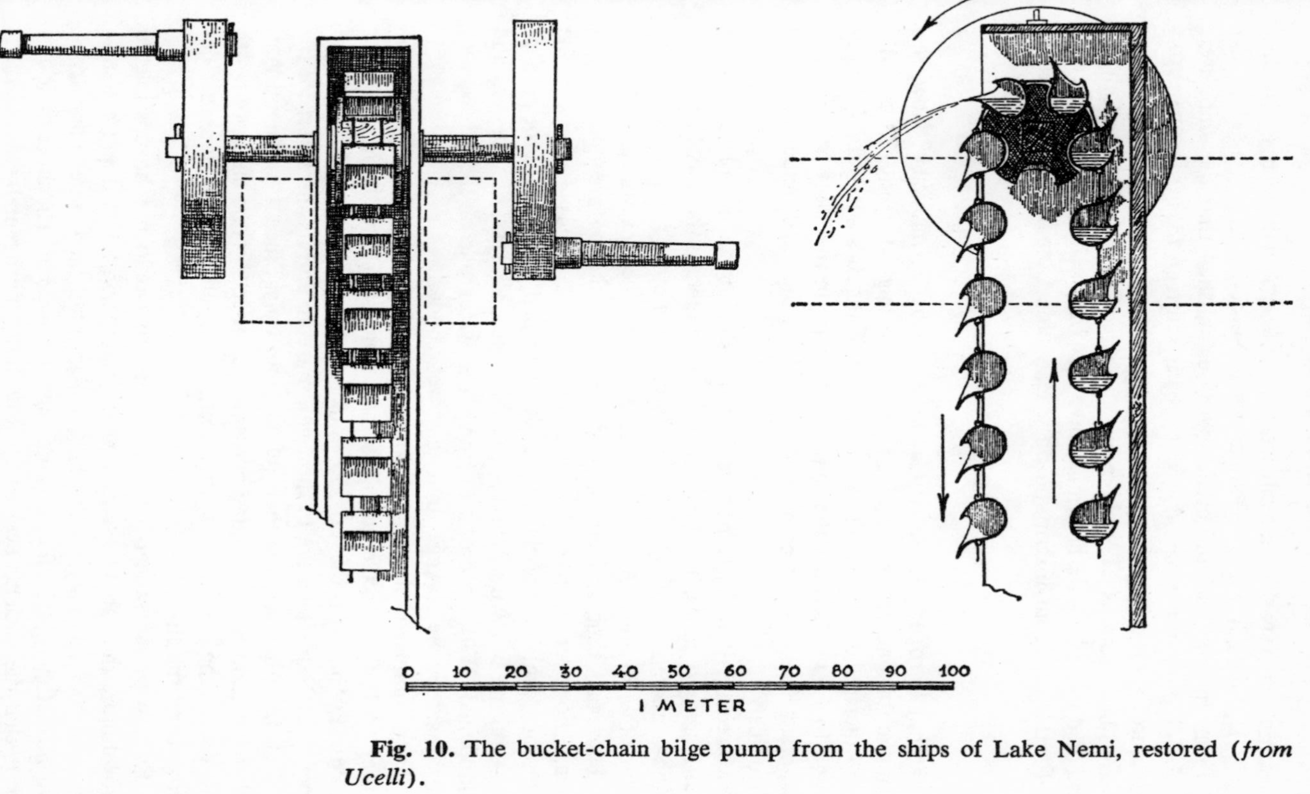

Fig. 10. The bucket-chain bilge pump from the ships of Lake Nemi, restored (*from Ucelli*).

sheathing, and with a movable stock of a type not seen again until +XVIII, when it was reinvented in the Netherlands.

From inscriptions, the archeologists established that the ships were probably built in the period +35 to +55. Thirteen tiles from the superstructure of one ship bore the stamp of Dama, a slave of Domitius Afer. This Afer was a noted orator, who served as consul under Caligula and died in +59; so the ship must have been built before +59.

Furthermore, an inscription on a lead pipe from the wrecks reads:

C. CAESARIS AVG. GERMANIC

meaning: "Property of Gaius Caesar Augustus Germanicus." Now, this was the full and real name of Caligula, who reigned from +37 to +41. But other indications point to the reign of Caligula's successor Claudius (+41 to +54). Perhaps the ships were begun in the short principate of the crazy Caligula but completed, or at least modified, under the pious Claudius.

Moreover, it is fairly certain that the ships were not pleasure craft but floating temples. They may have been erected to the ancestors of the Julio-Claudian families, or to the Egyptian mother-goddess Isis, or to some other divinity. All these hypotheses have their partisans.

The exact time and occasion of the ships' sinking are unknown. It seems, however, as though after the reign of Claudius–perhaps in the reign of Nero; perhaps during the civil war that followed Nero; perhaps not until the following century–the ships were stripped of valuables and abandoned. With no crews to pump the water out of the bilges, they slowly settled by leakage until they went under.

The remains of these marvelous ships, alas, were destroyed in the spring of 1944. The Germans, having previously shown their regard for *Kultur* by burning the 80,000 books and manuscripts of the Royal Society of Naples, set fire to the Nemi museum and its ships, just for the hell of it, in the course of their retreat from Italy. Since then the museum has been restored, and today one can see models of the ships, one-fifth the size of the originals, together with those relics of the ships not destroyed by the fire.

Even with the saucer-like Nemi ships, it would not be possible to crowd more than about twenty rowers–half facing forward and half aft–on each oar. Actually, in the Middle Ages, ten men were found to be the most that could efficiently be put on one oar. Therefore it is not likely that the ancients ever used more save experimentally.

The actual arrangement in large classical galleys–eighters or tenners and up—was probably that of a super diere or triere, with the oars in two or three banks and a varying number of rowers on each oar. Hence Demetrios Poliorketes' elevener, for instance, may have had oars in two banks, with six men on each upper oar and five on each lower, so that the number 'eleven" still means the number of files of rowers on each side. Philadelphos' thirtiers might have had three banks of ten-man oars or equivalent; or they might have been catamarans.

Philopator's fortier we have discussed. If four-bankers were tried, they would have given the same old trouble with keeping the oars in step as with four banks of one-man oars. Certainly, Philopator's ship was no great success. Plutarch says it "was for show, and not for service, scarcely differing from a fixed edifice ashore, and was not to be moved without extreme toil and peril."[35] As nothing is heard of the sluggish monster after Philopator's time, the chances are that his successors soon scrapped it.

Shipbuilding was only one of Archimedes' technical interests. He also built an astronomical instrument for demonstrating the movements of the heavenly bodies. His book on "sphere making,"[36] now lost, described this orrery. This was a "planetarium" in the original sense of the word, used before the Zeiss optical planetarium was invented to bring the wonders of the heavens to the masses at the cost of cricks in their necks. The older device is also called an "orrery" after the Fourth Earl of Orrery (*c.* 1700) who had one made for himself.

As nearly as we can tell from allusions to Archimedes' orrery, it was a machine in which, by turning a wheel, pointers, balls, or disks representing the sun, the moon, and the planets were made to revolve around a bronze ball representing the earth. The idea stemmed from the simple, non-mechanical bronze celestial globes of the earlier scientists Thales of Miletos and Eudoxos of Knidos, on which the fixed stars and the constellations were marked.

Later, the philosopher Poseidonios and the engineer Heron of Alexandria built similar orreries, though even fewer details are known about these devices. Students long thought that Archimedes' clockwork universe was either a myth or a premature invention that died without issue. Now, however, there is reason to suspect that it not only existed but also gave rise to a long line of progeny that comes right down to modern clocks. As nearly all trace of these early devices has perished, we are like a paleontologist who is trying to reconstruct the evolution of a whole family of dinosaurs from a few claws and teeth, widely scattered through geological time and space.

After Archimedes, the next fossil of this branch of engineering is the Antikythera machine. In 1900, a crew of Greek sponge divers discovered a mass of statues and other antiquities lying in 200 feet of water off the Aegean island of Antikythera, where a ship of classical times had foundered. Captain Kondos, with more public spirit than many such discoverers, reported his find to the Greek government and was engaged to bring up the relics. This he accomplished with great risk and effort.

Part of the haul consisted of several lumps of bronze, badly corroded and partly covered with limestone. Although at first sight the find seemed to be nothing but a bucketful of junk, a Greek archeologist recognized it as the remains of a mechanism.

Little by little during the following decades the thing was cleaned. In 1958 Derek Price, a young British historian of science, persuaded the Greek National Archeological Museum to let him study the remains.

The Antikythera machine turned out to be an astronomical computer. In its original form it was a wooden box a little over a foot high, looking something like a large eighteenth-century table clock.

In front, a hinged bronze plate, covered with inscriptions telling how to use the device, served as a door. Behind this door was another plate with a large circular hole through which the gearing behind could be seen. Around this hole was a dial with a slip ring on its rim.

Two graduated scales ran around this circle, one on the bronze plate and the other on the movable slip ring. The outer scale listed the signs of the zodiac; the inner one, the months of the year. The purpose of the slip ring was to adjust the device when (as happened under the Egyptian calendar, which lacked leap years) the months got out of phase with the seasons. Moreover, examination of the setting of the slip ring, together with other evidence, shows that the device must have been built just about the year –82, when Julius Caesar was a youth of eighteen.

The back of the machine had another bronze door and behind that a bronze plate with two dials, one above the other and each provided with several slip rings. Between the front and back dials was a mass of clockwork made of bronze cogwheels with simple triangular teeth.

A large four-spoked wheel was driven by a crown gear and a shaft, which projected through the side of the box and was doubtless furnished with a key or handwheel. This main wheel in turn drove twenty or more smaller gear wheels, some in complex planetary mountings. These gear wheels turned pointers on the dials to indicate the phases of the moon and the positions of the planets on different dates. A couple of minor breakages had been mended, showing that the device actually worked.

The ship bearing this astronomical computer, perhaps from Rhodes to Rome, sank about –65. Nothing more is heard about such devices for over a thousand years. Yet they must have been built, for about +1000 al-Birûni, an Iranian savant who traveled in India, described a similar machine. As in the Antikythera machine, al-Birûni's computer had 60° triangular teeth, and his gear wheels were mounted on axles with square shanks.

Much nonsense has been uttered about the lost secrets of ancient sages. There never was any prehistoric civilization having all the modern scientific knowledge and engineering techniques, as some cults would have us believe.

On the other hand, these astronomical computers represent an advanced, sophisticated line of development, actively pursued by a whole school of engineers, of which all but a few traces have disappeared. Just as the course and very existence of an underground river may be known only from a few places where wells and mine shafts have broken into it, so these few traces of clockwork show the evolution of an advanced form of mechanical engineering from ancient times on down, even though few traces of the engineers remain.

Hieron II, tyrannos of Syracuse, was for the most part a staunch ally of all-devouring Rome. But, after he died in –215, his successors went over to the Carthaginian side in the Second Punic War. As a result, Syracuse found itself besieged by the Romans, with Appius Claudius Pulcher heading the Roman army and Marcus Claudius Marcellus commanding the fleet.

Hieron had, some time before, persuaded his cousin Archimedes to serve as his general of ordnance and to prepare engines for the defense of the city. As a result, the Romans came up against the world's most advanced artillery. Pulcher attacked the land side with scaling ladders and penthouses. Soon, however, his forces reeled back from the storm of arrows, sling bullets, crossbow bolts, catapult darts, and catapult balls that tore through their ranks.

Marcellus, attacking from the seaward side, had made careful preparations. He had built four engines, each based upon a pair of galleys fastened together. On each platform over a pair of ships he erected a mast and a hoist. By means of this hoist, a device called a "harp" or *sambuca* was brought into use. This was a large ladder or staircase with sides and a roof to shield those using it. At its upper end was a gangway, on which four soldiers, protected by wicker shielding, fought off those who tried to stop the operation. The attackers maneuvered the sambuca into place so that the gangway lay atop the defenders' wall, and the soldiers could swarm into the defenses.

At least, that was the theory. While the approaching Romans quailed under the storm of missiles, a series of 600-pound leaden weights and stone balls smashed the scaling engines into kindling.

Plutarch says that these missiles were shot from catapults, but Polybios says that the weights were dropped from the ends of cranes that swung out over the wall as the engines neared it. Plutarch's tale is the more romantic, for such a catapult would be the biggest ever used in the ancient world.

But Polybios' dryly sober account is the more plausible. To judge from Polybios' narrative, Archimedes used an enlarged version of an existing weapon called the "dolphin," with which merchant ships were sometimes armed.

Whereas the mast on a galley was a light spar, made to be taken down and left ashore before a battle, a merchantman had a stout, permanent mast. Some merchantmen, operating in pirate-infested waters, had a kind of strongly built boom or yard attached to this mast. At the end of this yard was slung, by rope and pulleys, a heavy leaden weight. When a pirate laid alongside and its hairy, screeching horde prepared to board, the crew of the merchantman swung the dolphin out over the pirate's deck and released the rope. Down went the missile, to smash through the thin piratical deck as through papyrus. A later Greek writer (+VI) describes a similar device, mounted on a wheeled platform for use on land.

Another Archimedean trick was to lower a grapnel from a crane, catch the bow of a small Roman ship, and hoist until the ship was vertical and the crew tumbled into the blue Mediterranean. Then the rope was let run, dropping the ship to wallow awash and useless.

A story grew up in later centuries that Archimedes:

> . . . devised an hexagonal mirror, and at an appropriate distance from it set small quadrangular mirrors of the same type, which could be adjusted by metal plates and small hinges. This contrivance he set to catch the full rays of the sun at noon, both summer and winter, and eventually, by the reflection of the sun's rays in this, a fearsome fiery heat was kindled in the barges, and from the distance of an arrow's flight he reduced them to ashes.[37]

It is a colorful picture, but not likely. As a practical matter, Archimedes would have needed a much bigger set of mirrors than he could probably have made. People who have experimented with sun-powered steam engines have found that it takes 200 to 300 square feet of mirror to generate steam for a little one-horsepower engine. Moreover, Archimedes would have had to make the target ship hold still for some time for his burning glasses to take effect.

It is, however, possible that Archimedes did perform experiments with such mirrors. He might even have set some old hulk on fire and let people whisper about his fearful secret weapon. But the exact truth, as in so many ancient matters, has vanished down the voracious gullet of time.

Despite Archimedes' wonderful war engines, Marcellus captured Syracuse by a surprise attack on a weakly defended tower while the citizens were celebrating a religious festival. Marcellus, who was not without humanity and respect for Archimedes' genius, ordered his soldiers not to molest free Syracusans. It was impossible to forbid them to loot, for a general who tried to stop his soldiers from plundering a city captured after a hard-fought siege would have been instantly slain by his own men.

During the looting, a legionary approached the seventy-five-year-old Archimedes, who was drawing geometrical diagrams in the sand. Archimedes, preoccupied with his mathematics, cried: "Keep off, you!" and the angry soldier killed him. At least, that is the commonest story. There are others, and any or none may be true. Marcellus mourned Archimedes but took the latter's orrery as booty to Rome, where it could still be seen a century later.

The death of Archimedes, practically speaking, ends the story of Hellenistic engineering. It is not that Hellenistic engineering really stopped, albeit it produced no more such towering figures. Instead, it was absorbed into the story of Roman engineering, just as the Hellenistic kingdoms, one by one, were swallowed up by Rome.

THE EARLY ROMAN ENGINEERS

SIX

Roman engineering, like that of the earlier watershed empires, was based upon the intensive application of rather simple principles, with plenty of raw materials, cheap labor, and time. Rome possessed abundant supplies of brick, stone, and timber. Labor was especially cheap during the last century of the Republic and the first of the Principate, because the swift extension of the Empire had crowded Rome with thousands of slaves.

The Romans devoted more of their energy and capital to useful public works than did their predecessors. They built roads, harbor works, aqueducts, temples, forums, town halls, arenas, baths, and sewers. Although, like all ancient peoples, the Romans had little sympathy to spare for the underdog, they did like to see the masses prosper and enjoy themselves. A magnate, governor, or emperor might be a scoundrel in other respects, but he believed that he gained honor by presenting the people with some useful or entertaining public work. Under the Republic this was a method of bidding for votes; and, even after elections had ceased in the Principate (the technical name for the early Empire) the tradition lived on.

Therefore Frontinus, Trajan's water commissioner, is not being merely patriotic when, in describing Rome's aqueducts, he bursts out: "With such an array of indispensable structures carrying so many waters, compare, if you will, the idle Pyramids or the useless, though famous, works of the Greeks!"[1]

Modern writers often point out that the Romans contributed practically nothing to pure science: "To the scientific habit of mind which has made our present attempt at civilization possible and is rapidly making it impossible, no Roman ever contributed anything."[2] Still, although Rome produced almost no pure or research scientists, she brought forth many remarkable soldiers, statesmen, administrators, and jurists.

Moreover, in the applied sciences, Romans were very active. Roman architects and engineers made far-reaching advances towards improving the machinery of everyday life and reshaping the earth for man's convenience. In the story of civilization, these things have their own importance.

Although the Romans inherited some of their engineering ideas from the Etruscans, they adopted many more from the swarm of Greeks and orientals who came to Rome, voluntarily or as captives, in the last centuries of the Republic. One of the newcomers, for instance, was Hermodoros of Salamis (–II) who built temples in Rome to Mars and to Jupiter the Stayer.

Moreover, the Romans themselves actively developed new methods of building. This vigorous engineering activity continued from the early days of Roman expansion in –IV for nearly eight centuries, until the barbarians overran the western half of the Empire.

Roman engineering was mainly civil engineering: the building of roads, bridges, public buildings, and other permanent structures. A consul, senator, or other magistrate commanded the whole of such a governmental enterprise. Under him the *architectus* or engineer, in his turn, bossed a crew of minor technicians: *agrimensores* or surveyors, *libratores* or levelers, and others. In addition, private builders without special technical training practiced, for private landowners, the craft in which they had been reared.

The profession of *architectus* became so popular by –I that even the emperor Tiberius' son Drusus dabbled in it. Vitruvius complained "that an art of such magnificence is professed by persons without training and experience, by those who are ignorant not only of architecture but even of construction . . ." But then, Vitruvius wanted his ideal architect to "be a man of letters, a skillful draftsman, a mathematician, familiar with historical studies, a diligent student of philosophy, acquainted with music; not ignorant of medicine, learned in the responses of jurisconsults, familiar with astronomy and astronomical calculations."[3]

Although Vitruvius gives sound reasons for having all these skills—

for instance, a knowledge of music is useful in tuning catapults by striking the tension skeins–the difficulty is the same as in all professions in all ages. Vitruvius' requirements are a counsel of perfection, because nobody lives long enough to learn everything that might be useful to him.

The Etruscans, from whom the Roman engineers originally borrowed their technics, were the Romans' northern neighbors. Although the Etruscans were racially much like other Italians, their language, only partly deciphered, was a non-Indo-European tongue unrelated to any other known language. Nobody knows where they came from. The ancients thought they had come from Asia Minor, while some modern scholars believe that the Etruscans, like the Basques, were a pocket of pre-Indo-European speakers who had dwelt there before the horse-driving Aryans conquered the rest of Europe.

The Etruscans, for instance, may have invented the candle. While the taper, long known, was a string dipped only once or twice in melted wax, the candle was dipped many times to build up a thick solid rod of wax. The candle, used at first as a religious accessory rather than as a means of profane illumination, spread around the Mediterranean in Hellenistic times. It was known to Philon of Byzantium in –III. In addition, the Etruscans pioneered the making of dentures.

The larger Etruscan houses were built on the Near Eastern central-court plan, with a hole in the roof over the central court and a cistern below the hole to catch the rain water. Etruscan temples followed the Greek pattern, save that the Etruscans, even after they had substituted stone columns for wooden posts, continued to make their lintels and roof structures of wood. As wood is more efficient for long beams than stone, their temples had more widely spaced columns and broader eaves than those of Greece. Etruscan builders covered the wooden roof structure with an elaborate coating of terra cotta, or brick molded into ornamental forms.

The Etruscans differed from all their neighbors in the fact that they used the arch. We cannot tell, though, whether they got it through some unknown intermediary from Mesopotamia or invented it independently. They did invent a system of decorating a series of arches with a molding that ran from the base of one arch to the next (the impost) and then around the outer edge of each arch in a semicircle (the archivolt).

The Etruscans also conceived the bloodthirsty idea of making slaves, prisoners, and criminals fight each other as gladiators. The Romans took over all these culture traits and, by their conquests, spread them around the Mediterranean world. Rome began its expansion from a mere city state in –V and never stopped growing until Trajan completed his conquest of Dacia (modern Rômania) and Babylonia at the beginning of +II.

The early Roman house was built on the Etruscan plan. The *domus* of a prosperous early Roman bourgeois had about a dozen rooms, ranged around a partly roofed court. There was a square hole in the roof to let in the rain and a cistern beneath to catch it. This court was called the atrium, from *ater,* "black," because the walls were blackened by the smoke of cooking. Poor peasants, of course, continued to live in one-room huts as they always had.

When in Hellenistic times the Romans came under Greek influence, they added a second court on the Greek model. This court was called the *peristylum* or peristyle, from the Greek words meaning "surrounded by columns."

The practice of planting a flower garden in the peristyle also came into the classical world at this time. Such formal gardens were a Persian custom, unknown to the Greeks of the Golden Age, who cared little for the beauties of nature and who, if they planted anything, preferred something edible like onions or cabbages.

But the Hellenistic Alexandrines, living on a wretched and monotonous sand spit, discovered the beauties of nature and wrote sentimental poems about shepherds piping in an imaginary Arcadia of groves and waterfalls, nymphs and satyrs. The real Arcadia is a rugged, barren, inhospitable country, but few of the poets or their admirers had been there.[4]

As wealth and sophistication increased in the last two centuries of the Roman Republic, other types of dwelling house evolved. The rich built large country houses, called *villas.* While these were of many kinds, in the early decades of the Principate they evolved away from the traditional Mediterranean type.

The traditional house, which was closed in upon itself, now unfolded and opened outward, like a flower. The peace of the early Principate made it no longer necessary to build each house like a miniature fortress. A common type of villa consisted of a main building, in a solid block, with a pair of wings, one containing the bedrooms and the other the bathrooms. Instead of running around a closed court, the colonnade extended in a straight line along the south side of the house.

At the same time, the increase in population in Italian cities, and especially in Rome itself, made builders eager to find space for more tenants in a given area. An ancient city could not expand so easily as a modern one, because of the walls that surrounded it.

The walls themselves did not present a serious obstacle, but military considerations did. If the wall was to serve as an effective defense, a space had to be kept clear on both sides of it. On the inside, a space about 30 yards wide was kept open so that troops could move quickly along it, engines could be set up in it to shoot over the wall, and the foe could not easily ignite the nearest houses by shooting incendiary missiles into the city.

The wall itself, to be effective, had to be at least 30 feet high and 15 feet thick. The walls of Martyropolis in Armenia, 20 feet high and 4 thick, were considered useless against a determined attack, even without heavy siege machinery, because a numerous enemy could overrun them with scaling ladders alone.

The clear space outside the wall was about 200 yards wide. In wartime, this space was shaped into a series of obstacles to slow the approach of the besiegers. First, a series of ditches was dug. Philon of Byzantium recommends three, about 35 yards wide and as deep as the nature of the soil allows. The earth from these ditches was piled up on the spaces between them. But these areas were also rendered more difficult by barricades of tree branches stuck into the ground, with the ends whittled to points so that they looked like antlers. This was the ancient equivalent of a barbed-wire entanglement. Bells hung from cords on a line of stakes gave warning of the approach of the foe at night.

A form of tank trap was made by burying a lot of large empty jugs, their mouths stopped by moss, just under the surface of the soil. Infantry could walk over the jugs without effect; but, when heavy wheeled engines tried to cross the same ground, the jugs broke and the wheels sank in.

Because of these defenses, once an ancient city had taken form, no builder could add new houses around the edge of the existing built-up area. Once the cleared strip inside the wall was reached, the builder had to go more than an eighth of a mile, outside the defenses, to find another building lot. Moreover, lack of transportation checked any tendency towards suburban sprawl.

So Roman builders did what builders in Babylon and Tyre had done before them, and what modern builders have done in places like New York City, where surrounding waters prevent the city from growing outwards in all directions. They went up. Buildings rose to at least three stories in –III and five or more by –I. Augustus set a limit of 70 feet on the height of buildings, and Trajan reduced this to 60 for greater safety.

These big modern-looking brick apartment houses, filling whole blocks, were called *insulae* or "islands." Under the Principate, insulae became far commoner in Rome than private houses of the old domus type. A survey of Rome about +300 listed 1,797 domûs and 46,602–twenty-six times as many–insulae.

Insulae furnished accommodations of every grade, from dark little single rooms several flights up to luxurious duplex apartments on the more desired lower floors. Writers of the early Principate complain of jerry-built insulae, which are always either catching fire and frying their tenants or collapsing and burying them.[5] However, the actual remains of insulae that have come down, as at Ostia, indicate well and solidly built structures. But then, perhaps that is why these particular buildings have survived; all the flimsy ones fell down or were demolished in ancient times.

For that matter, a good Roman house was built in a sturdier and more substantial way than most of the houses you can buy today. Roman builders had not learned all the short cuts used by modern developers. The houses they turned out were quite as durable and as handsomely decorated as a modern house–although, lacking running water, electricity, oil heat, upholstered chairs, and substantial tables, they would seem to us a lot less comfortable to live in.

Windows had long been closed by wooden shutters. In Hellenistic times, builders experimented with translucent windowpane materials: oiled cloth, sheepskin, mica, horn, and gypsum shaved down to a thin pane. Some of these materials continued in use for many centuries; but, in the more opulent homes, they were slowly ousted by glass, which at last had become clear enough for the purpose.

The Phoenicians took over the art of glass-making from the Egyptians and improved it greatly. They learned how to make glass clear enough for windowpanes and also (–I) how to *blow* glass. This new method lowered the cost of glass drinking vessels until nearly everybody could own them: vessels of clear glass for the rich and of opaque glass for the poor.

In the early Principate these glass-making techniques were brought to Italy. A curious story is told about an incident said to have happened in the reign of Tiberius.[6] A man went to the Emperor with a scheme for making unbreakable, flexible glass and proved his claims by dropping a vessel on the floor without breakage. Thereupon the Emperor had the man's shop demolished (or, in another version, had the man beheaded) lest his product somehow destroy the value of gold and silver.

We know enough about glass today to be sure that the story as it stands is untrue. But it may have a basis of truth. Forbes thinks that some enterprising manufacturer, by introducing improved Syro-Phoenician methods of making glass vessels into Italy, competed so severely with the makers of metal vessels that the latter raised an outcry to the Emperor, who intervened on the side of the metalsmiths.

This attitude was common among ancient rulers. From ancient to modern times, every important invention has brought an outcry from those whose livelihood would be harmed by its adoption. Sovrans have often heeded such complaints. In the reign of Emperor Vespasianus, for instance:

> An engineer offered to haul some huge columns up to the Capitol at moderate expense by a simple mechanical contrivance, but Vespasian declined his services: "I must always ensure," he said, "that the working classes earn enough money to buy themselves food." Nevertheless, he paid the engineer a very handsome fee.[7]

Queen Elizabeth I of England felt the same way about a proposed knitting machine. While it is useless to blame those whose business was threatened for trying to save themselves, such opposition from vested interests has greatly slowed technical progress. Ancient rulers did not understand that real income is the consumer goods one can buy with what one earns, that the goods cannot be consumed unless they are produced, and that therefore whatever increases production increases real wealth. If the vested interests had always won out, we should still be living as people did in ancient Sumeria.

During the later years of the Roman Republic, glass began to be used to let light into houses. The first panes were little round skylights, the glass of which was too irregular and impure for true transparency. But during +I, glass windowpanes of the modern type appeared. By the end of that century, glass factories had become common in Italy and were spreading into Gaul.

Even the greenhouse was known. The agricultural writer Columella (+I) advised raising cucumbers in wheeled boxes covered by *speculares* –transparent panes, whether of gypsum, mica, or glass–and suggested that these boxes be rolled out of doors on sunny winter days.

Door locks also approached the modern type. There were several kinds. In the typical Roman lock, the bolt had a number of holes in it.

When the bolt was in the locked position, several pins in the lock dropped down into these holes from above, much as the pins do in a modern cylinder lock, and held the bolt in place.

The key had a set of prongs, at right angles to its shank (like the bristles of a toothbrush) and corresponding to the holes in the bolt. When the key was thrust into the keyhole, these prongs were inserted into the bolt from below, pushing the pins up out of the way. Then a sidewise pull on the key drew back the bolt. The holes and pins were differently arranged in different locks, and a duplicate key was called an "adulterous" key—a name for which I can think of at least two explanations. Locks of this Roman type are still used in Arabia.

The Romans also took great strides in house heating. They invented central indirect heating; or rather, they reinvented it. In 1954, at Beycesultan, Turkey, a British expedition dug up the palace of the king of Arzawa, a kingdom that flourished in southwestern Anatolia before −1200. Here, ducts beneath the floors suggest a central heating plant. Then nothing more is heard of this invention for over a thousand years.

This disappearance is not really surprising. For, in the ancient world, an invention could be lost and rediscovered several times. The numbers of engineers and inventors were extremely small compared to those of today. Inventors often tried to keep their inventions secret; in the absence of patent protection, there was no other way to stop rivals from copying the invention and reaping where the inventor had sown. Inevitably, some inventors took their secrets to the grave. And, what with the lack of encouragement of inventors, the lack of mass production to multiply specimens of inventions, and the lack of printing to disseminate descriptions of them, knowledge of an invention was likely to be very restricted.

Because of the small scale on which invention was carried on, ancient inventions led a more precarious life than do most modern ones. Some catastrophe, such as the sack of a city, could easily destroy the only specimens of an invention, along with the only men who knew how to make it; and there were no patent files to which a later seeker could refer.

This does not mean that we should, like the occultists, believe that the ancients had scientific and engineering knowledge beyond our own. They did not. But it did sometimes happen that a useful invention was made, lost, and rediscovered centuries later.

In the case of central heating, the rediscoverer was a Roman businessman, Gaius Sergius Orata, who lived near Naples. About −80, Orata, already successful at raising fish and oysters for the market, thought he could do even better if he could only keep his edible sea-creatures growing through the winter. Perhaps he was inspired by the sweat baths of Baiae, heated by volcanic steam.

At any rate, Sergius Orata built a series of tanks which, instead of being sunken in the earth, were propped up on little brick posts. The smoke and hot air from a fire at one side of a tank circulated through the space below the tank to warm it.

Not yet satisfied, Orata applied his invention to human comfort. As I have told you, country houses in his time were evolving into a form with a central hall and a pair of wings, one of which was devoted to bathing chambers. Orata bought country houses, equipped them with *balnae pensiles* or "raised bathrooms," heated by means of ducts under the floor, and resold them at a lusty profit. As a result, Orata became famous for his ingenuity, his business acuteness, his refined and luxurious taste, and the jovial life he led.

Some aristocrats sneered at Orata as one of the new-rich. Pliny accused him of avarice, but this was the standard resentment that members of the landowning Roman gentry felt towards self-made men who had gained their wealth in vulgar trade.

Lucius Licinius Crassus (not the triumvir, but an older member of the family) once represented Orata in a lawsuit, which indicated that Orata was engaged in a bit of sharp practice in buying and selling houses. Orata also occupied public waters with his fish and oyster ponds at Lake Lucrinus, just off the Bay of Naples, until a suit was brought to make him stop. This time Crassus remarked that the lawyer for the defense "was mistaken in saying that keeping Orata away from Lake Lucrinus would deprive him of oysters; for, if he was prevented from catching them there, he would find them on the roofs of his houses."[8]

After Orata, builders applied his system, called a *hypocaust,* to whole buildings. Under the Empire, Romans in the northern provinces built hypocaust houses to keep winter at bay. Central heating died out with the fall of Rome and was not revived until modern times. The last-known hypocaust, strangely, was in the house of a +XI Viking earl of Orkney, the formidable Thorfinn the Mighty, on the isle of Birsay.

Under the Principate, Roman houses became more and more like those of today. The atrium disappeared; then the columns vanished from around the peristyle. Finally the peristyle itself dwindled and disappeared, as the house in one solid block was found to be easier to heat and made more efficient use of ground space than the old Mediterranean courtyard house.

For construction, the Romans used wood, stone, clay, mud brick, baked brick, and mortar. The wood and mud brick have disappeared, as is natural in a country with wet winters. For stone, the Romans first made use of tufa, a soft tan or brown volcanic rock. Then they began to use the harder *lapis gabinus* and *lapis albanus* (now called *sperone* and *peperino* respectively) formed by the action of water on a mixture of volcanic ash, gravel, and sand.

Later still, during the Principate, they exploited *lapis tiburtinus*—modern "travertine," named for the town of Tibur or Tivoli. This is a hard, handsome limestone; but, unlike the volcanic rocks mentioned, it is not fire-resistant. Under heat it crumbles into powder.

Baked brick, though long known, likewise did not come into general use until the late Republic. According to Vitruvius (–I), mud brick had been made illegal for house walls in the city of Rome.

Earlier builders often used baked clay in thin slabs called tiles. Roman bricks came in many shapes and sizes; but the Romans preferred long, wide bricks only an inch and a half thick. They had learned that such shapes were less likely to warp or crack in drying than thicker forms.

They also used lime mortar from the earliest times. But, for public buildings, they preferred to trim their stones as accurately as possible and depend upon a close fit and iron cramps to hold the structure together.

About –III, Roman builders made an important discovery. Near Vesuvius and elsewhere in Italy were deposits of a sandy volcanic ash which, when added to lime mortar, made a cement that dried out to rocklike hardness and even hardened under water. They called this stuff *pulvis puteolanus* from Puteoli,[9] where it occurred in huge beds. By mixing this cement with sand and gravel they made concrete.

Builders had experimented with mortars and plasters for many centuries; witness the concrete bed of Sennacherib's aqueduct. But here for the first time was a completely satisfactory waterproof concrete, which formed a synthetic rock as hard as most natural rocks. In fact, samples of Roman concrete that have come down to modern times in buildings, conduits, and the like are harder than many natural rocks would be after so many centuries of exposure.

The Roman builders did not at first fully realize the possibilities of their new material. They used it in a small, tentative, nervous way. After all, how did they know it would not soon crumble like common plaster?

But, as the strength and durability of the new substance became plain, builders used it more and more freely, until it became the typical building material for large structures under the Principate.

While we do not think of the Romans as great innovators, their builders deserve credit for one thing. Having stone, brick, and concrete to work with, they doggedly tried out almost every possible combination of these materials.

In the early days, the Romans built stone walls in the form called *opus quadratum*—"squared work." This was simple ashlar masonry with all the stones trimmed to a rectangular form of the same size. As a rule, not all the stones had their long axes parallel. To strengthen the wall, in some courses the alternate blocks, called "headers," had their long axes at right angles to the face of the wall, so that they extended into the interior of the wall, or clear through it, and helped to bond the whole together. The Greeks and their predecessors had used a similar system.

Another system was *opus incertum* or "random work." The stones were still trimmed to a rectangular shape, but were of different sizes, so that there was no continuous course of stones. Such a wall was more trouble to put together but stronger when finished.

Builders used similar systems in erecting walls of brick—first of mud brick, later of baked brick. But then came the discovery of concrete. At first concrete was used as little more than a superior mortar. As its properties became known, however, builders used more and more of it and less and less of the stonework or brickwork that it was supposed to be holding together.

Soon, walls were made with thin facings of stone or brick and a thick filling of concrete, spaced out with gravel, stones, and bits of broken brick and tile. As the builders did not have powered cement mixers, they could not prepare their concrete in the enormous quantities we do.

Instead, they laid a layer of it either between wooden forms or between facings of stone or brick already assembled. Then they pressed the filling material down into it. Then they laid another layer of concrete; and so on.

In the late Republic and early Principate, walls became essentially masses of concrete, in whose surfaces stones or bricks had been inserted. These bits of stone or brick added nothing to the strength of the wall, as the concrete was stronger than the facing materials to begin with. Often they could not possibly have added to its beauty either, because the wall, facing material and all, was then covered by a layer of stucco or by marble slabs.

You might ask: Why, if the Romans had such good concrete, did they not simply build structures of pure concrete, as we often do, without these useless and not very beautiful facings? We do not know, but perhaps this was an effect of Roman conservatism. A customer who ordered a wall expected to see a surface of solid stone or brick and would have felt uncomfortable without it, regardless of the wall's interior. So the builders obliged.

In the era of concrete walls, the old terms for wall construction took on new meanings. *Opus testaceum* meant a concrete wall in which had been inserted wedge-shaped pieces of brick, with their apices inward and their bases exposed to view. These wedges were so closely packed that, to the eye of the beholder, the wall appeared to be of solid brick.

Opus incertum acquired a new meaning, namely: a wall in which irregular pieces of stone had been inserted into the face of the concrete and then, when the concrete hardened, had been chiseled down flush with the concrete surface.[10]

Opus reticulatum or "netlike work" was a more refined version of this system. Little square pyramids of stone were inserted in the wall, point first, so that the bases presented a lozenge or diamond pattern of square stones set closely in the concrete. This form of construction, though rather costly and not so firm as the others, and though often hidden by a coat of stucco, attained a great vogue through –I and +I and then died out.

Cheap working-class houses, the upper floors of domûs, and the partition walls of the top stories of insulae were made of *opus craticum* or "basket work." The wall was a mere trellis of lath or cane, covered with mortar or plaster. Though neither fireproof nor waterproof, such a wall was light and cheap.

In the late Empire, builders used mixed constructions, courses of brick alternating with those of stone. Then, after the fall of the Western Empire, builders were driven to re-use bricks from old fallen buildings, or whatever they could get.

Although Italy has good marble, the Romans of the Republic were slow to take up its use, even after they had become familiar with Greek building methods. When Licinius Crassus in the –90s adorned his house on the Palatine with six 12-foot columns of Attic marble, his contemporaries scoffed at him as the "Palatine Venus."[11]

Later in that same century, marble became popular for public buildings, although the Romans preferred to use it in thin slabs for facing, purely for decoration, instead of for main structural elements. Hence late in his reign Augustus boasted: "I found Rome a city of mud bricks; I leave her clothed in marble."[12]

The Romans not only developed new materials of construction and new combinations of old materials; they also evolved new architectural forms. Some, like the arch and the vault, they took over from the Etruscans and improved; some were original with them. And they combined these elements in an endless variety of ways, giving large buildings a spaciousness and adaptability to their functions far beyond anything the world had seen.

In the architect's never-ending fight against gravity, the arch was found to be useful in many ways. For instance, when concrete construction came in, arches could be cast in concrete. However, to erect an arch of any kind, the builder must set up a scaffold, called *centering,* whose upper surface corresponds to the inner surface of the arch. This centering holds up the wedge-shaped stones or bricks of the arch (the *voussoirs*) until all are in place. Then the centering can be taken away, and the arch will hold itself up. With an arch of concrete, a very massive centering was needed to hold up the weight of all that wet cement.

So the Romans put up a light centering and erected over that an arch of thin bricks. These bricks, which could support themselves once they were all in place, in turn served as centering for the concrete, thus making construction easier. To lighten the superincumbent mass of concrete, builders inserted empty jars or blocks of porous pumice stone into the work at points where no great stress was to be resisted.

Another development was the straight arch as a substitute for lintels of wood or stone over doors and windows. A straight arch is made of tapering voussoirs just as a curved arch is, but they are arranged in a single horizontal line over the opening.

However, the straight arch has much less strength than a semicircular arch. So, to protect straight arches, Roman builders developed another trick: the relieving arch. This was an arc of bricks (usually less than a full semicircle) imbedded in the wall over the area to be protected. It took some of the load of the overlying wall and distributed it to the sides of the opening, where the continuous wall could withstand it.

The original arches were merely round tops to openings in walls, taking the place of lintels. But it was learned that arches could be used as independent units, supported on piers or columns. Diocletian's palace at Spalatum[13] (+III) has arches mounted on columns, and this system became common in medieval architecture.

If an arch is prolonged along its central axis, it becomes a vault. The Etruscans used vaults but only, it seems, for such modest structures as culverts, drains, and gates. The Romans used the semicircular vault, commonly called the "barrel vault," to roof over their huge public buildings.

In earlier times, the uncluttered width of a chamber was limited by the length of timbers that could be obtained to hold up the roof. In a country like Mesopotamia, where the only native timbers were palm trunks, a room could not be much more than fifteen or twenty feet wide unless columns or piers were set up between the walls, as in the Hypostyle Hall of Rameses II and the audience halls at Persepolis, to carry the load. With vaulting, however, unobstructed halls 80 or 90 feet wide became possible.

Furthermore, vaulted halls did not have to be single semicylindrical structures. In +I the Romans learned that one vault could cross another at right angles and the two vaults would still stand up. The inner surfaces of these vaults intersected along elliptical lines called groins. Such a structure is called a groined vault or cross vault. Such a cross vault could roof a large square area and be supported wholly by piers or columns at the corners.

Another method of roofing such an area was by means of the cloister vault, which may be described as a square dome. The Etruscans had, like the Mycenaean Greeks, built tombs in the form of corbelled domes; but the Romans began to build both the true dome and the cloister vault about –I.

Like the arch, the true dome is made of stones or bricks tapered downwards so that, once the last one is in place, the whole structure supports itself. In fact, a dome need not be completed all the way up. It stands as firmly if a circular hole is left in the top.

Large domes, however, brought up another problem. A circular dome must be supported all the way round. If you put a dome on top of a square hall, some of the rim of the dome will rest on nothing but air, either inside or outside the square. Hence you must either make your hall circular or make its walls enormously thick, so that the lower edge of the dome shall not stray off its support.

Roman architects solved this problem in late +I. When the emperor Domitian rebuilt a palace on the Palatine Hill (now ruined), the builders joined square chambers to domes by means of *pendentives*. A pendentive is a triangular piece of masonry, leaning inwards from the upper corners of the chamber, to form a transition between the square ground plan and the circular base of the dome. Although a pendentive may be a flat triangle, the best design makes it, like the dome, a section of a sphere. The use of pendentives to support huge domes became a major feature of Byzantine architecture.

All these arches, domes, and vaults posed an additional problem. An arch does not simply press straight down upon the columns, piers, or sections of wall on which it rests. It also presses outwards; just as when a man stands spraddle-legged, his feet push outwards along the ground as well as down into it.

Vaults and domes produce similar outward stresses. Hence, if such a structure is erected over a chamber whose walls are only thick enough to support the straight vertical load it receives from the roof, the outward thrust of the arch, vault, or dome will push the walls outwards. And down will come the whole structure.

One cure for this state of affairs is to make the walls thicker. But no builder wants to use far more material in his structure than is really needed.

So the Romans of the Principate (+II) learned that a wall could be greatly strengthened against overthrow, without much adding to its total bulk, by vertical thickenings at intervals, either inside or outside. The principle is the same as that of stiffening cardboard or sheet iron by corrugating it. Therefore, late Roman vaulted and domed buildings had walls with numerous niches or recesses. The thick parts between the recesses are called buttresses. Later, in Byzantine and medieval times, buttressing was developed to an extraordinary degree.

Roman builders used wooden scaffolding much like that employed by Italian builders today. To economize on scaffolding, builders put up walls with recesses to receive the ends of timbers, or with projecting corbels to which scaffolding could be made fast. When the building was finished, these projections could be carved into ornamental shapes or trimmed down flush with the wall face. If you look closely at the picture of the Pont du Gard (Plate XIII) you will see a number of such projections.

In their methods of construction, the Romans far outdistanced the builders of the ancient watershed empires. The Romans developed methods of erecting huge, well-constructed buildings in a fraction of the time and with far less expense than in former centuries. For the first time in history, we begin to see an appreciation of the advantages of efficiency.

In building their cities, the Romans did much what the Greeks had done before them. In a new city built from the ground up, like Thamagudi[14] in North Africa, the town was well planned and laid out on a gridiron scheme. An old city, notably Rome itself, which had started off as a jumble of huts in a tangle of alleys, kept on growing without any organization despite the efforts of reforming sovrans to straighten it out. Making a virtue of necessity, some people in classical and medieval times insisted that narrow, crooked alleys were a good thing, because they kept "unhealthy winds" from sweeping through the city.

Most of the streets of Old Rome were not over sixteen feet wide—gloomy alleys flanked by the towering cliffs of insulae. As no ancient city had zoning laws, buildings of all kinds—mansions, hovels, insulae, taverns, temples, workshops, and warehouses—were all jumbled together.

Street traffic was so dense that Julius Caesar ordered wheeled vehicles (with certain exceptions) to move in the city streets at night only. This law, renewed from time to time while the Western Empire lasted, thinned the daytime crush of traffic. But light sleepers were kept awake all night by the rumble of cartwheels, the shouts of the carters, and the screech of ungreased axles.

In building large edifices, the Romans excelled in secular rather than religious buildings. Their temples were essentially Hellenistic post-and-lintel structures in the ornate Corinthian mode, differing little except in size from those of the Attalids and the Ptolemies.

The most spectacular of these that still survive are a group of huge temples, which the emperors of +II and +III put up in Syria. For the site, they chose an ancient Syro-Phoenician religious center. They called the place Heliopolis, the City of the Sun, but it has now regained its old Semitic name of Ba'albakk. The Romans tolerantly identified the local ba'alim with their own Jupiter, Bacchus, and Venus, so that every man could worship freely.

For the main temple, that of Jupiter, the Romans quarried columns from the granite of Aswân in distant Egypt. A Roman innovation was to make temple columns of one solid piece, instead of a stack of stone drums. In the case of Ba'albakk, the whole fifty-four columns, 65 feet long and 7½ feet thick, were rafted down the Nile, shipped across the southeast corner of the capricious Mediterranean, and hauled thirty-odd miles over the mountainous Lebanese hinterland to Heliopolis. Perhaps the Romans used Chersiphron's method of inclosing the columns in wooden cradles and rolling them on their sides.

Under Rome, theater architecture took great strides as a result of the government's policy of giving the people of the city of Rome bread and circuses.[15] The bread was a dole of grain to poor citizens, at first (–II) below cost and later, at the instance of the demagogue P. Clodius Pulcher (–58), free. This dole of food enabled the Romans to live while working only 206 days a year and to spend the remaining 159 days at shows and festivals. The circuses were periodic public entertainments on a scale theretofore unimagined.

The rulers of the old watershed empires had never thought it their duty to amuse their subjects. The obligation, if any, was the other way round. But the Ptolemies and the Seleucids, with their huge parades featuring polar bears and other oddities, tried to reconcile their subjects to their rule and to divert any thoughts of revolt by staging shows for them. Roman politicians of the Republic and later the emperors themselves carried this policy to the point of lunacy. Ninety-three days of the year were given over to spectacles.

Plays were given in semicircular theaters of the Greek type. Originally these were of wood, but wooden theaters had a deadly habit of collapsing. Although permanent theaters of stone had long been forbidden as a decadent Greek idea, Gnaeus Pompeius Magnus, "Pompey the Great," built one in –55 in imitation of one he had seen in the Aegean.

Plays, however, appealed only to a limited upper-class audience. The masses preferred racing and "games." Chariot races were staged in race courses ("circuses" in the literal sense) a quarter of a mile or more long, with a long island, the *spina,* down the middle. In the late Principate there were four or five circuses in Rome, the biggest being the 700-yard-long Circus Maximus. This structure, begun by Julius Caesar and completed by Trajan, seated over a quarter of a million spectators.

The favorite mass entertainment, however, was the "games." Becoming popular in –II, these comprised fights between gladiators; the killing of animals; fights between animals; and the killing of people by animals and by other ingenious tortures. For instance, a play in which the hero was burned alive at the end would be produced with a victim really burned alive. The victims were condemned malefactors, prisoners of war, and sometimes, in the case of gladiatorial combats, volunteers.

The scale and cost of these games was fantastic. Pompeius Magnus opened his theater with a show that included the killing of 500 lions and most of 18 elephants. When Titus opened the Colosseum in +80, 5,000 beasts were killed in one day; 11,000 were slain in a series of games given by Trajan in +107. In the last centuries of the Empire, when the Western emperors were at their wits' ends for tax money to hire enough soldiers to keep out the barbarians, they continued to lavish their wealth on the games.

To furnish victims for such spectacles, the Moroccan variety of the African elephant was exterminated and the white rhinoceros, once common throughout the Sudan, was reduced to a few survivors. Nearly all Romans enjoyed this "good, red-blooded sport." Even the timid and scholarly emperor Claudius was an avid gladiatorial fan. He often arrived before dawn and skipped his lunch in order not to miss a drop of gore, though he would have run like a rabbit from a blade bared against him.

Not satisfied with duels, battles, and massacres, the emperors dug artificial lakes in which naval battles engaging thousands of men were staged. This was considered a convenient way to get rid of prisoners of war. For, to the Romans as to the Japanese of the Second World War, a prisoner had no rights. He could have fought to the death, couldn't he?

The Romans' excuse for gladiatorialism was that an imperial people had to be inured to the sight of bloodshed to teach them courage and contempt for death and to toughen them for the task of keeping the lesser breeds in order. The very ones who advanced these reasons objected to Greek athletic meets because of their nudity. But this Roman theory did not work. When in +II and +III sadistic mobs were howling for more and more blood, the Italian people were becoming completely demilitarized, until as soldiers they were worth no more than so many sheep.

The Christian emperors stopped gladiatorialism but not public massacre. Beginning with Constantine (early +IV) they had magicians, heretics, and other infidels publicly tortured, crucified, and fed to the lions with all the gusto that the pagans had previously exercised on the Christians. Pope Leo I, "the Great," (+V) who dissuaded Attila the Hun from sacking Rome but had less success with Gaiseric the Vandal, persecuted Manichaeans and other heretics and heartily approved of death for unbelievers. Nowadays people are no longer slain for religious heresy, but in many parts of the world they are killed for economic deviationism. So far have we come.

However, our concern here is not with the rights and wrongs of gladiatorialism but with the structures to which it gave rise. After several experiments, the amphitheater or bowl was found to be the most practical building for such shows. There had to be a high barrier around the arena to keep the victims, human and animal, from climbing up among the spectators to give them a first-hand taste of the blood for which they screeched.

After the main permanent amphitheater was destroyed by fire in +64, Vespasian began and Titus finished a mighty new amphitheater. Originally called the Flavian Amphitheater, it came in the Middle Ages to be called the Colosseum. This great bowl, of which the greater part still stands despite Renaissance quarrying, was about 600 feet long and 175 feet high. The arena could be flooded for small naumachiae or sea battles. The Colosseum remained the largest structure of its kind until the Yale Bowl was completed in 1914.

Under the late Principate, the Colosseum, one other amphitheater, and two reservoirs for naumachiae were all in use. Most of the other cities of the Empire also had amphitheaters, though the Greeks bitterly resisted the introduction of gladiatorialism into their country.

A pleasanter side of Roman building is seen in their public baths. Bathhouses in the form of small private establishments were old in the Mediterranean world. By +I there were several hundred of these in Rome alone. As the ancients did not often have running water in their houses and therefore did but little washing at home, such establishments were more necessary to them than to us.

The Hellenistic kings began to build large public bathhouses to which people were admitted for a nominal fee. Polybios described the jovial, brilliant, and eccentric Antiochos IV, king of Syria (–II), bathing in the public baths with the people of Antioch. When he was being anointed, somebody remarked:

"Lucky fellows, you kings, to be able to use such sweet-smelling stuff!"

Next day, Antiochos had a jar of the ointment unexpectedly poured over the man's head. A general roughhouse ensued, with everybody, including the naked king, falling down where the ointment had made the marble slippery and scrambling up again with roars of laughter.

In –II, public baths were introduced into Rome, reaching full flower in +I. They combined the original purpose of the private bathhouse—that of getting people clean—with some features of the Greek gymnasium. Around a central hall were ranged chambers for steam baths, hot baths, tepid baths, and cold baths; also exercise rooms, game courts, gardens, and even libraries.

Seneca, who once took lodgings over a public bath, complained of the racket: the whack of the masseur's hand, the grunts of the gymnast as he swung his dumbbells, the splashing of the swimmer, the roars of the man who sang as he bathed, the yelps of the man who was having his armpits depilitated, and the cries of sellers of sausage, cakes, and other goodies. Under the Principate, men and women bathed together save when some bluenose like Trajan or Marcus Aurelius ruled against it.

The first of the large, luxurious public baths in Rome was built by Augustus' minister Agrippa. Thereafter a long succession of emperors built bigger and bigger baths, each striving to outdo his predecessor. The ruins of two of the largest, the so-called Baths of Caracalla[16] and of Diocletian, still dominate the city of Rome. The latter had a main hall

200 by 80 feet in plan and 120 feet high, cross-vaulted at the top.

About –310 an ambitious Roman politician, Appius Claudius Crassus, was elected censor. As the ancients did not have our sense of clear-cut departmental specialization, they were apt to assign almost any duty to any magistrate, whether or not it was related to the duties he already had. Hence the censors of the Republic not only conducted the census but also guarded public morals, decided who should be Senators, and had charge of public lands and public works.

As a politician of democratic leanings, Appius Claudius Crassus distinguished himself by lavish public works. At least, they were considered lavish in austere Republican Rome. He exercised his power over membership in the Senate in a way that his critics considered extremely arbitrary, admitting to that body plebeians and even, to the horror of the landed aristocracy, the sons of freedmen.

Crassus built the Aqua Appia, the first of Rome's great web of aqueducts. Then followed a splendid road southeast from Rome. First it ran for sixty miles in a straight line to Anxur, later called Tarracina, on the coast. Thence it followed the Tyrrhenian coast to Capua, northeast of Naples. The Via Appia, about 115 miles long, passed through the heart of the Latin country, which had been united to Rome for over two centuries. Crassus undertook to build an all-weather road by paving his Appian Way. But, like the Achaemenid kings, he probably paved only the parts near cities, where traffic was thickest. The rest was paved much later.

Avid of glory as well as full of constructive engineering ideas, Crassus tricked the other censor into resigning his post ahead of time and "by various subterfuges, is reported to have extended the term of his censorship, until he should complete both the Way and the aqueduct."[17] Thus he made sure that both public works should be named for himself alone. A monumentally stubborn and self-willed man, he used a legal quibble to argue that the law setting the term at a year and a half had been superseded.

Despite failing eyesight, Crassus went on to a distinguished career as consul, praetor, senator, orator, writer, and poet. In fact, he was just about the first real man of letters that Rome produced.

About eighty years later another popular leader, Gaius Flaminius, executed two more costly public works. One was the Flaminian Circus, Rome's first permanent race course. The other was the Via Flaminia, which extended from Rome across Italy to Ariminium (modern Rimini). Crassus and Flaminius probably paid for their works by the sale of public lands. A few years after building his road, Flaminius, commanding a Roman army as consul, fell in the disastrous battle against Hannibal, at Lake Trasimenus (–207).

Meanwhile the Appian Way had been extended to Tarentum and Brindisium in the heel of the Italian boot. Other roads were added and existing ones paved until Italy was covered with a fine all-weather network.

When Octavianus ruled Italy as a triumvir, before he became the Emperor Augustus, he ordered the architect Lucius Cocceius Auctus to cure the traffic bottlenecks around Naples, caused by the spurs of the Apennines that here run down to the sea. Auctus drove two tunnels, one at Cumae and the other between Naples and Puteoli, about 10 feet wide and varying in height from 9 to 70 feet. Seneca, who once used the latter, bitterly complained of the dust and darkness, although the tunnels much shortened the route.

Excellent roads were also extended into the provinces. Trajan built a road in Arabia, from Aqabah to Bostra, which was not only paved but divided into two lanes by a row of raised stones.

Among the conquered peoples of the Empire, the Gauls, being skilled in making carts and chariots, already had roads of a sort. But these were mere tracks, maintained only to the extent of cutting down bushes and trees that grew up in them. Not being paved, the Gallic roads were worn deeper and deeper by traffic until they became sunken roads, like the legendary road that was once supposed to have ruined the charge of Napoleon's Guard at Waterloo. Some of these roads can still be traced by shallow ditches across the face of France. Roman roads, on the other hand, being paved and embanked, survive in the form of low ridges.

One of the most important later Roman roads was that which, leaving the Italian network at Tergeste (modern Trieste) extended down the Dalmatian coast to Dyrrhachium[18] and thence, as the Egnatian Way, across the Macedonian mountains to Thessalonica and Byzantium. In later times, when the Empire had two capitals, one at Rome and one at Constantinople, this road became the vital link over which the emperors of East and West marched, sometimes to support each other against the invading barbarians but more often to attack each other.

To understand Roman road construction, we must remember that Roman roads were intended primarily to enable the army to move swiftly, and that the Roman army consisted, until the late Empire, mainly of heavy infantry. So the road builders were more concerned with providing a firm footing for marching soldiers than a soft surface for the hooves of animals. Therefore the more important roads were hard-paved, although most secondary and provincial roads were merely graveled. Alongside of the paved roads, the practical Roman engineers often placed one or two unpaved strips for those who preferred a softer surface.

As routes for infantry, again, Roman roads were kept as straight as the contour of the land and the presence of existing structures allowed. Romans preferred that a road should go over a hill rather than wind around one, even if this meant grades of as much as 20 per cent. This is perilously steep for vehicles, especially in the days before brakes, when the only way to check the vehicle's descent was to lash one wheel so it could not turn.

Depressions and valleys in the path of the road were filled or bridged, but the Romans did not like cuttings to reduce the grade in passing over a ridge. They liked to place as much of the road as they could on an embankment above the level of the surrounding land. This kept the road clean of mud, guarded it against blockage by snow in winter, and made it hard for a foe to ambush the marching legions.

Such a road system supported not only the marches of the legions but also a brisk civilian traffic afoot, mounted, and on wheels. Like the Achaemenid and Hellenistic kings before them, the rulers of Rome maintained an official postal service. There were relay stations at intervals of ten or twelve miles and larger stations, with inns, smithies, and other facilities, at intervals of thirty to forty miles. The governmental mail carts and horses were for the use of governmental officials and military officers only, though the drivers could probably be bribed to carry a private passenger. The emperor issued a pass called a *diploma* to authorized users of the postal service and to cronies whom he wished to favor.

Under the Principate, private companies also carried mail and parcels and rented animals to travelers. The one travel facility to which the Romans never attained, and which did not appear anywhere before +XVII, was that of the scheduled common carrier: the vehicle running from here to there at definite times and open to all comers.

To lay out a route for a road, Roman surveyors used a few simple instruments for establishing horizontal lines and right angles. For the former purpose they employed levels of several kinds (*libra, libella, chorobates*). These worked either by a plumb-bob hanging at right angles to the line of sight, or by a water level.

For setting up right angles, the Romans used a *groma*. This was a pair of boards fastened together to make a right-angled cross, mounted horizontally atop a post or stand. Plumb lines, hanging from the four ends of the crosspieces, made it possible to level the instrument, and the surveyor sighted along the crosspieces. At Pompeii, the blacksmith Verus kept a groma in his smithy and presumably did surveying on the side.

A Roman paved road was a massive structure that has been well compared to a wall lying on its side. The makers began by digging a trench several feet deep. If the earth was still soft at that depth, they often drove piles to strengthen it. On the surface so prepared, the road was built up in layers.

The precise nature of the road depended on the material to be had locally and the importance of the route. A Roman poet, Publius Papinius Statius, describes the procedure; and if you don't think road-building a poetic subject, you don't know Roman poets. He is celebrating a new and shorter road built by Domitian (late +I) near Naples:

Here the belated one, borne on an axle,
Formerly clutched at the pole as it tottered;
Mud malignant sucked down the wheels, while
Folk in the midst of the meadows of Latium
Shuddered at evils like those of seafaring.
Nor was the ongoing rapid, but sluggish, as
Echoless ruts impeded the carriage,
While the exhausted, high-yoked animal
Staggered, protesting the onerous burden.
Howso, a journey that once took till sunset
Now is completed in scarcely two hours.
Not through the heavens, ye fliers, more swiftly
Wing ye, nor cleave ye the waters, ye vessels.
First comes the task of preparing the ditches,
Marking the borders and, deeply as needed,
Delving into the earth's interior;
Then, with other stuff filling the furrows,
Making a base for the crown of the roadway
Lest the soil sink, or deceptive foundations
Furnish the flagstones with treacherous bedding;
Then to secure the roadway with cobbles
Close-packed, and also ubiquitous wedges.
How many hands are working together!
Some fell the forest while some denude mountains,
Some smooth boulders and balks with iron;
Others with sand that is heated, and earthy
Tufa, assemble the stones of the structure.
Some with labor drain pools ever thirsty;
Some lead the rivulets far to the distance.[19]

As Statius indicates, the many hands of the Roman road builders built their roads to last. A major, fully paved highway in Italy usually consisted of four or five layers, altogether about 4 feet thick and 6 to 20 feet wide. At the bottom was a layer of sand, mortar, or both. Then came a layer of flat squared stones set in cement or mortar. On top of this was a stratum of gravel set in clay or concrete. Then came a layer of rolled sand concrete. On top of all was laid a crowned pavement of large many-sided blocks of hard rock set in concrete and dressed on their upper surfaces.

In places where there were no good materials for making concrete, and where the Romans still wanted a strong paved road, they sometimes set the surface blocks in molten lead. Few traces of such costly road-building methods remain, because as soon as Roman authority lapsed, the local people pried up their roads in order to use or sell the lead.

Parts of some Roman roads still remain in much their original condition. Today you can drive over the Old Appian Way, extending southeast of Rome. The Italians have laid down a layer of asphalt over the imperial flagstones, but there are gaps in the asphalt where you go bumping over the same stones that rang to the tread of the legions and the triumphs of the Caesars.

The Roman theory of road building was that if the road was made solid enough to begin with, it would not need much maintenance. Hence a fully paved Roman road, under normal wear, lasted eighty or a hundred years before it had to be renewed. Only in the present century, with the advent of heavy automobile and truck traffic, have nations begun to return to road-building methods like the Romans'.

As with the Persians and Macedonians before them, the Romans conquered a great empire less because of any peculiar national virtues than because they had, at that time, a better army than anybody with whom they fought. This army was important in shaping not only the history of Rome but also the development of Roman engineering.

In its earliest stages, in –VI and –V, the Roman army was much like that of the Greek city-states. It consisted of a number of large, solid blocks of spearmen, divided into classes according to how much the soldiers could afford to buy in arms and armor. Those of the richest class bore a panoply much like that of the Greek *hoplitai*.

Little by little, this army evolved into the form it attained in –I, at the end of the Republic. At that time the infantryman carried a pair of javelins for throwing instead of a pike for thrusting. At close quarters he fought with a two-foot broadsword.

Legions were divided into maniples of 200 men, which could maneuver independently. The phalanx of Philip and Alexander had had this flexible system of subdivision, too; but the Hellenistic kings had let the phalanx degenerate into an unwieldy solid mass, which could hardly maneuver at all. The general could only aim it, set it in motion, and hope.

During the days of the Republic, the legions—that is, the Roman army proper—consisted entirely of Roman citizens, armed as heavy infantry. Light infantry, cavalry, and missile troops were always soldiers from the allied or dependent states, or from the provinces, and were called "auxiliaries."

As Roman rule expanded, Roman citizenship was also extended, until in +212 citizenship was conferred upon all free provincials. Thereafter auxiliaries were recruited from Germans, Alans, and other barbarians from beyond the frontiers. The practice of recruiting legionaries from the border provinces in preference to those in the interior finally made the Roman Empire into a molluscan organism. It had a hard shell surrounding a soft interior of provinces whose people had become demilitarized by long peace.

The Roman army depended less upon brilliant generalship than on the warlike spirit of the people from whom it was recruited; on a sound and flexible system of infantry drill and maneuver; and most of all on stern discipline and rigorous training. Soldiers practiced with swords and shields twice as heavy as those they fought with, so that in battle their weapons seemed light and easy to wield. Discipline was harsh. The centurion Lucilius, killed in a mutiny under Tiberius, bore the nickname "Fetch-another,"[20] because he was always breaking his vine stick on his soldiers' backs and shouting for another.

When not fighting, the legionaries were often kept busy on public works. They built roads, dug canals and drainage ditches, made bricks, and even mined silver. They did not always do such work willingly. Under Tiberius they mutinied against road building. The energetic reformer Probus became emperor at a time (+III) when the legions had become demoralized by idleness. When he put some of them to work draining bogs and planting vineyards at Sirmium,[21] they killed him.

A strange instance of such "keeping the boys busy" came to light in 1961, when a cache of about a million iron nails, 6 to 16 inches long and weighing a total of seven tons, was unearthed at Inchtuthill, Scotland. The Roman army built a fortified camp here, at its most northerly outpost in Britain, in +83 and abandoned it six years later. Before evacuating the camp they buried the nails in a pit, which they carefully hid.

The likeliest explanation for the manufacture of seven tons of nails in distant Scotland is that the Romans expected to use them in a chain of wooden fortified camps along the neck of land between the Firths of Forth and Clyde. This was actually done by the Antonine Wall half a century later; but this proved impractical. The Antonine Wall was occupied for only fifteen years, beginning in +143; then abandoned, reoccupied, and abandoned again. Evidently, when the order came through in +89 to abandon the fortress, the local commander, lacking means of transporting seven tons of nails, had them buried to keep the wild Caledonians from making weapons of the iron.

Roman military operations made good use of engineering techniques. Roman armies, on ending a march, set up a fortified camp on a standard, square plan, which remained the same regardless of the terrain. A Greek army adapted the form of its camp to the ground, but a Roman army adapted the ground to the form of the camp, no matter how much digging this entailed.

Whereas the Greek soldier, though a great fighter, disdained digging as slaves' labor, the legionary was expected to work as hard with his spade as with his sword. When Marcus Licinius Crassus (later the political partner of Caesar and Pompeius) campaigned against Spartacus' army of slaves and gladiators, he trapped them by digging a 30-mile ditch, 15 feet wide and 15 feet deep, clear across the toe of the Italian boot, and backing the ditch with a wall. Spartacus' army escaped by filling up the ditch on a snowy night. Soon, however, the revolting slaves were destroyed when the men became overconfident and forced Spartacus to lead them into battle against his will.

To cross streams, Roman armies did not have to depend upon fords, rafts, or water wings made of inflated goatskins. Instead, their engineers threw bridges across the rivers. Caesar built one–probably a pontoon bridge–across the Saône in a day, and another, a quarter-mile long, across the Rhine near Coblenz in ten days. As he described it, he

> . . . caused pairs of balks a foot and a half thick, sharpened a little way from the base and measured to suit the depth of the river, to be coupled together at an interval of two feet. These he lowered into the river by means of rafts, and set fast, and drove home by rammers; not, like piles, straight up and down, but leaning forward at a uniform slope, so that they inclined in the direction of the stream. Opposite to these, again, were planted two balks coupled in the same fashion, at a distance of forty feet from base to base of each pair, slanted against the force and onrush of the stream. These pairs of balks had two-foot transoms let into them atop, filling the interval at which they were coupled, and were kept apart by a pair of braces on the outer side

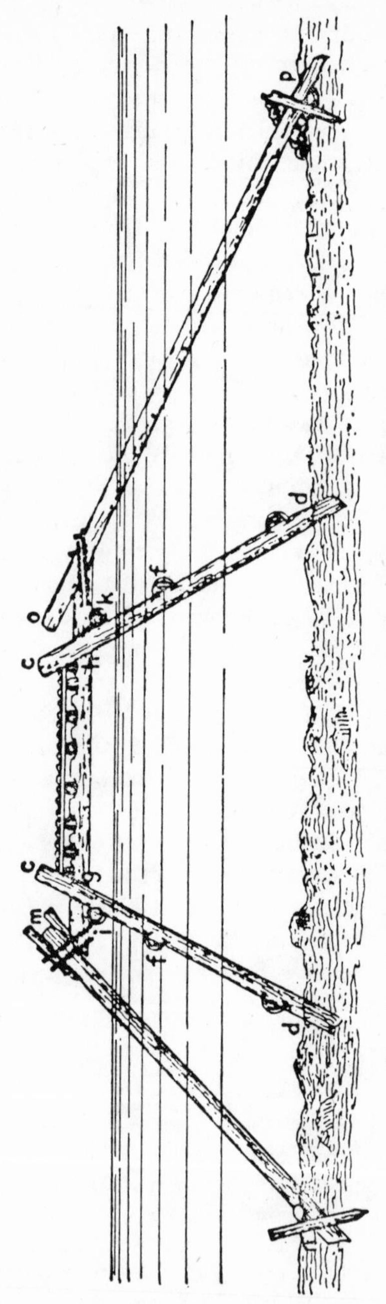

Fig. 11. Cross-section of Caesar's bridge across the Rhine (*from Merckel*).

at each end. So, as they were held apart and contrariwise clamped together, the stability of the structure was so great and its character such that, the greater the force and thrust of the water, the tighter were the balks held in lock. These trestles were interconnected by timber laid over at right angles, and floored with long poles and wattlework. And further, piles were driven in aslant on the side facing down stream, thrust out below like a buttress and close joined with the whole structure, so as to take the force of the stream; and others likewise at a little distance above the bridge, so that if trunks of trees, or vessels, were launched by the natives to break down the structure, these fenders might lessen the force of such shocks, and prevent them from damaging the bridge.[22]

Back in late –VI, according to the half-legendary history that has come down from those times, a dynasty of Etruscan kings ruled the small city-state of Rome. One of these kings, the usurping Lucius Tarquinius Priscus, called "the Proud," so provoked the Romans by his oppression that they drove him out. King Tarquinius toured the Etruscan courts, trying to rally support against the spread of this wicked new idea of throwing out kings and setting up republics. He succeeded in convincing his kinsman Porsena, king of Clusium, eighty miles north of Rome. So:

Lars Porsena of Clusium
By the Nine Gods he swore
That the great house of Tarquin
Should suffer wrong no more.
By the Nine Gods he swore it,
And named a trysting day,
And bade his messengers ride forth
East and west and south and north,
To summon his array.[23]

As the Etruscan army neared the north bank of the Tiber, the Roman countryfolk stampeded into the city. All seemed lost when:

Then out spake brave Horatius,
The Captain of the Gate:
"To every man upon this earth
Death cometh soon or late.
And how can man die better
Than facing fearful odds,
For the ashes of his fathers,
And the temples of his Gods . . ."

Horatius urged the Romans to break down the bridge while he and two others held the northern end, for "In yon strait path a thousand / May well be stopped by three." As the Romans attacked the bridge with ax and crowbar,

> Meanwhile the Tuscan army,
> Right glorious to behold,
> Came flashing back the noonday light,
> Rank behind rank, like surges bright
> Of a broad sea of gold.
> Four hundred trumpets sounded
> A peal of warlike glee,
> As that great host with measured tread,
> And spears advanced, and ensigns spread,
> Rolled slowly towards the bridge's head,
> Where stood the dauntless Three.

When the earth before the dauntless Three was heaped with corpses, the bridge crashed down. Although Horatius' comrades dashed back before it fell, Horatius had to save himself by swimming. But what about this bridge?

Being made of wood and not more than ten feet wide, it was called the Pons Sublicius or "bridge of piles." An earlier and even more legendary king, Ancus Marcius, had built it. For religious reasons he had made it entirely of wood, without any iron, and placed the bridge in charge of the high priest, who thus came to be called the Pontifex or "bridge maker." This sobriquet, shortened to "Pontiff," has been inherited by the Pope. At a yearly ceremony, dummies made of rushes were thrown from this bridge into the Tiber, probably as a vestige of an earlier custom of annual human sacrifice to the god of the river.

The fallen Pons Sublicius was soon replaced and, with successive repairs, survived down to the end of the Republic. Then a bridge of stone and wood took its place until a flood destroyed it in +69. The ruined piers in the Tiber now called the *avanzi del Ponte Sublico* are more probably the remains of a bridge built by Probus (+III).

In the meantime, as cross-Tiber traffic grew, the Romans built other bridges. The Pons Aemilius began as a structure with stone piers and a wooden deck, like Nabopolassar's bridge at Babylon. Later stone arches were installed and a stone paving took the place of the wooden deck. With repair and rebuilding this bridge was kept in service until 1598, when it was finally abandoned.

The two short bridges connecting the Isola Tiberina, the island in the

Photograph by the University Museum, University of Pennsylvania

Plate I. The step pyramid of King Joṣer at Ṣaqqâra, Egypt.

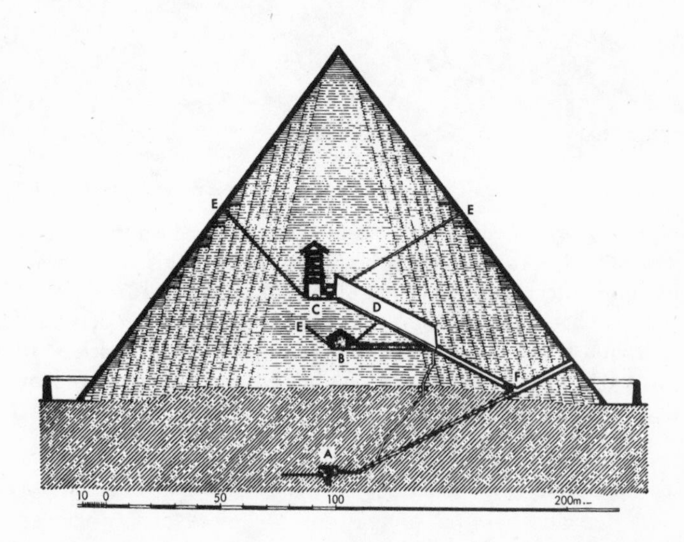

Plate II. The Great Pyramid of King Khufu at Giza. A–abandoned rock chamber; B–unfinished chamber ("Queen's Chamber"); C–burial chamber ("King's Chamber"); D–Grand Gallery; E–ventilation shafts; F–granite plug.

Photograph of a relief in the British Museum

Plate III. Assyrian siegecraft. At the right, King Ashurnasirpal (reigned –884 to –860) shoots arrows at a besieged city while his armor-bearer holds up a shield to protect him. To the left of the king, a belfry attacks the wall with a ram while archers shoot from the top of the tower. The defenders have looped a chain around the ram and strive to pull it up to make it useless, while the Assyrians try to pull it down again. Farther left, a cutaway view shows engineers digging a tunnel under the walls. Still farther left, others attack the wall with crowbars.

Plate IV. The Ishtar Gate of Babylon, as reconstructed in the Vorderasiatisches Museum in Berlin.

Photograph by the Metropolitan Museum of Art (New York) bequest of Levi Hale Willard, 1883

Plate V. A model of the Parthenon (the New Hekatompedon of Athena Polias) of Athens.

(after Payne-Gallwey)

Plate VI. Torsion dart thrower.

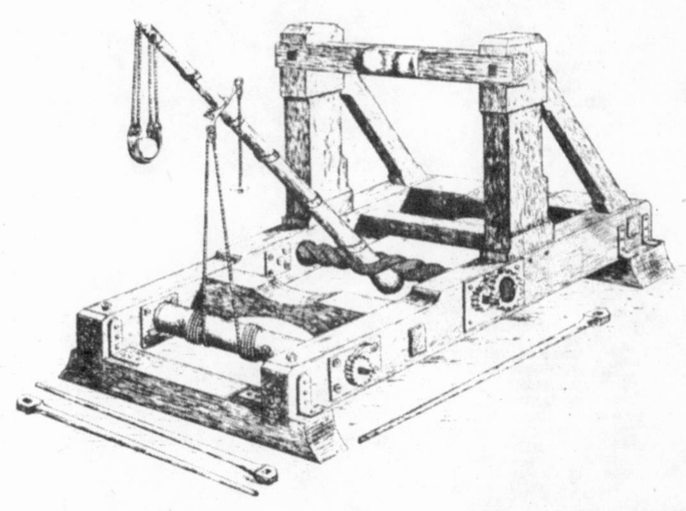

(after Payne-Gallwey)

Plate VII. Onager or one-armed torsion stone thrower.

(*after Payne-Gallwey*)

Plate VIII. Trebuchet or counterweight stone thrower.

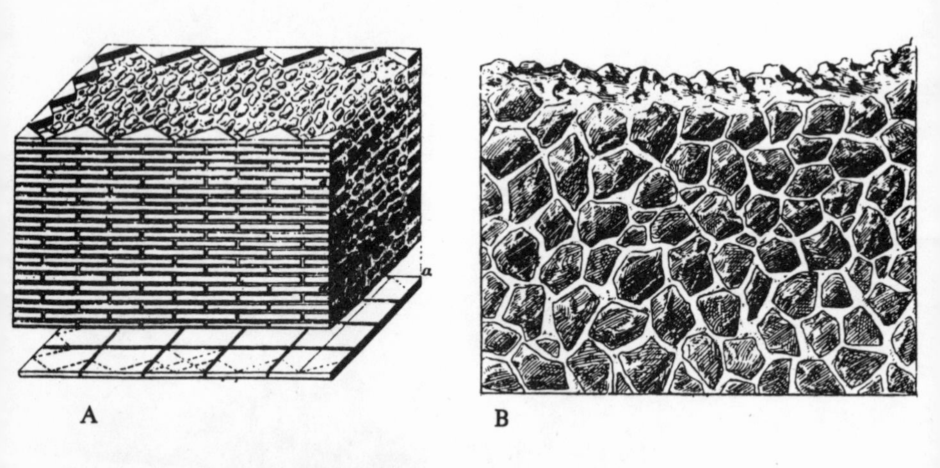

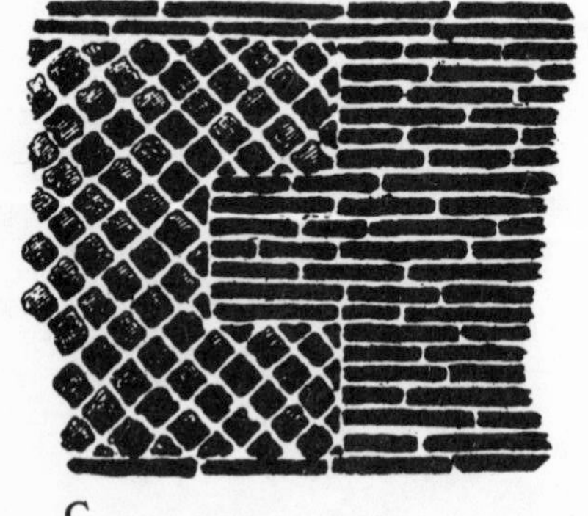

C

Plate IX. Roman methods of wall construction: A–concrete wall faced with brick; B–*opus incertum;* C–*opus reticulatum* (on the left)

(*from Schreiber*)

Drawing by G. G. Woodward © The Illustrated London News

Plate X. The Pharos of Alexandria, restored.

Photograph by the University Museum, University of Pennsylvania

Plate XI. The relief on the funerary monument of the Haterii, in the Lateran Museum of Rome, showing a crane powered by a treadwheel.

Tiber, with the mainland—the Pons Cestius to the west and the Pons Fabricius (named for its architect) to the east—are still in service, carrying swarms of Fiats and motor scooters during the rush hour. So are Hadrian's Pons Aelius, leading to the Castel Sant' Angelo, once Hadrian's tomb; the Pons Mulvius, north of the city, where Nero reveled and Constantine slew his rival Maxentius; and the Pons Aurelius. These bridges can carry modern traffic because the Romans, not knowing the stresses in their bridges and hence being unable to calculate safety factors, built them much stronger than they had to be to bear the loads of Roman times.

The other bridges that stood in Imperial times (Neronianus, Agrippae, Aemilius, Sublicius, and Probi) have all fallen. Nor are the surviving bridges in their original form. All have been repeatedly repaired as they began to crack and crumble; often one or more arches have been replaced. Since the names by which these bridges are known have changed through the centuries, the bridges of Rome afford wide scope for scholarly disputes as to which is which.[24]

Roman stone bridges were all built on one simple plan. Each consisted of one or more semicircular arches of stone, which was preferred to concrete for large arches because of the difficulty of building forms to hold so much concrete. Such bridges were put up in great numbers in Italy and in the provinces. Two outstanding ones still survive in Spain near the Portuguese frontier: Lacer's bridge over the Tagus at Alcántara, 600 feet long and 175 high; and the bridge over the Guadiana at Mérida, 2,500 feet long and over 30 high.

The number of spans that have lasted nearly 2,000 years shows that this bridge design is very solid and substantial. However, it meets the same difficulty that Nabopolassar's bridge did. The thick piers seriously reduce the cross-sectional area of the river, so that they produce the effect of a dam with holes in it. Because the water is held back, there is a noticeable drop in the level of the river between the upstream and the downstream sides of the bridge. Hence the water rushes under the bridge with augmented speed.

Now, the power of moving water to carry silt, sand, gravel, and stones varies steeply with the velocity. Hence the swift flow between the piers of a bridge scours away the bottom until the undermined piers totter and fall.

In sinking their piers, the Romans used methods like those employed today. To form a cofferdam around the site of the pier, they drove piles into the river bed with a simple muscle-powered pile driver like that used by Caesar in bridging the Rhine. Then they pumped the water out of the dam and dug down into the mud until they reached bedrock, or until the leakage of water into the cofferdam thwarted their efforts. Then they built the pier of stone and strengthened it by piling rocks around its base.

Because of the limitations of muscle power, however, the Romans could not build their foundations so solidly as we do. Therefore scour still threatened their foundations, especially during spring floods.

There were two possible ways to try to strengthen such a bridge. One was to build more or larger piers. But this reduced the flow of water still further and aggravated the problem of scour. The other was to use fewer and larger arches, to hold back the river less. Following this plan, the Romans built some very large arches, such as that of the bridge over the Nar near Narnia, where the main arch has a span of 139 feet.

However, with semicircular arches, each arch must be exactly half as high as its span is wide. So, if the span is enlarged, the height rises accordingly. This is all right where the river flows between high banks, as at Narnia. But, if the banks are low and flat, the bridge becomes much higher in the middle than at the ends. Hence it must have either long approaches or steep grades. The Romans built some big single-arched bridges with stairs at the ends, which made them unusable by vehicles. No solution was found to this problem until Renaissance architects discovered (as the Chinese already had) that arches could be in the form of a segment of a circle, or a semi-ellipse, so that the span could be increased without raising the deck.

As you spiral down to the Roman airport, your eye is caught by a straight reddish-brown streak across the flat green Campagna. It is an aqueduct—or rather the remains of an aqueduct, for there are many gaps in it—looking just as it did in your high school history textbook. The aqueducts of Rome have been so extensively pictured that the term "ancient Rome" at once conjures up a picture of such a row of arches.

The Romans were, of course, not the first folk to build aqueducts. You remember King Sennacherib's aqueducts and Polykrates' aqueduct with its tunnel at Samos. The Phoenicians also constructed aqueducts, and about −180 Hellenistic Pergamon built a water system that brought water from a nearby mountain under a pressure of 16 to 20 atmospheres by means of pipes, which passed through hollowed stones.

The Roman aqueducts were distinguished from the earlier ones mainly by their size and number. Sections of their arcades are still to be seen, not only near Rome, but also elsewhere in Italy and in France,

Spain, North Africa, Greece, and Asia Minor. Parts of some of them, as at Rome, Athens, and Segovia, are still in use or at least were within recent decades.

The arcades or aqueduct bridges were all built on a simple pattern. On a row of tall piers, of stone or brick, rose a series of small round arches. Above these lay the actual water channel of concrete, with an arched or gabled roof above it. When an aqueduct crossed an exceptionally deep gorge, as at Segovia, Tarragona, or Smyrna, two or even three rows of arches were erected one atop the other. The famous Pont du Gard at Nîmes (Roman Nemausus) has three superimposed arcades. The footbridge that runs alongside the lowest arcade, with arches parallel to those of the aqueduct, was added in 1743.

As the arcades are the most conspicuous part of an aqueduct, we tend to think of "aqueduct" as meaning "arcade." In fact, however, the arcades formed only a small part of the whole system. The rest of an aqueduct took the form of conduits and tunnels. When Rome's web of aqueducts was nearly completed in the early Principate, all the aqueducts together totaled about 260 miles, of which only 30, or one-ninth, were on arches. The actual length of the arcades was even less, because in some places two or three water channels shared the same arcade.

The many Roman aqueducts put up in other parts of the Empire were built on similar principles. In +152 the Roman engineer Nonius Datus sent in the following report about the conditions he found when he inspected an aqueduct under construction at Saldae, Algeria:

> I found everybody sad and despondent. They had given up all hopes that the opposite sections of the tunnel would meet, because each section had already been excavated beyond the middle of the mountain. As always happens in these cases, the fault was attributed to me, the engineer, as though I had not taken all precautions to ensure the success of the work. What could I have done better? For I began by surveying and taking the levels of the mountain, I drew plans and sections of the whole work, which plans I handed over to Petronius Celer, the Governor of Mauretania; and to take extra precaution, I summoned the contractor and his workmen and began the excavation in their presence with the help of two gangs of experienced veterans, namely, a detachment of marine infantry and a detachment of Alpine troops. What more could I have done? After four years' absence, expecting every day to hear the good tidings of the water at Saldae, I arrive; the contractor and his assistants had made blunder upon blunder. In each section of the tunnel they had diverged from the straight line, each towards the right, and had I waited a little longer before coming, Saldae would have possessed two tunnels instead of one![25]

The answer to Datus' question about what more could he have done is, of course, "Stay on the job and don't wander off for four years at a stretch." But then, perhaps poor Datus was ordered to other projects by his government and had no choice in the matter.

Why did the Romans build so many aqueducts, when other peoples got along without them and we do not build such arcades to carry our water? Well, a city that is built on a river, as most of the world's cities are, can usually manage for drinking water by scooping it out of the river. But even in ancient times, when men were ignorant of bacteria, they knew that spring water is better than water from a river. Moreover, if the river lies in a deep bed between high banks, as the Tiber does at Rome, a lot of muscle power must be spent in hauling water up to street level. Finally, the Hellenistic and Roman ideals of civilized living called for far more fountains, baths, and gardens than had been customary before, and these required more water.

The Romans had to build their aqueducts on elevated structures because the water flowed in an open channel instead of in an underground pipe as with us. Flowing in an open channel, the water depended on gravity to move it all the way from the source to the point of distribution. Therefore the channel had to have a slight and fairly constant down grade of two or three feet to the mile.

As Rome stands on seven hills amid a wide flat plain, and most of the sources were in a spur of the Apennines fifteen miles to the east, the streams had to be carried across the Campagna on stilts to arrive at a level high enough to be useful. The builders followed a natural ridge across the plain as far as they could, but thence the water channel had to take to the air.

Why did the Romans not use pressure pipe all the way? By this time, men knew how to make pipe of bronze, lead, wood, tile, and concrete.

However, the art of pipe making was still young, and the difficulty of making good pipe increases with the size. All the materials the Romans had to work with have their shortcomings. Bronze makes a fine strong pipe but is hard to work and costly, so that sections left unguarded were liable to be stolen. Wood rots and splits; while tile and concrete, though durable, have but little strength in tension and so cannot withstand much pressure from inside. Moreover, the Romans liked to keep as much of their aqueducts as they could above ground, where leaks could be easily seen and repaired.

Most Roman piping was in fact made of the remaining substance, lead. Leaden pipes were made by rolling a sheet of lead into a cylinder and soldering the edges. Small lead pipes, with an oval or elliptical cross-section, carried water from distributing points at the outlet of each aqueduct to the places where the water was to be used. To avoid excessive pressures, risers and reservoirs were installed in such pipe lines. These kept the pressure constant on the same principle as that of Ktesibios' water clocks.

However, extensive use of water from lead pipe is bad for the health, for the lead goes into solution in the form of poisonous salts. The Romans were not entirely ignorant of this fact. Vitruvius wrote:

. . . water is much more wholesome from earthenware pipes than from lead pipes. For it seems to be made injurious by lead, because white lead paint is produced from it; and this is said to be harmful to the human body. . . . We can take example by the workers in lead who have complexions affected by pallor. For when, in casting, the lead receives the current of air, the fumes from it occupy the members of the body, and burning them thereupon, rob the limbs of the virtues of the blood. Therefore it seems that water should not be brought in lead pipes if we desire to have it wholesome.[26]

This was, however, but one of many sound suggestions, scattered through ancient literature, that were ignored and forgotten. Lead poisoning was not properly diagnosed until Benjamin Franklin wrote a letter about it in 1768.

The ancients used lead not only for piping but also for kitchen utensils. Furthermore, leaden vessels were used in making wine, and the stoppers on wine jars were sealed with a cement containing red lead. As a result, some students think that lead poisoning was very common in ancient times, and that some cases of alleged poisoning of ancient notables by their enemies were really cases of lead poisoning, or else of food poisoning caused by careless methods of handling food–a hazard to which travelers in tropical countries are still subjected.

How about the high-pressure pipe used at Pergamon, which I mentioned earlier? We do not know what the pipe was made of, but from the head of water it sustained it was probably metal. Since this water system was not duplicated elsewhere in ancient times, the Pergamenes may have found that maintaining it against leakage and theft was more trouble than it was worth.

Although most Roman aqueducts took the form of gravity-powered open channels, the Romans did know the principle of the inverted siphon–that is, a U-shaped pipe higher at the intake than at the outlet. They sometimes used an inverted siphon instead of aqueduct bridges to cross deep valleys. At Arles such a siphon crossed the Rhone by means of pipes laid in the river bed.

At Lyon the aqueduct crossed three valleys, so deep that an open-channel system would have required arcades 215 to 300 feet high, while a system of piping only would have developed pressures at the bottoms of the inverted siphons of nine to ten atmospheres. To avoid both difficulties, the Romans built inverted siphons, the lower parts of which were carried across the rivers on arcades 50 to 60 feet high. Moreover, the water was carried, not by one large pipe, but by a number of small pipes, the better to resist the pressure. In crossing the river Garon, the number of pipes was increased from nine to eighteen, each $6\frac{5}{8}$ inches in diameter.

The first of the aqueducts of the city of Rome was the Aqua Appia, built in –312 by the same Appius Claudius Crassus who built the Appian Way. It was made entirely of stone and was nearly all underground. It carried water from the Anio, a tributary of the Tiber. Although it began only seven miles from Rome, the total length of its winding course was ten miles. Nearly all traces of this aqueduct, including its one small 100-yard arcade, have now disappeared.

Forty years later the censors Dentatus and Cursor built a second aqueduct, tapping the Anio higher up, thirty miles east of Rome. Hence this aqueduct was about four times as long as the Appia. Originally called the Aqua Anio, it was later known as the Anio Vetus or "Old Anio" to distinguish it from the subsequent Anio Novus.

During the next three centuries, seven more aqueducts were built to service Rome. The later ones were built mainly of concrete. The Claudia, which we saw from the air, is the best preserved. The longest of these aqueducts was the 58.4-mile Marcia, built by the praetor Marcius Rex around –145. Its water was considered the best. On the contrary, the water of the Alsietina–which, unlike the others, came down from the north to service, not Rome proper, but the Ianiculum region west of the Tiber–tasted so bad that it was used mainly for watering gardens and for filling the artificial lake that Augustus had dug for naumachiae.

All the other aqueducts came from the south and east. Near Rome, the Anio Novus and the Claudia shared the same arcade; likewise, the Marcia, the Tepula, and the Iulia shared another arcade. When two or more aqueducts occupied one arcade, their channels were placed one above the other. The Romans considered the water of each aqueduct best for certain purposes; hence they rarely mixed them.

Arriving at the city, each aqueduct ended in an extensive distributing system. First the water flowed into one or more tanks to let mud and pebbles settle out. Thence the water was piped to a tower called a *castellum* or "little castle," although, because the water in the aqueduct was not under pressure, the water level in the castellum could not be higher than in the aqueduct from which it came.

From the castellum the water flowed into several smaller tanks, whence leaden pipes distributed it to fountains, baths, industrial establishments, and private users. Insulae often had a running-water supply, but on the ground floor only, because the pressure was not great enough to supply the upper stories.

Towards private users, the Roman government had trouble in finding a policy. According to the early Roman theory of government, chief magistrates were expected to build public works. For money they used cash from the sale of public land, or the spoils of foreign wars or, when both these sources failed, their own fortunes. During the late Republic, some of the fierce foreign wars and the savage looting of the provinces by Roman officials were probably the result of efforts to get funds, in the only way they knew, for building the public works they felt obliged to furnish.

Moreover, these public works were supposed to be given to the Roman people for their free use and enjoyment. The grateful people, in their turn, would put up statues of their benefactor. Poets would recite panegyrics to give the official that sense of glory, pride, and self-esteem that to a noble Roman of the time was more than life itself.

Although the original Roman theory of water supply did not contemplate any private users, leading citizens, and later the emperors and their favorites, managed to have lines run to their houses, where bronze faucets controlled the flow almost as well as modern taps. Although the rules forbade private water for anybody else, other Romans saw no reason why they should not have this convenience, too. So they either bribed the water officials to run pipes to their houses or secretly bored holes into the water channels and tanks and laid their own pipes.

Under the Principate, the water commissioners seem to have given up trying to distinguish between "eminent citizens" entitled to free private running water and the rest of the people. They charged fees for private water pipes, though we do not know the size of the fees or how often they were collected.

Private users, however, still had to have an imperial document entitling them to the water, so private running water remained the privilege of a few. Other folk were expected to take their jars to the public fountains as in the days of yore. The Romans never did quite grasp the idea that such a public utility could be made to support itself by offering service to all in return for a fixed and rational system of charges.

The unit used in measuring water was a *calix* or standard nozzle.[27] The standard calix was the *quinaria,* a length of bronze pipe 1¼ digits (= 0.728 modern inches) in diameter and 12 digits (= 8.75 inches) long, connecting the distributing tank to the user's pipeline.

Users of larger calices were charged in a rough proportion to the cross-sectional area of their nozzles, as nearly as the Romans could calculate these areas with their awkward system of numerals. These charges were made on the assumption that doubling the cross-sectional area would double the flow, when in fact it would more than double it.

The Romans also knew that the flow of water through an orifice is greater if the hydraulic head or water pressure is higher. But they did not know how much greater. So, not having water meters, they could not adjust their charges accordingly.

Most of what we know of these matters comes from a book on the aqueducts of Rome, which Trajan's water commissioner, Sextus Julius Frontinus, wrote around +100. Frontinus was a typical Roman civil servant of the better type, who had served as consul and as governor of Britain. As water commissioner he bossed a small staff of engineers, surveyors, and clerks, and a crew of 700 governmental slaves, including inspectors, foremen, masons, plumbers, and plasterers.

Frontinus bitterly complained of the frauds that had taken place under his predecessors. Users employed calices larger in diameter or shorter than the standard, so that they got more water than they paid for. Or, when they installed a new calix, they left the old one in place and illegally drew water from both. Frontinus was shocked to find secret, illegal pipes running to irrigated fields, shops, and even whorehouses.

The landowners of the countryside, over whose property the aqueducts ran, were even worse. They not only stole the water but also illegally planted trees and erected tombs right next to the aqueducts, damaging the foundations of the piers.

Many students, from Frontinus on down, have sought to calculate the total amount of water conveyed to Rome by the nine aqueducts. Frontinus tells us that the total supply was the equivalent of 14,018 quinariae. Unfortunately, Frontinus' "quinaria" is a measure, not of the volume of flow of a stream, but its cross-sectional area. And the cross-sectional area does us little good if we do not know how fast the water flows.

Still, for want of effective water meters, this system of measuring water supply long continued in use; it was employed in Paris as late as the 1850s. When large discrepancies appeared in Frontinus' figures, he thought that these were due entirely to theft of water and leakage. In fact, however, they were also due to his crude methods of reckoning.

Modern estimates of the total volume delivered to Rome runs from 85 to 317 million gallons a day. Whatever the maximum delivery of all the aqueducts in working order, the volume was much reduced in practice. Besides theft and leakage, one or more aqueducts were often out of service for repairs. So the actual flow may have sometimes been no more than half the theoretical maximum.

Neither can we accurately estimate the amount of water available per person, because the population of Imperial Rome is not closely known, either. Estimates vary from 200,000 to 1,600,000. The most reasonable figure, I think, is around a million. At that, Rome at its greatest was much larger than any of the great cities that had gone before it. As nearly as we can estimate, Babylon, Nineveh, Athens, Syracuse, Carthage, Alexandria, Antioch, Capua, and Republican Rome had all, at their height, harbored somewhere from 250,000 to 500,000 people. Probably, larger cities were impractical because of the difficulty of bringing food from a distance to feed their populations. Roman roads made it possible to import food more cheaply and therefore to concentrate more people in one metropolis. Hence Imperial Rome and–later, for similar reasons–medieval Constantinople, Baghdad, Anuradhapura in Ceylon, and Hangchow in China all approached or exceeded the million mark.

Some have claimed that each Roman had two or three times as much water at his disposal as the inhabitants of most modern cities. But, besides the losses already mentioned, most of the water was used in fountains and baths. Hence it was consumed on a constant-flow system, not on a demand system like that in a modern private house with its valves and taps. Therefore much of the water was wasted.

About all we can say is that the average Roman's water supply was *comparable* to ours. Anybody who needed water could get it, even if he had to haul his own jugful to the topmost story of his insula.

Aside from the abuse of the water system by grafters and water stealers, the system suffered from natural causes, which made its upkeep a heavy responsibility to a conscientious bureaucrat like Frontinus. The water channels of the aqueducts were always cracking and leaking. Although Frontinus complains of sloppy workmanship, this cracking probably could not be helped. Some was caused by the settling of the piers of the arcades.

Moreover, the Romans did not understand thermal expansion. Hence the expansion and contraction of a straight concrete channel several miles long, between a hot summer day and a cold winter night, was enough to crack the cement. As the water was heavily charged with mineral salts, the leakages built up thick limestone concretions around the piers, like the forms you can see at Mammoth Hot Springs in Yellowstone Park.

After Frontinus died about +103 and was succeeded by the younger Plinius, two more aqueducts were built in Rome: the Traiana and the Alexandrina. After Constantine founded a second capital in the East at Byzantium (renamed Constantinopolis, +330) and Honorius shifted the western capital from Rome to Ravenna (+404), dwindling Rome no longer needed more water.

When the Goths besieged Rome in +537, they cut the aqueducts. Although the damage was soon repaired, the shrinking city found it harder and harder to keep up the system. Because of the cracking of the water channel, an aqueduct had only to be neglected for a few years before it stopped giving water.

Although the aqueducts were fitfully repaired during the dark centuries that followed, all of them finally failed about +X. Thereafter the people of Rome went back to the yellow Tiber for their water. For several centuries the shrunken and impoverished population, ruled by a murderous group of "nobles" who were merely successful gangsters, could not support so vast a water-supply system. Furthermore, when Christianity did not regard cleanliness as positively sinful, it placed it down low in the scale of virtues.

The popes of the Renaissance, in rebuilding and beautifying Rome, began the renewal of the water supply. Nicholas V started restoration of Aqua Virgo in 1453. In the next century, Sixtus V built the Acqua Felice in place of the ruined Alexandrina. And in +XVII, Paul V built the Acqua Paolina in place of the ruined Traiana. During this work, many arches of the old arcades were demolished for their masonry.

The headstrong and energetic Sixtus V was especially ruthless in this respect. Like Mehmet Ali in nineteenth-century Egypt, he considered himself a modernist. This meant that he was not at all sentimental about monuments of antiquity and had no compunction about tearing them down for his own projects. He completely demolished, for its stone, the Septizonium, an ornamental façade that Septimius Severus had built on the Palatine Hill in +203.

Just as the ancients built aqueducts to bring water to their cities, they built drains and sewers to take away water that they did not want. Although, among ancient cities, Rome was pre-eminent for its sewer system, it was not the first to have one. Drains had been built at Kalakh (modern Nimrud, Iraq) and elsewhere. Sometimes they were simply storm drains to prevent the flooding of streets in cloudbursts; you can see such a drain today in the ruins of Kameiros on the island of Rhodes. Sometimes they carried off waters used in religious rites, as at Jerusalem. And sometimes, as in the palace at Knossos, they disposed of human waste.

Thus the royal latrine in the palace of Sargon II at Dûr Sharrukîn had, beside each seat, a jar of water and a clay dipper by which the user could flush the appliance after use. At Knossos, on the other hand, the Minos depended on a constant-flow system for flushing. The appliance was built over a channel through which water constantly ran. Latrine drains were generally kept separate from other drains to keep noisome gases from rising through all the system's inlets.

The sewers of Rome started back in the days of the Tarquins, when a ditch was dug to drain the swampy land between the seven hills. According to legend, Tarquinius Priscus, the Proud, worked the common folk so hard on this project that many killed themselves to escape the never-ending drudgery. Tarquinius stopped this wave of suicide by crucifying the bodies of all who slew themselves. The thought that their mortal remains would suffer such ignominy nerved the Romans to keep on living. However, as other ancient authors tell a similar anecdote about the despondent virgins of Miletos, the story should not be given much weight.

The first sewer followed the course of an existing stream and had a stone-lined channel to carry off storm waters. Successive generations enlarged and improved this drain, covered it over with a stone barrel vault, and led affluents into it. It became the great Cloaca Maxima, through which, before the water level rose in recent centuries, it was possible to row a boat. The oldest parts of the present Cloaca date back to –III, and much of the original masonry was replaced in Imperial times by concrete.

The Cloaca Maxima remained primarily a storm-sewer system with numerous openings in the streets to drain off rain water. The grammarian Krates of Mallos, visiting Rome on a diplomatic mission (–II), broke a leg by stepping into one of these openings. Krates aroused the admiration of the Romans by doggedly continuing his lectures throughout his convalescence.

By the time of the Principate, a number of large public latrines had been built and connected with the sewer system. Used water from baths and industrial establishments was channeled to flush these appliances.

A modern man, entering such a structure, would be struck by the complete lack of privacy and by the costly mosaics and marbles. So lavish was the decoration of these establishments that when one was dug up at Puteoli about a century ago, the archeologists at first mistook it for a temple.

Although this latrine system was a great advance over conditions of primitive times, it was still a long way from modern sanitation. Most insulae had their own latrines, nearly always on the ground floor and often connected with the sewer system. A few (at least at Ostia, Herculaneum, and Pompeii) had sewer connections to their upper stories. But vast numbers of Romans lived either at an awkward distance from the latrines or on the upper floors of insulae not so equipped. These either carried their sewage to cesspools or, laws to the contrary notwithstanding, threw it out the window. So Rome, despite its splendid sewers, was still pervaded by the bouquet of all pre-industrial cities.

Withal, the Roman sanitary system was advanced enough so that when, in 1842, a British Royal Commission was appointed to consider ways of improving the health of the people of London, the commission included in its report a description of the sanitary arrangements of the Colosseum and of the Roman amphitheater at Verona. They were better than anything Britain could boast at the time.

Another kind of drainage system was designed to dry up lakes and swamps. Southeast of Rome, between the Via Appia and the sea, sprawls a 40-mile stretch of low, flat land, covered in a wild state by a patchwork of forest and swamp. It is called the Pomptine Marshes. From early Republican days on, Rome looked at this land with speculative eyes. Romans wondered if, by drainage, this land could not be made into good wheat-growing country.

For nearly a thousand years, Roman leaders attempted to reclaim these marshes. The more vigorous of them—Appius Claudius Crassus, Augustus, Trajan, and Theodoric the Goth—made inroads into this malarial flatland. But, as soon as their efforts relaxed, the marshes went back to their pristine state. Only in recent decades has much of this refractory fen been reclaimed.

Central Italy also has a number of lakes without outlets. Some lie in the craters of extinct volcanoes. Whereas the level of a normal lake is held to close limits by the outlet, the level of these landlocked lakes varied widely with the seasons. They filled up in winter and spring and dried out to bogs and mud flats in summer and autumn. Hence these lakes were not much good for fishing, pasturage, or grain crops.

As far back as early –IV, some Roman magistrate undertook to drain Lake Albanus, fourteen miles southeast of Rome. He drove a tunnel a mile and a third long, to lead off the water. This tunnel still serves its purpose, and the water from the lake is used for irrigation. About a century later, another politician executed a similar work at Lake Velinus, 43 miles north of Rome.

The most ambitious drainage project of this kind, however, was undertaken in the +40s by the emperor Claudius. This was the drainage of Lake Fucinus, high in the Apennines at the center of Italy. Whereas the previous lakes had been only a couple of miles long, Lake Fucinus was ten miles in length.

The freedman Narcissus, Claudius' secretary and the most powerful man in the Empire, had charge of the project, on which thirty thousand men worked for eleven years. Along the course of the 3.5-mile tunnel, several dozen shafts were sunk from the surface, so that the workers could descend to the cuttings.

To celebrate the opening, Claudius prepared a great naumachia on the lake. Nineteen thousand convicts manned two fleets of triemes and quadriremes, the "Rhodian" and the "Sicilian" fleets, while soldiers on rafts surrounded them, lest some crew run their ship ashore and make a dash for the hills.

When the gladiators gave the conventional cry: "Hail, Caesar! They salute you, who are about to die!" Claudius replied: "Or not, as the case may be." Although this was just one of the scholar-emperor's little jokes, the fighters thereupon refused to fight, declaring that the emperor had pardoned them. It took threats of massacre to get the battle underway, and the combat continued until a satisfactory quantity of blood had been shed.

The first attempt to lower the lake level did not work well, because the tunnel did not tap the lake at a low enough point. Claudius had the work improved and staged a second grand opening, with more gladiators. He also spread a banquet near the outflow of the tunnel.

When the floodgates were opened, the volume of the stream was much greater than expected. It swept away part of the banquet and some of the banqueters. Thereupon a noisy quarrel arose between Narcissus and Claudius' fourth wife, the fiendish Agrippina, who was also his niece and the mother of Nero by an earlier marriage. Agrippina screamed that the work had gone awry because of the graft that Narcissus had made from the project. The gouty Narcissus, enraged, "was not silent, but inveighed against the domineering temper of her sex, and her extravagant ambition."[28]

Soon thereafter Agrippina, fearing that Narcissus would turn Claudius against her, poisoned the Emperor with toadstools. As soon as Nero was on the throne, she had Narcissus jailed. Narcissus either was killed or slew himself.

The tunnel soon became blocked and went out of use. Although Trajan and Hadrian reopened it, control of Lake Fucinus was not made secure until a modern tunnel was completed by Prince Torlonia in 1876.

Perhaps Narcissus had indeed been grafting. We must remember, however, that our two main sources for that period, Suetonius and Tacitus, abominated the Julio-Claudian emperors. Their sympathies lay with the Senate–that is, the landowning aristocracy–with which these emperors often came in conflict. Therefore they tended to magnify the emperors' faults, attenuate their virtues, and condemn their associates. Although some of the Julio-Claudians were pretty appalling, they may not fully deserve the judgment: "An arch-dissembler was succeeded by a madman, and a fool by a monster."[29] If they had been, it is hard to see how the Empire could have survived them. On the other hand, for lack of evidence, we cannot tell how much better the Julio-Claudians were than their gruesome literary portraits.

Because the land was mountainous and the rivers short, canals did not play a large role in classical Italy. One canal, however, paralleled the Appian Way between Forum Apii, 43 miles from Rome, and Tarracina. Forum Apii was probably founded as a stopping place by Appius Claudius Crassus. Travelers from Rome often left the road at Forum Apii to take an overnight canal-boat ride to Tarracina, where they took to mules and carriages again. Horace tells of such a journey:

Presently night began to spread her shadows over the earth and to scatter the stars across the heavens. A babel of voices arose, slaves abusing boatmen and boatmen abusing slaves. "Make a stop here!" "That's plenty. You've got three hundred on board, now." We wasted a whole hour paying fares and harnessing the mule. The cursed mosquitoes and frogs in the marsh made sleep impossible, while a boatman and a passenger, soused with flat wine, rivaled one another in singing to their absent mistresses. Finally the passengers became exhausted and dropped off to sleep while the lazy boatman tied the halter to a rock, turned out the mule to graze, and lay on his back and snored. It was already dawn before we noticed that the craft was not moving. Then a hotheaded fellow, one of the passengers, jumped from the boat, cut himself a willow cudgel, and clubbed the mule and boatman over the head and sides. Even at that we only landed at ten o'clock to wash our hands and faces in your holy water, Feronia.[30]

Under the Principate, harbors were improved all over the Empire. According to doubtful traditions, King Ancus Marcius, alleged builder of Horatius' bridge, founded a Roman colony at the mouth of the Tiber (–VII) and called the settlement Ostia, "entrance." During the Punic Wars this became a naval station. Then, as the city of Rome expanded and needed more grain from abroad, Ostia became the main port for the trans-shipment of this grain.

By the time of Augustus, silting had impaired the value of Ostia's harbor. The shore line had moved seaward, and the harbor bar was becoming a menace, either because the water over it was shallower or because the grain ships were larger.

As a port, Ostia competed with the Bay of Naples and its several excellent natural harbors. In –36, Augustus added a great new harbor at Baiae, on the west shore of the bay. The harbors of the Bay of Naples, especially Puteoli, still received most of the more valuable and less bulky goods bound for Rome. This freight was moved to the capital over the Appian Way.

Bulk goods, especially grain, came to Ostia. If it came in a small enough ship, the ship went on up the river to Rome–rowing if it had oars, being towed if it did not. Larger ships unloaded their cargoes at Ostia, whence it was taken to Rome in barges pulled up the river by slaves or by oxen, or was carried over the Ostian Way. Very large ships had to anchor outside the bar and send their cargoes ashore in lighters. Many of the captains of these ships therefore preferred to unload at Puteoli, despite the high costs of land transportation.

Claudius wanted to enable large ships to enter the harbor in winter and tie up safely, so that, if the summer's importation of grain fell short, ships could continue to bring in food throughout the winter. He therefore built a new harbor, called Portus, across the mouth of the Tiber from Ostia.

Two curving breakwaters extended out until they formed an almost complete semicircle. A man-made island near the opening shielded the harbor from swells and supported a lighthouse. To form a foundation for the island, Claudius had a special ship, which had brought an obelisk from Egypt, filled with concrete and sunk.

Despite these precautions, a violent storm in Nero's reign (+62) sank 200 ships in the harbor. Therefore Trajan improved the facilities still further. He dredged out a hexagonal inner harbor, with a slip for small craft to tie up.

Similar harbors were built elsewhere in the Mediterranean. Between –20 and –10, Rome's ally Herod the Great, king of Judaea, built a new city, Caesarea, on the harborless coast of Palestine. The semicircular breakwater of Caesarea's man-made harbor had its monumental opening on the northern side, because the strongest blows in that region came from the south.

Puteoli had a breakwater of original design. It consisted of two parallel arcades with a space between them. The arches of one arcade were in line with the piers of the other, so that no wave could sweep straight through. Yet the openings in the arcades lessened the impact of the waves on the structure, somewhat as the holes in a panel of soundproofing material absorb sound waves.

Most Roman harbors had lighthouses. While their design was somewhat like that of the Pharos, they were smaller and less ornate. A typical Roman lighthouse consisted of several stories, round or polygonal, each smaller than the story below it. The light at Ostia had seven stories; that at Boulogne, twelve.

The Boulogne lighthouse was repaired by Charlemagne in +811 and was fortified and equipped with cannon by the English in the 1550s. In the 1640s, as a result of reckless quarrying and neglect by the magistrates of Boulogne, the cliff whereon it stood collapsed, bringing down the lighthouse with it. Remains of Roman lighthouses still stand at Fréjus (Forum Iulii) and at Dover.

And how about the ships that were guided by these lighthouses into the Roman harbors? During the Punic Wars, the Romans, like their Carthaginian foes, used as their standard warship the Hellenistic pentere or quinquireme, with a single row of five-man oars on each side. In these wars the Romans made one technical advance: mounting a flying bridge on the bow of a ship, with a projecting spike to catch the foe's deck when the bridge, called a *corvus* or "crow," was dropped on the other ship for boarding. By thus converting naval battles from ramming contests into shipboard infantry battles, the Romans were able to inflict a series of smashing defeats on the skillful seamen of Carthage.

Later, the Romans made some use of giant galleys of the Hellenistic type. The evolution of these super-galleys ceased with the battle of Actium (–31) where Marcus Antonius employed some of them against Octavianus. His flagship was a decireme or tenner. But disaffection broke up Antonius' fleet in the midst of the battle, and Antonius and Cleopatra had to flee with such ships as remained loyal to them.

Once Octavianus, under the name of Caesar Augustus, ruled the entire Mediterranean, there was nobody left for these big costly ships to fight. To keep down pirates, triremes and biremes were adequate. So, for the next five centuries, the Roman navy consisted of these small galleys. The favorite type was the Liburnian, a bireme based upon a design formerly used by Illyrian pirates.

Larger galleys were revived after the fall of the Western Empire, in the wars between the Byzantine Empire–the eastern rump of the Roman Empire–and the Arabs. The Byzantine battleship was called a *dromon*. This was a bireme with one man to each lower oar and, depending upon the size, one, two, or three men to each upper oar.

On the other hand, the design of merchant ships advanced under Rome. Grain freighters were often over 100 feet long and carried cargoes of over 1,000 tons, compared to 180 tons for the *Mayflower*. They were broad, tubby, solidly built ships, which wallowed along at about six knots under one large square sail. To this sail, however, was added an artemon or foresail, a triangular topsail or raffee sail, and sometimes a small mizzen sail on a third mast at the stern. These small sails at bow and stern were intended less for speeding up the ship than for making it easier to maneuver. In harbors these ships were moved by tugs, which were large rowboats like a modern lifeboat.

An important change, almost certainly of Greek invention, also took place in rigging. The fore-and-aft sail appeared, making it possible to sail much closer to the wind than before. When a fisherman named Alexandros of Miletos died near Athens, a tombstone was put up showing Alexandros steering his fishing smack. Alexandros' sail is similar to the triangular lateen sail of later centuries, except that the sail is squared off at the forward end instead of coming to a point. A sail of this type is called a short-luffed lug.

Other Greeks of about the same time had tombstones showing fore-and-aft sails of another type, the spritsail. The spritsail is fastened to the mast at its upper and lower forward corners. A diagonal spar from the base of the mast to the upper after corner spreads the sail, and the lower after corner is belayed to the deck by the sheet.

As these tombstones are not dated, their exact time has to be guessed. Casson, who has been prowling the Mediterranean in search of them, is sure that some go back at least to +I and that one is from –II. Moreover, a coin of Nero's time shows several ships with short-luffed lug or lateen sails in the harbor of Ostia.

For several centuries after their invention, sails of these types appear to have been used only on small boats such as fishing craft. It was especially important for coastal craft to be able to sail whithersoever they wished, to avoid being blown ashore. After the fall of the West Roman Empire, the lateen sail became popular not only in the Mediterranean but also around the shores of the Indian Ocean. For a thousand years, through the Dark and Middle Ages, most ships in the Mediterranean had lateen sails.

Perhaps you wonder why, once shipbuilders had fore-and-aft sails, they did not abandon square rig altogether. The answer is that square rig has many advantages for long-distance, deep-water sailing. In crossing wide seas and oceans, the captain can take advantage of prevailing wind belts so that during most of the voyage he has the wind astern. And, with the wind astern, the square rig gives a greater area of sail and makes more efficient use of it than the fore-and-aft.

Moreover, possession of a fore-and-aft sail does not, by itself, mean that it will be used for sailing close-hauled. The sailing ships of the Arabian Sea, of the kind called "dhows," have long used the short-luffed lugsail. But they do not sail close-hauled or tack. Instead, their masters wait for the seasonal monsoon winds to blow whither they wish to go; for, as they say: "No one but a madman or a Christian would sail to windward."[31] It is not the custom, and anyway their sails are probably too baggy to make such sailing practical.

THE LATER ROMAN ENGINEERS

SEVEN

In the final century of the Republic, Rome expanded swiftly, gobbling up all the other nations of the Inner Sea. At the same time the nation was convulsed by ferocious civil wars among rival politicians, who butchered their opponents by tens of thousands and piled pyramids of heads in the Forum. Nobody heeded the Roman Constitution any more; an ambitious man cared less for the rules of political advancement than for control of troops by which he could impose his will.

In the –40s, the most gifted of these politician-adventurers, Gaius Julius Caesar, crushed his opponents, the party of Pompeius. Less than a year after he had attained supreme power in Rome, however, Caesar was murdered (–44) by diehard republicans. After another round of civil war, Caesar's great-nephew succeeded to the murdered dictator's power. Originally named Gaius Octavius, this youth changed his name, as customary, to Gaius Julius Caesar Octavianus[1] when his great-uncle adopted him in –45.

In –27, having in his turn liquidated his enemies, Octavianus took the name or title of Augustus, together with a number of Republican offices. Although in theory the constitution of the Republic remained in force, in fact the Republic had ended and the Empire had begun. The title of the early emperors was Princeps or "first citizen," so that the early Empire is properly called the Principate.

Augustus' forty-year reign was a time of unprecedented peace and prosperity. Trade routes stitched the Mediterranean. Goods and gifted men poured into Rome, which became the world's most magnificent city, while Roman roads, law, and citizenship were bit by bit extended to the provinces. Beyond the frontiers, the barbarians were not yet a serious threat, although the Germans wiped out one Roman army that strayed too far into their somber forests.

Slavery reached vast proportions. In Italy, the number of slaves approached or perhaps even exceeded that of free men. A proposal in the Senate to make the slaves wear distinctive dress was hastily squelched when somebody pointed out that "It would be dangerous to show the wretches how numerous they really were."[2]

In later centuries, as the Empire ceased to expand, one of the main sources of slaves–conquest–dried up, and the freeing of slaves was easy, common, and socially approved. Hence the number of slaves declined, while the law gave them more and more protection. At the same time, the lot of the poor free worker worsened as more restrictions were clamped upon him. Consequently, at the end of the Western Empire, there was little to choose between the lot of the slave and that of the free worker.

The wealth and peace of the early Empire, the lust of the first emperors for glory-by-building, and the abundance of cheap labor combined to foster a great surge in the construction of public works and the practice of engineering.

The most eminent engineer and builder of the Roman world, after Appius Claudius Crassus, was Marcus Vipsanius Agrippa (–63 to –12). Born into an obscure Roman family, he studied at Apollonia, a Greek city on the Adriatic coast opposite the heel of Italy. With him studied Gaius Octavius, the future Augustus.

Agrippa and Octavius became lifelong friends. When Julius Caesar, fighting his last campaign against the surviving Pompeians in Spain, sent for his adopted son, he included Agrippa in the invitation. In Spain, Agrippa learned the art of military command.

Back in Rome the following spring, Caesar was murdered, and Octavianus (as he was now called) came into power as one of the Second Triumvirate. Young Agrippa, now praetor, won victories in Gaul and Germany. As consul in –37 he commanded Octavianus' fleet against the forces of Sextus Pompeius in the waters around Sicily.

Here Agrippa showed his technical flair. First, he had to prepare a lot of untrained soldiers and sailors to fight a formidable foe. To overcome this handicap, he made Lake Lucrinus–Orata's old oyster-hunting ground–into a training area.

Shallow Lake Lucrinus was separated from the Bay of Naples by a narrow neck of land. Half a mile inland from Lake Lucrinus lay the deeper Lake Avernus in the crater of an extinct volcano. Agrippa joined these lakes with each other and with the bay, using Lake Avernus as a storm-proof anchorage and the other lake for exercises. Few traces now are left of Agrippa's Portus Iulius, as he named this complex, because an earthquake and eruption in 1538 changed the lay of the land.

Agrippa also invented two devices to give him a military advantage. One was a collapsible tower for missile troops, which could be quickly raised from the deck when a ship neared an enemy. The other was a grapnel that could be shot from a catapult, to catch another ship and pull it close for boarding.

When Sextus Pompeius and his pirates were cleaned up, Agrippa became aedile in Rome and began his notable public works. He repaired the older aqueducts, built two new ones, and further improved the water works by constructing 130 water-distributing stations, 300 large cisterns, and 500 fountains. He even took a boat ride through the Cloaca Maxima –the great sewer–to direct its renovation.

In –31, Agrippa took part in Octavianus' war against Marcus Antonius. Having reduced Antonius' forces to hunger by capturing their naval stations and blockading them in Western Greece, he commanded the whole Octavianist fleet at the decisive battle of Actium. Octavianus, who knew that he was no military genius, watched from a distance.

Having become rich from the property of political enemies, which Octavianus confiscated and gave to his friend, Agrippa spent his own money as well as the state's on public works. He built the first public bath in Rome, the forerunner of those immense bath halls erected by Diocletian and Caracalla. This was his own property, but when he died he left it to Augustus (as Octavianus now called himself) with a hint that it ought to belong to the people. So the Princeps turned the edifice over to the state.

Agrippa also built another bridge across the Tiber, a series of temples and porticoes, and a hall for counting votes. As the emperors soon stopped holding elections, this last building became a theater. Agrippa also built a naval headquarters building and insulae for the masses to live in.

The most celebrated of all Agrippa's constructions was the Temple of All Gods–the Pantheon–in Rome. It consisted of two main parts: a rectangular portico and, behind the portico, a large rotunda.

It is hard to say how much of the present building is Agrippa's original, or even how much it looks like the original. The Pantheon was damaged by fire in the reign of Titus, repaired by Domitian, damaged by fire under Trajan, drastically rebuilt by Hadrian in the +120s, further repaired by Septimius Severus and Caracalla, turned into a Christian church in +608, and stripped of its gold-plated tiles by the Byzantine emperor Constans II, who was slain by Saracen pirates on his way back to Constantinople. During the Middle Ages it was robbed of its marble facings, all the statues that once occupied the niches of the rotunda having already disappeared.

The last big depredation occurred in 1625, when Pope Urban VIII (Maffeo Barberini) took the bronze girders that held up the ceiling of the portico. He melted them up to cast eighty cannon. These he mounted around Hadrian's tomb, which earlier popes had already turned into a private fortress under the name of the Castel Sant' Angelo.

Urban's rape aroused no little comment, even in an age when the despoiling of ancient monuments was common. A wag of the time said:

Quod non fecerunt barbari fecerunt Barberini

or, "What the barbarians did not do, the Barberini have done." Another writer defended the action on the ground that it was a "worthier destiny . . . that such noble material should keep off the enemies of the Church rather than the rain."[3]

Any historian of technology would agree with the first comment; for these were the only all-metal girders ever known to have been made in ancient times. We cannot tell, now, whether they formed part of Agrippa's original structure or were put in during Hadrian's rebuilding. In any case, Pope Urban replaced the bronzen structure with one of wood.

The front of the portico, at least, may be original, for it still bears the inscription:

M · AGRIPPA · L · F · COS · TERTIVM · FECIT

meaning: "Marcus Agrippa, son of Lucius, made (this) in this third consulship."

The rest of Agrippa's life was spent on military and diplomatic missions. He pacified the Gauls in −19 and found time to build four great roads in Gaul. He also furnished the town of Nemausus (modern Nîmes) with a graceful temple, a public bath, an arena, and an aqueduct. All still stand except the bath, which was partly demolished for military reasons in 1577. The aqueduct includes the celebrated Pont du Gard. The temple is now known as the Maison Carrée or Square House. At Thomas Jefferson's urging, it was taken as a model for the Virginia State House at Richmond. Jefferson mistakenly thought it had something to do with the Roman Republic, which he admired.

Agrippa went on to Spain, crushed rebellious tribes, and built roads. At Augusta Emerita (modern Mérida) he erected temples, baths, a circus, a theater, and a naumachia, of which the last two are still in good condition. Visiting Syria in –15, he built a bath and other structures at Antioch. After further travels and missions, robust and energetic though he seemed, he died suddenly, probably of gout, at 51.

Although Agrippa ranks with Rameses II and Nebuchadrezzar II as a builder, we cannot tell much about his personality. For one thing, the attention of ancient historians was glued to monarchs and generals to the neglect of other folk. For another, Agrippa was such a modest man, in an age when paranoid self-aggrandizement was considered normal behavior, that he refused some of the triumphs Augustus offered him.

We can infer that Agrippa was sober, hard-driving, honest by the standards of the time, devoted to his building projects, but personally not very ambitious. His technical judgment seems to have been sound. In other words, he was the perfect executive engineer. He looked the part, too: heavy-set but rather handsome in a beetle-browed, beak-nosed, jut-jawed way.

Augustus, on the other hand, was a cold, crafty, merciless little man who learned to play to perfection the kindly role of father of his country. Augustus made the shrewdest of his many clever moves when he attached Agrippa to himself. For he came near to falling more than once and might well have done so without Agrippa's staunch help.

The more tightly to bind Agrippa to him, Augustus persuaded Agrippa to divorce his first wife and marry Augustus' niece Marcella, then to divorce Marcella and wed his daughter, the promiscuous Julia. This was probably not so painful as it sounds, because, to most upper-class Romans, marriage was more a matter of business than it is with us. They traded wives back and forth as liberally as movie stars do today. There is reason to think that Augustus planned to name Agrippa his successor, for it probably never occurred to the frail Princeps that he would outlive the lusty Agrippa by a quarter of a century.

A few years before Agrippa's death, Augustus' stepson Tiberius had married Agrippa's eldest daughter Vipsania. After Agrippa's death, Tiberius' mother Livia prevailed upon him to divorce Vipsania and marry the amorous Julia, daughter of Augustus. As Tiberius loved Vipsania, the experience soured him for life, and he became a morose and miserly emperor. Agrippa left several other children, most of whom came to violent ends. A son of one of these children became the emperor Caligula, while one of Caligula's sisters was the dreadful Agrippina, mother of Nero.

The geography and autobiography that Agrippa wrote are lost, but we are lucky to have a work by a contemporary and colleague of his, the architect Marcus Vitruvius. Very little is known about Vitruvius, save that he had worked for the state as an artillery engineer, that he built a basilica or town hall at Fanum, and that he described himself as an ugly little old man.

We must think of Vitruvius as a writer on comparatively early Roman architecture. He borrowed most of the historical parts of his treatise from his Greek predecessors, and he discusses the building methods of Hellenistic and Roman Republican times. In his day, the most famous Roman buildings, such as the Colosseum and Hadrian's villa, had not yet been built, so he could not deal with them. Despite certain shortcomings, his treatise is one of our main sources of information on ancient art, architecture, and technics. Although many other such treatises once existed, time has spared Vitruvius' *De Architectura* alone.

Of Vitruvius' ten "books," the first deals with the qualifications of an architect, architecture in general, and town planning. The second tells of building materials, the use of which it traces from prehistoric times; and it also describes the Roman methods of using masonry, brick, and concrete. The third book is about temples; it derives the proportions of their parts from the proportions of the human body. The fourth tells of the three orders: Doric, Ionic, and Corinthian. The fifth describes public buildings of other kinds–basilicas, theaters, baths, and so on–and discusses acoustics and the wave theory of sound.

The sixth book tells about dwelling houses, while the seventh goes into interior decoration, with much detail on plaster, paint, and mosaics. Book VIII is on water supply: springs, aqueducts, wells, cisterns, and so forth. The ninth deals with geometry, astronomy, measuring, and the design of water clocks. The tenth and last expounds mechanics. In this Vitruvius includes hoisting devices, pumps, water wheels and mills, the water organ, and a geared taximeter to measure the distance a carriage has gone. He also devotes several chapters to catapults, tortoises, sambucae, belfries, and other engines of war.

Vitruvius' work dropped out of sight in the Middle Ages. But, after a

Painting by Zeno Diemer, Wasserleitungen im alten Rom, *in the Deutsches Museum in Munich*

Plate XII. The aqueducts of Rome, restored. The Anio Novus and the Claudia in the left foreground, combined in a single structure; the Marcia, Tepula, and Iulia, also borne by one single arcade, on the right.

(from Schreiber)

Plate XIII. The Pont du Gard, a three-level Roman aqueduct bridge near Nîmes, France.

(from Schreiber)

Plate XIV. Trajan's bridge over the Danube, with the emperor sacrificing in the foreground. A relief on Trajan's Column in Rome

Plate XV. The great clock of Gaza, as described by Procopius.

(from Diels)

Plate XVI. The Church of Santa Sophia in Constantinople (modern Istanbul).

Photograph by Ewing Galloway

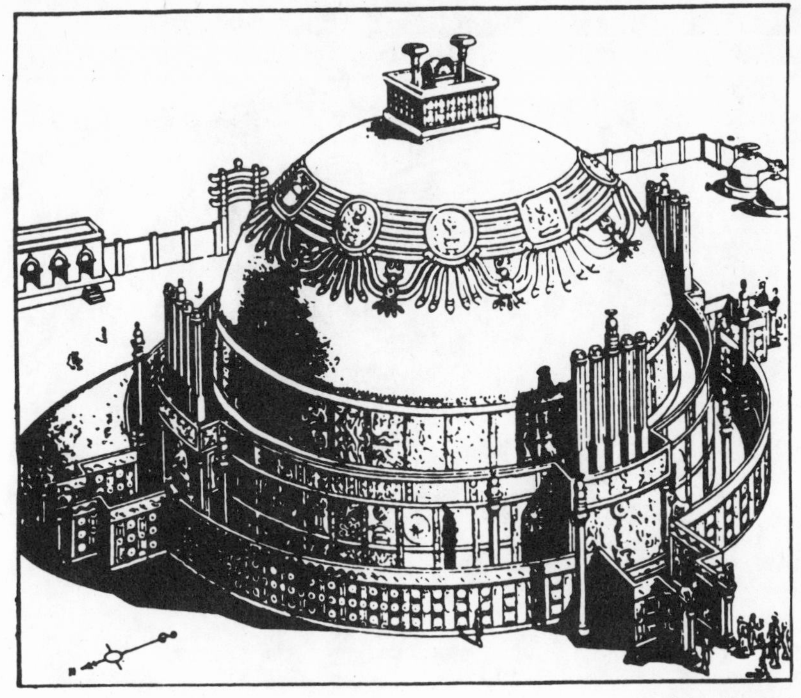

(after Brown)

Plate XVII. The stupa of Amrâvatî, restored.

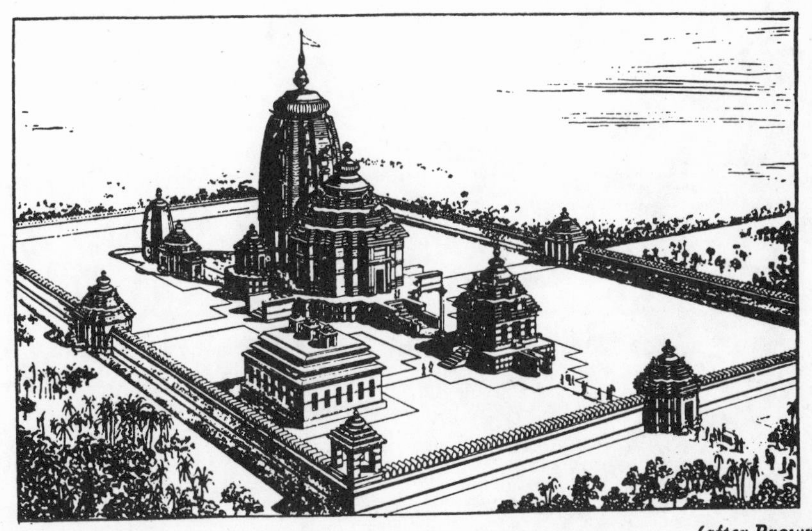

(after Brown)

Plate XVIII. The Temple of the Sun at Konârak, Orissâ, India, restored.

Drawing by John Christiansen in Needham, Wang, and Price

Plate XIX. Su Sung's astronomical clock tower at Kaifêng, restored.

Plate XX. The east end of the Cathedral of Notre Dame de Paris, showing flying buttresses.

Photograph by Philip G. Cavanaugh

Plate XXI. A reversible overshot water wheel for hoisting water out of a mine, as depicted in Georgius Agricola's *De Re Metallica* (1556).

Plate XXII. A post windmill, as depicted in Agostino Ramelli's *Le Diverse et Artificiose Machine* (1588).

(from J. J. von Wallhausen: Art Militaire à Cheval, *1616)*

Plate XXIII. Light v. heavy cavalry in early +XVII. The lancer on the left wears a three-quarter suit of armor and carries a lance, a sword, and a pair of horse pistols. The rider on the right is armored only with helmet and cuirass and armed with sword and pistols.

manuscript copy of it was rediscovered in +XV and Vitruvius' writings again became well known, architects came to look upon him as an infallible authority, much as medieval schoolmen regarded Aristotle.

After the aged Augustus died, his successor, the somber and thrifty Tiberius, did little in the way of public works, save to build a temple to Augustus and restore Pompeius' theater.

Succeeding Tiberius, the mad Caligula started the aqueduct later known as the Claudian. Instead of concrete and brick, which had become usual for such works, he ordered it made of the costlier stone. He also had the Isthmus of Corinth surveyed with the idea of cutting a canal across it. Otherwise he indulged in such freaks as assembling a bridge of boats across the Bay of Naples, like Xerxes' bridge across the Hellespont, solely to stage parades led by himself in fancy dress.

Claudius, intelligent and well-meaning if prematurely senile, was a more vigorous builder. He finished the languishing Claudian aqueduct. I have told about his drainage of Lake Fucinus. When a grain shortage caused the Roman mob to pelt Claudius with bread crusts, he speedily built the harbor of Portus, next to Ostia, to secure the food supply.

The Julio-Claudian emperor with the most creative engineering ideas, however, was Nero. Usually thought of as a monster, Nero was the most gifted, artistic, and versatile of the line, as well as the most contradictory and the last.

As a Hellenophile, Nero, despite his personal fondness for murder, deplored gladiatorialism and tried to wean the Roman public away from it. When he gave "games," he refused to allow anybody to be killed and even made hundreds of Roman gentlemen appear in the arena to see what it felt like. He also encouraged plays, concerts, and ballets as a substitute for the gory national spectator sport.

Nero was intensely serious about his own artistic ambitions. While his voice was too weak for good professional singing, his poetry is said to have been not bad. When he appeared in an artistic contest, he tried to see that the judges judged his performance fairly. But the judges, knowing the spoiled, capricious, and violent temper of their Princeps, took no chances; they gave him the prize regardless of the merits of the performance.

All this cultural propaganda had no effect on the Romans, who went right back to blood and guts. Nero's artistry they despised, as this was not the sort of thing a Roman gentleman did. They did not especially mind his having his mother murdered; most agreed that Agrippina deserved what she got. But it incensed them that their emperor should so demean the Empire as to play the lyre and sing in public.

In +64, a conflagration burned the greater part of Rome. Gone were many structures that had come down from the days of the kings and the early Republic. Gone, too, were countless art treasures plundered from the Greek lands.

During the fire, men were seen running about with torches, spreading the fire and defying anybody to stop them. Some said that these were Nero's agents, but this is unlikely. Others blamed the Christians, and Nero massacred numbers of Christians as punishment. It is not impossible that the incendiaries were in fact Christians, because Christianity then included many wild-eyed fanatics who went about crying that the world was about to end. But perhaps the arsonists were merely slaves, venting their hatred of their masters and of Rome, or ruffians out for loot.

Nero energetically directed the fire fighting. Afterwards he took prompt and vigorous action to succor the people, collect contributions for their relief, import an emergency store of food, and rebuild the city.

In rebuilding, Nero used the services of a pair of able architects, Severus and Celer. This time the streets were laid out on an orderly gridiron plan, with wide avenues and open spaces in place of the former tangle of crooked alleys. Some of the avenues of modern Rome still follow those of Imperial times; thus the Via del Corso is the old Via Lata or Broad Way. The new insulae were limited in height and required to use a certain amount of fireproof construction. They were also provided with balconies to help in fighting fires, while the water works were extended to make more water available for this purpose.

Nero did not forget to reserve for himself a large burned-over tract, stretching from the Palatine Hill to the Oppian Mount. Here he built an enormous palace, the Golden House, with a mile-long colonnade and a statute of himself, by the sculptor Zenodoros, as big as the Colossus of Rhodes but, as a result of advancing methods of construction, completed in a fraction of the time. Vespasianus later turned this figure into a statue of Helios by putting a crown of solar rays, like those of the Rhodian Helios, on the statue's head. Then the tireless Hadrian, with the help of the architect Decrianus, had the features reworked to look less like Nero's and, by means of twenty-four elephants, moved the statue to a new site.

Severus and Celer persuaded Nero to an even more daring scheme. This was to dig a 160-mile canal along the Italian coast from Ostia to Lake Avernus. There, by means of the channels cut by Agrippa, this canal would communicate with the Bay of Naples. Nero mobilized thousands of convicts to do the work. Digging began, but at Nero's death the project was dropped.

Later historians cited this canal as just one more example of Nero's megalomania. Actually, it was a brilliant idea. By means of this canal, heavy grain freighters would have been able to sail safely to a point close to Rome. As it was, they had either to stop at Puteoli, which raised the cost of goods by requiring their transportation by land for 160 miles, or to anchor at Ostia with its dangerous bar and cramped harbors.

For most of the distance, the canal was quite feasible; you recall that one already existed parallel to the Appian Way. Southeast of Tarracina the spurs of the Apennines, which come close to the sea in several places, might have given the engineers some trouble, but the project was by no means absurd.

In +66, Nero made a grand tour of Greece. While trying to show the skeptical Greeks what a fine artist and cultured Hellene he was, he revived the idea of cutting a canal across the Isthmus of Corinth. This plan had intrigued not only his uncle Caligula but also several other eminent predecessors like Julius Caesar.

Again Nero mobilized convicts, including 6,000 Jews captured in the Jewish War, just beginning. He himself swung the first mattock and carried off the first basketful of dirt, as Ashurbanipal had done in his day. In +67, however, rumors of plots and revolts at home sent him back to Italy and the digging stopped.

This canal could perhaps have been completed, to the advantage of impoverished Greece. In 1881 a French company, finding Nero's old route the best, undertook to dig the canal. After they gave up, a Greek company finished the job in 1893. If it took twelve years with modern machinery, you can see what a job it would have been for the Romans. The actual quantity of dirt and rock to be moved was so vast that several emperors in succession would have had to work at it.

Back in Italy, Nero did not long survive, having alienated the Senate by his murders of Senators, the army by his pacifism, and all the other Romans by his affectation of Greek culture. When revolt burst out in Gaul and Spain, the bewildered Nero lost what wits he had but kept to the last his interest in technical matters. Having summoned the leading citizens of Rome to discuss the emergency, "after a brief discussion of the Gallic situation, he devoted the remainder of the session to demonstrating a new type of water-organ, and explaining the mechanical complexities of several different models. He even remarked that he would have them installed in the Theater 'if Vindex [one of the rebellious generals] had no objection.'"

A few weeks later, Nero had stabbed himself, murmuring: "What an artist dies in me!"[4] Whatever Nero's artistic merits, it looks as though the world also lost a slightly mad but naturally gifted engineer that day.

Three emperors quickly followed Nero but came to violent ends. Then building began again under the competent Vespasianus and continued under his sons Titus and Domitianus. Making use of cross-vaulting on an enormous scale, they built the sinister but awesome Colosseum, as well as temples, baths, and other public buildings.

After Domitianus was murdered in +96, the elderly Nerva reigned for two years and died, having chosen as his successor Marcus Ulpius Traianus–Trajan, as we call him. Trajan, an upper-class provincial of mixed Italian-Spanish descent, was one of the ablest Principes. Under him, the Empire reached its greatest extent. Nerva, Trajan, and the three who followed them (Hadrian, Antoninus Pius, and Marcus Aurelius) are sometimes called the Five Good Emperors; the period in which they reigned, +96 to +180, was the most prosperous time that Rome was ever to know. The Romans themselves became better behaved, too. They toned down the wild excesses of the preceding two centuries and recovered something of their former dignity and sobriety.

Trajan, a mighty builder, employed Apollodoros of Damascus as his architect. The Romanized Spaniard and the Hellenized Syrian adorned Rome with its finest forum. It lies, abutting the earlier forum of Augustus, 150 yards northwest of the egregious pseudoclassical nineteenth-century monument to Vittorio Emanuele II.

Entering the forum through a triumphal arch at the south end, one found oneself in a paved plaza about 125 yards long and 100 wide, with an equestrian statue of Trajan at the center, colonnades around the sides, and adjacent buildings that included a library and a market. Athwart the forum at the north end of the plaza stood the Basilica Ulpia, a hall for public business. Covering an area 100 by 500 feet, the Basilica Ulpia must have looked much like the present church of Saint Paul Outside the Walls at Rome.

If one passed through or around the Basilica Ulpia, one reached a smaller open space north of this building, dominated by Trajan's 100-foot column, which still exists. A spiral band of reliefs, illustrating Trajan's campaign against the Dacians, runs around this column to the top. A statue of Saint Paul has taken the place of that of the emperor atop the column.

Finally, at the north end of the forum rose Trajan's personal temple, famed for size and richness. Two churches now occupy the site.

Elsewhere, Trajan and Apollodoros erected a bath and dug a

naumachia. Another of Trajan's executives, Gaius Julius Lacer, erected one of the most impressive surviving Roman bridges near Alcántara in Spain. On it he proudly inscribed:

PONTEM PERPETVI MANSVRVM IN SAECVLA

meaning: "I have created a bridge that shall last for the ages."[5]

Their Roman contemporaries were even more impressed by the bridge that Apollodoros built for Trajan across the Danube. The purpose of this bridge was to enable the Emperor to conquer the Dacians, who lived in what is now Rômania. The bridge had twenty piers of squared stone, 150 feet high, 60 feet wide, and 170 feet apart. Furthermore, these piers were set up where the river was deep and swift, with a muddy bottom.

A relief on Trajan's column shows this bridge. On the stone piers were built a series of wooden arches, and these in turn supported a wooden deck. If you will look at Plate XIV, you will see that, if the sculptor who carved the relief was accurate, Apollodoros used diagonal bracing in the wooden part of the bridge. As far as I know, this is the first definite example of the truss, which depends for its strength on the rigidity of three beams fastened together to form a triangle.

Trajan nominated as his successor his younger cousin Publius Aelius Hadrianus, who turned out to be the most brilliant of all Roman emperors. Hadrian was a writer, poet, wit, musician, artist, architect, critic, philanthropist, lawyer, diplomat, general, and supreme executive. Thinking that the Empire had overextended itself, Hadrian gave up Dacia and removed the superstructure of the Danube bridge to discourage the Dacians from raiding into Moesia.

At first Apollodoros worked under Hadrian with another architect, Decrianus. After the latter had moved Nero's statue by means of his twenty-four elephants, Apollodoros planned a similar statue of the moon goddess. But poor Apollodoros seems to have been too outspoken for his own good. Hadrian:

> . . . first banished and later put to death Apollodoros, the architect, who had built the various creations of Trajan in Rome—the forum, the odeum, and the gymnasium. The reason assigned was that he had been guilty of some misdemeanor; but the true reason was that once when Trajan was consulting him on some point about the buildings he had said to Hadrian, who had interrupted with some remark: "Be off, and draw your gourds. You don't understand these matters." (It chanced that Hadrian at the time was pluming himself upon such a drawing.) When he became emperor, therefore, he re-

membered this slight and would not endure the man's freedom of speech. He sent him the plan of the temple of Venus and Roma by way of showing him that a great work could be accomplished without his aid, and asked Apollodoros whether the proposed structure was satisfactory. The architect in his reply stated, first, in regard to the temple, that it ought to have been built on high ground and that the earth should have been excavated beneath it, so that it might have stood out more conspicuously on the Sacred Way from its higher position, and might also have accommodated the machines in its basement, so that they could be put together unobserved and brought into the theater without anyone's being aware of them beforehand. Secondly, in regard to the statues, he said that they had been made too tall for the height of the cella. "For now," he said, "if the goddesses wish to get up and go out, they will be unable to do so." When he wrote this so bluntly to Hadrian, the emperor was both vexed and exceedingly grieved because he had fallen into a mistake that could not be righted, and he restrained neither his anger nor his grief, but slew the man.[6]

It does not sound like Hadrian, who was usually genial and tolerant of criticism. Moreover, Dio, who tells the tale, had something of Suetonius' scandalmongering attitude. And the story is not even self-consistent.[7] But that is all we know of Apollodoros' fate.

Hadrian was the mightiest builder of all. A list of towns and cities in which he erected one or more buildings would include every place of importance in the Empire. If you see a Roman ruin anywhere around the Mediterranean and do not know its provenance, there is always an excellent chance that it dates from Hadrian's reign.

To take a few examples, Hadrian completed the great Olypieion, or temple of Olympian Zeus, with which the Athenians had been struggling for nearly seven centuries. Fifteen of its 104 fifty-six-foot Corinthian columns still stand. Another was blown over by a gale in 1852 and lies, its drums leaning slantwise against one another like fallen dominoes. And it was Hadrian who raised a wall across the north of Britain to keep out the wild Picts.

About +130, Hadrian also built at Tiburtina (modern Tivoli) a villa that was practically a small city. It stretched out over seven miles of rolling land and included replicas of the most interesting monuments that Hadrian had seen in his travels.

So vast and populous was Hadrian's villa that Hadrian was hard put to it to find a little seclusion. To solve this problem, he dug a pond in the center of one of the villa's many courtyards. In the center of this lake rose a tiny island on which stood a marble pavilion containing a single room. When the emperor crossed the few feet of bridge to his island, nobody–not even a Parthian ambassador with an ultimatum–might disturb him.

Hadrian's most celebrated work was the rebuilding of the Pantheon. In place of Agrippa's building, Hadrian erected a rotunda capped by a huge dome of concrete, 144 feet in diameter, with a circular opening or eye in its center. This dome still stands, soaring high above the patterned marble pavement of the interior. Barring earthquakes and nuclear wars, it may stand for many centuries yet, showing the world what Roman engineers could do.

None of the Roman emperors after Hadrian was outstanding as an engineer. Some did build spectacular buildings, like the baths of Caracalla and Diocletian and the circus of Maxentius. The later emperors, however, became too involved in fighting off barbarian raids, suppressing revolts, and keeping their own soldiers from murdering them to have much time for building.

Roman engineering, though vigorous and progressive for many centuries, was almost entirely civil engineering: building baths, bridges, lighthouses, and other static structures. The Romans never bothered much with mechanics, save where hoists and similar devices helped them to erect their public works. The reason is easy enough to see. Roman society was so organized that a leader could get glory by building a bath or other public work. But to tinker with gearing, as the Alexandrian engineers did, would have been deemed merely eccentric.

Meanwhile, however, other technical developments, not directly connected with Rome and its government, had been taking place around the Mediterranean and in adjacent lands.

One such non-Roman development was the wine press. In early times, the juice was squeezed from grapes by trampling them. Later, the grapes were put in a bag and squeezed. The press consisted of a beam pivoted at one end. When the wine makers chinned themselves on the other end of the beam, or hung weights on it, or pulled it down by a block and tackle, the bag of grapes was crushed.

In Roman times a better press appeared: the screw press, with a capstan for turning the screw. The screw had been invented some time in –IV or –III. But screws were never very common in antiquity. There were no screw-cutting machines, except for a simple device, described by the engineer Heron of Alexandria (+I), for cutting female screw threads. Therefore all screw threads, or at least all male screw threads, had to be laboriously cut and filed by hand.

Plinius, writing in the +70s, says that the screw press was a Greek invention, made within a century of his own time. In the ruins of Herculaneum was found a clothes press that worked on the same principle.

As Plinius is one of our main sources on ancient technology, he merits a few words. Gaius Plinius Secundus (+23 to +79), called "Pliny the Elder" to distinguish him from his literary nephew of the same name, was another of those keen, indomitable, and indefatigable Roman civil servants of the stamp of Agrippa and Frontinus, who kept the Empire going despite the Caligulas and Neros. He served as a cavalry officer in Germany, practiced as a lawyer, held various colonial posts, and ended his life as an admiral. A man of boundless curiosity and limitless energy, he had a slave read to him or take dictation while he bathed or ate, lest he waste a minute.

The only surviving one of his seven works, his *Natural History,* is the oldest existing encyclopedia. It was intended as a description of "nature," meaning everything not man-made or artificial. Plinius, however, included several categories of man-made things like inventions and works of art. His coverage is enormous but very superficial; he collected facts as a magpie does nest ornaments.

Although Plinius scoffed at Greek writers (whom he nevertheless used for sources) as credulous liars, he had little critical sense himself. Hence, while he tried to distinguish between fact and legend, he never came near to doing so. However, we can be thankful for the many facts he preserved, without being forced to believe, as he did, that an elephant could be taught to write, that the tooth of a hyena caught when the moon is in Gemini is a good ghost repellant, and that there dwelt in India a race of men without mouths, who subsisted by smelling flowers.

Even so, Plinius was sometimes right when his critics were wrong. Once, the most available version of Plinius in English was that which Bostock and Riley translated a little over a hundred years ago, and which was published in London. In this edition, where Plinius says: "Indeed it is generally admitted that all water is more wholesome when it has been boiled,"[8] the translators state in a footnote: "This is not at all the opinion at the present day." Only a decade after this note was published, the discoveries of Pasteur about the bacterial nature of disease and of Lister about asepsis proved that Plinius had been right after all.

Plinius died as he lived, seeking more facts. In +79 Vesuvius erupted, burying Pompeii under a rain of lapilli–lumps of volcanic ash–and Herculaneum under a landslide of boiling mud. Plinius, stationed with the fleet at the western tip of the Bay of Naples, ordered out the ships to rescue the fugitives. He himself prepared to take a ship across the bay to Stabiae, close to the volcano, and asked his nephew if he would like to come along. He received an astonishing answer; as the younger Plinius tells it:

"I replied that I would rather study; and, as it happened, he had himself given me a theme for composition."[9]

Whoever heard of a youth of seventeen preferring to study under those conditions? It's denarii to doughnuts that the boy was simply scared.

At Stabiae, Plinius puffed his way up to the villa of a friend, whence he had a good view of the eruption. He would not have missed such a phenomenon for anything. Unperturbed, he bathed, dined, and napped until the rain of lapilli became so threatening that all resolved to flee.

To avoid being brained by lumps of lava, Plinius, with perfect self-possession, directed those with him to tie pillows over their heads with scarves. So armored they set out. But, while waiting to embark, Plinius suddenly fell dead. He was fat, asthmatic, and in his fifties, and all this scrambling about through a rain of volcanic débris had been too much for his heart.

You will recall from the story of Hellenistic engineering that Philon of Byzantium (–II) mentioned water wheels. He describes overshot wheels, which are merely parts of his puppet shows, to move his figurines and make his little bronze birds seem to twitter. But he also describes a practical undershot water wheel with a circle of paddles and a bucket chain for raising water.

Since Philon hints that water-wheel designs other than his own were in use, he was probably not the original inventor of the water wheel; but we cannot tell what improvements, if any, he effected. Ancient engineering writers, in describing a machine, often neglected to distinguish between their own inventions and those of others, or between a mechanism that had actually been built and made to work and one that existed on papyrus only. Therefore we cannot often tell whether they invented a particular device or merely described what they had seen or heard of.

In any case, towards the end of the Roman Republic, the water wheel came into general use. It was adapted to two tasks, which for thousands of years had used up many millions of wearisome hours of human labor. One was raising water. Swapes, bucket chains, and the Archimedean screw were already known for some time before it occurred to anybody to use the force of the current to lift the water.

Several wheel-shaped devices were developed for hoisting water. One, because of its shape, is called the *tympanum* or drum. It has holes around the outside to let water in, while the inside is divided by spiral partitions. As the tympanum turns, the water in each section is raised until it runs out a hole in the hub. Such drums are still used in Egypt.

Another kind of water hoist has a series of buckets around the rim. These buckets pick up water at the bottom of their journey and dump it out at the top.

All these water-hoisting devices could be worked by human or animal power, and many still are. But, some time in –II or early –I, they were combined with a paddle wheel by which the energy of the stream was transferred to the hoist. The combination of an undershot water wheel with a water-hoisting wheel is called a *noria,* from the Arabic *na'ûrah.* Vitruvius described a tympanum, a bucket wheel, a bucket chain, and an Archimedean screw, all powered either by treadwheels or by paddle wheels.

The other task to which running water was put was milling. Throughout the ancient empires, bread was the principal food. To make it, wheat or barley grain had to be ground into flour. At first the grain was painfully pounded with a pestle in a mortar, as you can still see done in Central Africa.

At a later time, the grain was ground between two flat stones, one of which was pushed back and forth over the other. With such a mill, one person–usually a slave girl–could grind each day only enough grain to make bread for eight people. Hence, in a large household, several such women would have to spend their entire day at the weary task of pushing and pulling the upper millstone. The two ever-present sounds of ancient households were the clack of the loom and the grate of the hand mill.

In classical times, several improvements were made in this simple millstone. One was the lever mill, in which the upper stone was fastened to a beam, pivoted at one end and free to move at the other. The operator, pushing and pulling on the free end of the beam, worked the upper stone back and forth over an arc-shaped course. There were also roller mills, but the roller proved unsuitable for grain, although it worked well in crushing olives for their oil.

A further advance was to make the millstones circular, so that the upper stone was pivoted at its center on the lower and could be turned completely around. Spokes projecting from the upper stone enabled the operator to rotate the upper stone either back and forth or, as later became customary, round and round.

The history of the rotary hand mill or quern is very obscure. The earliest rotary mills may go back before –1000, but the people who made them saw no reason to inscribe dates on them.

Hellenistic and Roman rotary mills finally took the form of the hourglass mill. While the lower stone was conical, the upper was cylindrical. This cylindrical stone was hollowed out to fit the cone below and had a similar cone-shaped hollow above, into which the grain was fed. An adjustable wooden pivot made it possible to regulate the clearance between the stones and thus to grind the grain to the desired fineness. The hourglass mills at Pompeii were huge affairs 6½ feet high, weighing hundreds of pounds.

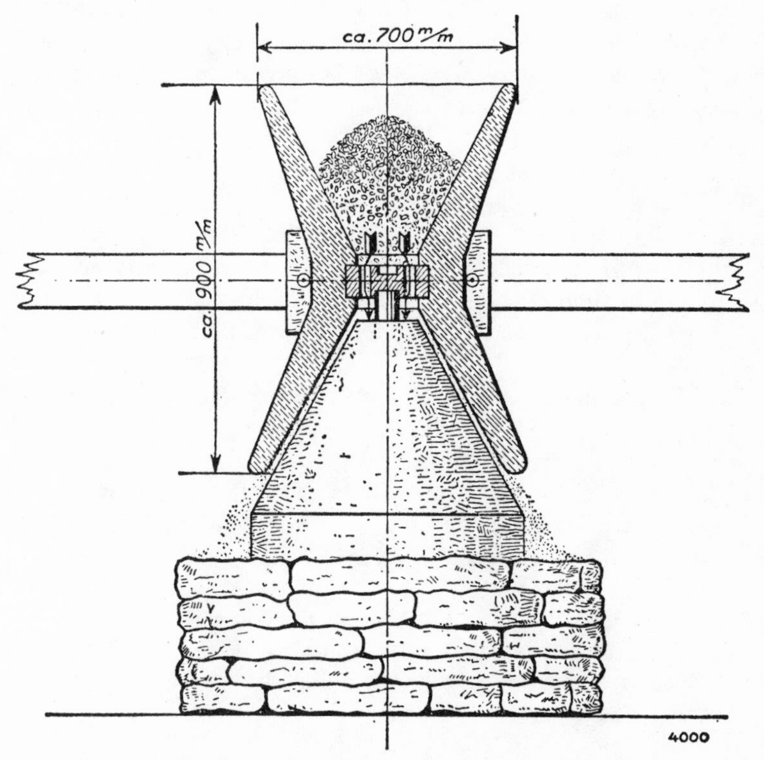

Fig. 12. Cross-section of a Pompeiian hourglass mill, restored.

Capstan spokes, projecting horizontally, made it possible to turn the upper stone. These stones were usually turned either by an ass walking the end of a boom around, or by slaves or free workers of the poorest class, pushing the capstan bars. Apuleius (+II) describes these unfortunates:

Good gods, what a sort of poor slaves were there; some had their skin bruised all over black and blue, some had their backs striped with lashes and were but covered rather than clothed with torn rags, some had their members hidden by a narrow cloth, all wore such ragged clouts that you might perceive through them all their naked bodies, some were marked and burned in the forehead with hot irons, some had their hair half clipped, some had shackles on their legs, ugly and ill favored, some could scarce see, their eyes and faces so black and dim with smoke, their eyelids all cankered with the darkness of that reeking place, half blind and sprinkled black and white with dirty flour marks like boxers who fight together befouled with sand.[10]

Apuleius was one of the very few Romans to show any concern over the plight of these wretches. When you hear somebody apologizing for slavery, whether in the old American South or in modern Arabia, and assuring you that the slave was not really so badly off, remember Apuleius' words.

One would think that an ass, being stronger than a man and cheaper to feed, would have been preferred to even the cheapest slaves. However, classical harness was so inefficient that it strangled the animal when it tried to pull really hard, and the boom to which the ass was attached was so short that the beast wasted much of its effort in turning. Therefore the cost of powering a mill by slaves or by asses was about the same.

Two further developments grew out of the use of the rotary grist mill, one economic and the other technical. With the rotary mill it became profitable to mill grain professionally on a large scale. Thus the professional miller, first seen in Egypt about −1500, appeared all over the civilized world in Hellenistic times. Close behind him came the professional baker, first recorded in Rome about −170:

There were no bakers at Rome down to the war with King Perseus [of Macedonia], over 580 years after the foundation of the city. The citizens used to make bread themselves, and this was especially the task of the women, as it is even now in most nations.[11]

In classical times, the business of milling and baking were usually combined in one shop. However, small domestic hand mills continued in use, even in the most civilized lands, for many centuries.

The technical development that sprang from the rotary mill was the invention of the crank. The ancient world knew the capstan and the windlass, which may be described as wheels with spokes but no rims, the former having its shaft vertical and the latter its shaft horizontal. In each case, however, the spokes by which men turned the mechanism radiated out from the hub.

By providing one of these spokes with a handle at right angles to the spoke and parallel to the shaft, a crank is made. For many applications, crank motion is a more efficient way of transmitting power than the capstan and the windlass.

Crank motion first appeared in querns, among people who still ground their grain at home. The quern maker drilled a vertical hole near the edge of the upper stone and stuck a peg in the hole. The operator could then spin the upper stone round and round more easily than she could with radial spokes, where she had to change her grip with every revolution. In fact, she could turn out about a bushel of flour a day–twelve times as much as before. The first such mills may go back to –IX in Syria, and crank-operated querns appeared here and there in the civilized world in Hellenistic times.

It took many centuries, though, for mechanics to adapt the crank mechanism to machines of other kinds. The only positive evidence of such cranks, before the Middle Ages, comes from the bilge pump of Lake Nemi (+I), the medical writings of Oreibasios (+IV), and the sketch in the Utrecht Psalter (+IX), whereof I told you in Chapter Five.

The next stride towards the age of power machinery was to combine the rotary mill with the water wheel, making a water mill. The earliest allusions to the water mill go back to early –I. Strabon mentions that: "It was at Kabeira that the palace of Mithridates [the Great of Pontus, –132 to –63] was built, and also the water mill . . ."[12] And a poet, Antipatros of Thessalonika, wrote:

Hold back your hands from the mill, O maids of the grindstone; slumber
Longer, e'en though the crowing of cocks announces the morning.
Demeter's ordered her nymphs to perform your hands' former labors.
Down on the top of the wheel, the spirits of water are leaping,
Turning the axle and with it the spokes of the wheel that is whirling,
Therewith spinning the heavy and hollow Nisyrian millstones.[13]

There are two simple ways to make a water mill. Obviously, the ordinary paddle wheel cannot be directly connected to a millstone, because the paddle wheel has its rim vertical and its shaft horizontal, whereas the millstone has its rim horizontal and its shaft vertical.

One kind of water mill, the horizontal mill, has a vertical shaft with horizontal paddles on the lower end. The shaft passes up through a hole in the lower millstone and is fastened to the upper millstone by a wooden crosspiece. Wheel, shaft, and upper millstone revolve together as water is squirted from a trough or nozzle at the paddles.

Although it is cheap and easy to build, a mill of this kind has no great power. Since the millstone turns slowly, such a mill will just about grind enough flour for one family. Moreover, it requires a small amount of water at high velocity. Therefore it is useful only in mountainous regions, where small, steep, swift streams are found.

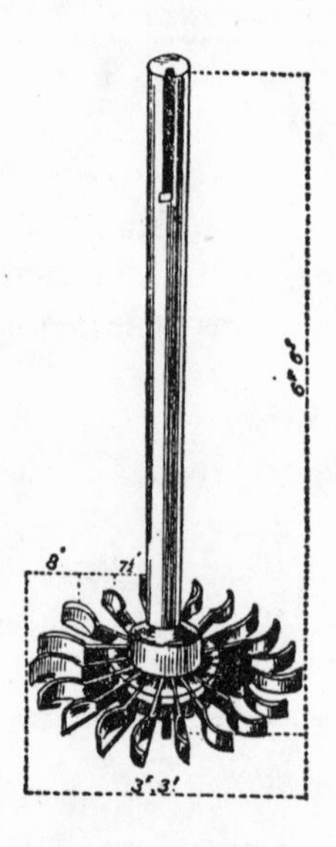

Fig. 13. Wheel and shaft of a medieval horizontal water wheel found at Drumtullogh, Ireland (*from the* Ulster Journal of Archaeology, *1856, No. 6*).

In another form of mill, the stream works an ordinary paddle wheel of the noria type. The horizontal shaft of this wheel is connected to the vertical shaft of the mill by gears at right angles. Vitruvius describes the geared mill, which we call the Vitruvian mill in his honor:

> Mill wheels are turned on the same principle, except that at one end of the axle a toothed drum is fixed. This is placed vertically on its edge and turns with the water wheel. Adjoining this larger wheel there is a second toothed wheel which is placed horizontally by which it is gripped. Thus the teeth of the drum which is on the axle, by driving the teeth of the horizontal drum, cause the grindstones to revolve. In the machine a hopper is suspended and supplies the grain, and by the same revolution the flour is produced.[14]

To make such gears work required an advance in engineering over that needed for the cogwheels of Ktesibios' water clocks and the Antikythera machine. In the clocks, the gears had to transmit but little load other than that of the friction of the machine itself. In the mill, on the other hand, the gears had to transmit a heavy load at high speed for long periods of time, so that any weaknesses in the design would at once

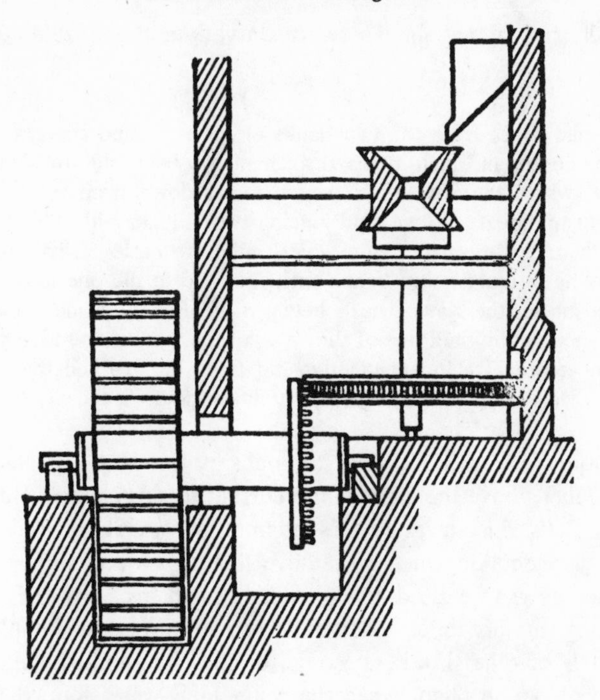

Fig. 14. Diagram of a Vitruvian water mill; undershot water wheel on the left, millstone on the upper right (*from T. Beck:* Beiträge zür Geschichte des Maschinenbaues, *1900*).

transpire. It took centuries of experiment with water mills to find the most efficient gear positions, sizes, and ratios.

By +IV, water power in Gaul had been applied not only to milling but also to sawing, as is shown by the verses of the Gallo-Roman poet, Decimus Magnus Ausonius:

Thee, swift Celbis, and thee Erubis, for marble illustrious,
Hurry with servile haste your waters forthwith to mingle:
Celbis renowned for glorious fishes, whereas its companion
Whirling the millstones that grind the grain with rapid rotations, and
Shriekingly drawing the saws through the glassy masses of marble,
Hears from one bank to the other a din that is strident and ceaseless.[15]

In late Roman times, the city of Rome got its flour from a battery of water mills on the Ianiculum Hill, where the Vatican now stands. In the siege of +537, Wittigis' Goths broke the aqueducts and thus put these mills out of action. To avert starvation, the Byzantine general Belisarius:

> . . . fastened ropes from the two banks of the river and stretched them as tight as he could, and then attached to them two boats side by side and two feet apart, where the flow of the water comes down from the arch of the bridge with the greatest force, and placing two mills on either boat, he hung between them the mechanism by which mills are customarily turned. And below this he fastened other boats, each attached to the one next behind in order, and he set the water-wheels between them in the same manner for a great distance. So by the force of the river all the wheels, one after the other, were made to revolve independently, and thus they worked the mills with which they were connected and ground sufficient flour for the city.[16]

The Goths attempted to wreck the mills by throwing logs and corpses into the Tiber above the city for the current to carry down. But the resourceful Belisarius thwarted this attempt by stretching chains across the river and detailing men to remove floating objects from the water as fast as they were carried down against the chains.

Evidently, at this time, water mills had more or less completely replaced mills of other kinds at Rome, or the city would not have been in such a desperate plight when the water mills were stopped. Furthermore, Procopius' tale of Belisarius' floating mills is one of the very few ancient accounts of the actual birth of a new invention.

A paddle wheel of the noria type, like that used in the Vitruvian mill, is called an undershot wheel. Although inefficient, it is useful where the stream is large but sluggish. A more efficient water wheel is the overshot wheel, where water is poured into buckets at the top and spilled out at the bottom. For an overshot wheel, however, there must be a fall of water at least equal to the height of the wheel.

Water wheels of all three types—horizontal, undershot, and overshot —may have come into practical use in late –II or early –I. However, the undershot wheel—the paddle wheel in reverse—is the only full-sized wheel whereof a description has survived from pre-Christian times. A mill of about +200 near Arelate (modern Arles, France) was powered by sixteen wheels, probably overshot. Also, an Irish manuscript, the *Book of Senchas Mor,* tells of horizontal water mills in Ireland in the time of Saint Patrick (+V).

Probably the overshot and horizontal wheels were not described until some time after they were invented, because they were only used in mountainous country. Therefore they escaped the notice of literary men, most of whom wrote in the gleaming cities of broad plains and level seacoasts, where such mills were not to be found.

In Roman times, a number of new materials and substances came into use. Window glass we know about. Wire cable appeared; a 15-foot length of bronze wire cable, an inch in circumference, was found at Pompeii.

Another discovery was brass, the alloy of copper and zinc. If we can trust the pseudo-Aristotelian treatise *On Wonderful Things Heard*–a collection of jottings by a later member of the Lyceum, some factual and others silly–brass was first made by the Mossynoikoi, a people of Asia Minor:

They say that the copper of the Mossynoikoi is shiny and white, not because tin is mixed with it, but because a certain earth is combined and molten with it.[17]

The "certain earth" was not metallic zinc. This metal was not discovered until +XVI, when Paracelsus named it *Zinken*. The "earth" was the ores zinc carbonate and zinc silicate, both called "calamine." In addition, brass was also made with cadmia or zinc oxide, which sometimes occurs as a deposit on the top of a copper-smelting furnace, mixed with other metallic oxides evaporated and recondensed near the vents.

Good brass is as strong as bronze, easier to work, and prettier. If the zinc comprises about 20 per cent of the total, the brass looks much like gold. Hence it was long regarded as a precious metal; Darius the Great treasured a brazen cup. When Plato wished to show the wealth of his fictional Atlanteans, he credited them with lavish use of *oreichalkôn,* "mountain copper," by which he probably meant brass.

Knowledge of brass became part of a body of doctrine that arose in the eastern Mediterranean, especially in Alexandria, in late Hellenistic and Roman times. This body of doctrine, afterwards called by the Arabic name of *alchemy,* was the forerunner of chemistry. It had four main sources:

Firstly, the old Egyptian art of making imitations of gems and precious metals for jewelry. When colored glass was first substituted for gemstones, this was not a "cheap" imitation; for, when the making of glass was in its infancy, glass was a rare and precious substance, and the glass gems were almost as costly as natural ones.

It was the same with the metals. To the ancients, whose ideas about elements were very different from ours, a metal that looked like gold was considered a kind of gold. Brass was one expensive *ersatz* gold; another was an alloy of silver, arsenic, and sulfur. Egyptian craftsmen also learned to plate or coat the cheaper metals with costlier ones.

In fact, some oriental craftsmen may have gone further than that. In 1936, at Khujut Rabu'a near Baghdad, was found a small pottery jar, about 5½ inches high and 3⅛ inches in diameter. Inside was a cylinder of thin copper, closed by an asphalt plug. Inside the cylinder was a rusted iron rod.

Similar jars, without the metal cylinders and rods, had been found at the ruins of Seleucia-on-the-Tigris, twenty-four miles below Baghdad. Three larger jars containing respectively ten copper cylinders, ten iron rods, and ten asphalt plugs, not yet assembled as in the Khujut Rabu'a jar, turned up at Ctesiphon, across the Tigris from Seleucia.[18]

All these objects date approximately from Roman Imperial times. The only use that anybody has been able to conceive for them is as battery cells for electroplating small metal objects with gold.

It has also transpired that the silversmiths of Baghdad, within the present century, used a similar apparatus for gold-plating their works. Those who knew about this method long assumed that the silversmiths had learned their electroplating from western sources. Although archeologists are not yet agreed about the mysterious jars, we must at least consider the strange possibility that electroplating was discovered in Iraq in ancient times; that, despite the ravages of the Mongols in +XIII, this technique survived down to the present century; and that, nevertheless, it failed to spread to other lands, presumably because the metal workers kept it secret.

A second source of alchemy consists of the Greek speculations about the nature of matter. The most influential ideas were Empedokles' theory that all matter was made of four elements—earth, air, fire, and water—and Aristotle's assertion that these elements could change, one into the other.

The third source of alchemy was an idea developed by the Daoist (or Taoist) philosophers of China, that a man's life could be prolonged indefinitely by an elixir made of rare substances like powdered gold. One Chinese emperor is said to have died as a result of drinking such a concoction. This idea seeped over the trade routes of Central Asia in late Hellenistic or early Roman imperial times, to be picked up by the early alchemists and added to their stock of lore.

The final source of alchemy was a cloudy mass of mystical, magical doctrine—a conglomeration of myth and legend, cosmic speculation and petty superstition, the pretensions of priests and the mummeries of medicine men, which in early societies passes for divine wisdom. A great surge of supernaturalism, which took place under the Roman Empire, spilled over into science and technology. It helped to spread and dignify astrology, and it contaminated the budding science of chemistry until it could hardly be recognized as a science at all.

As with astronomy and astrology, nobody in those days distinguished between alchemy and chemistry. We know enough today to tell the science from the pseudo-science, but the ancients drew no such distinction. To them there was just one science, most of whose practitioners insisted that prayers, incantations, and the use of bizarre ingredients and unintelligible directions were all necessary parts of sound scientific procedure.

Some early alchemists wrote treatises that survive. One, that of Bolos of Mendes, may go back as far as –300. The others, probably, all belong to the Christian era. Some were written under the name of Hermes Trismegistus, a combination of the Greek messenger-god Hermes with the ibis-headed Egyptian god of wisdom, Deḥuti or Thoth, transformed by legend into a mortal king of Egypt who reigned before the Flood and wrote 36,525 books on alchemy.

Other works were attributed to divinities like Isis and Agathodaimon, or to Moses, King Solomon, Ostanes, Demokritos, and Cleopatra, none of whom had anything to do with alchemy. However, names of real alchemists like Komarios and Mary the Jewess also appear. Mary the Jewess seems to have made significant chemical discoveries and to have invented the still. At least, she was the first to describe one. This invention in turn led to the discovery of alcohol in the days of the Caliphate.

While some of these early alchemists made practical discoveries, most of them soon wandered off into the search for picturesque nonentities, especially the Philosopher's Stone. This substance, usually described as a red powder, could transform a million times its own weight of base metal into gold or, if its quality was not quite up to specifications, into silver. It could be dissolved in alcohol to make the *elixir vitae,* the Elixir of Life, whereby the alchemist could cure all ills, rejuvenate his aging body, and prolong his life. Some sought the alkahest or universal solvent; though a few had the wit to ask: if you found it, what would you keep it in?

Moreover, the intense supernaturalism of late Roman times caused the alchemists so to confound their chemical procedures with magical rites that discoveries on the material plane became almost impossible.

Besides, a lust for secrecy and a love of mystification led them to write in an obscure and shapeless symbolism. One of these works, about the tail-biting serpent Ouroboros (an old Egyptian magical symbol) says:

A serpent is stretched out guarding the temple. Let his conqueror begin by sacrifice, then skin him, and after having removed his flesh to the very bones, make a stepping stone of it to enter the temple. Mount upon it and you will find the object sought. For the priest, at first a man of copper, has changed his color and nature and become a man of silver; a few days later, if you wish, you will find him changed into a man of gold.[19]

Alchemy flourished in the Byzantine Empire, the Caliphate, and medieval Europe. Its obscurity became more opaque and its magical content larger until one school of modern students has held that the alchemists were not after vulgar gold at all, but spiritual enlightenment, union with God, or some such lofty goal.

Doubtless some alchemists were of this sort. Some were plain fakers and swindlers, while others were professional organizers of secret societies, seeking neither Philosopher's Stones nor spiritual perfection but yearly dues. These last, under the name of Rosicrucians, we still have with us.

Most of the alchemists, however, really did busy themselves with retorts and alembics. They did try to make gold, until they ran out of money or gullible backers, or perished from breathing too much mercury vapor.

Scientific chemistry did not separate itself from alchemy until +XVI and +XVII. Chemistry was slower than most sciences to free itself from pseudo-scientific associations. The reason is that the laws of chemistry are so complex and interdependent that, in formulating these laws, one must hit upon the right scheme almost all at once, instead of proceeding in normal scientific fashion from the simpler problems to the more complex.

Moreover, chemistry concerns the behavior of atoms and molecules. Since one cannot see atoms and molecules, one must infer their existence and behavior from other facts, and it takes much trial and error to get such inferences right. Still, despite the lack of sound science in most alchemy, it was, perhaps, a stage through which chemistry had to go, just as an appalling child sometimes grows up to become an admirable adult.

Besides the works of Vitruvius and Plinius, several engineering treatises in Greek have come down from Roman times. One is a book on mechanics by Athenaios (called *Mêchanikos* to distinguish him from the better-known Athenaios of Naukratis, author of the literary miscellany *Deipnosophistai,* "The Dining Professors"). Athenaios the Mechanic used some of the same sources as Vitruvius, as a comparison of their works makes clear. As neither mentions the other, they may well have been contemporaries.

Athenaios discusses siege engines: sambucae, flying bridges, rams, tortoises, belfries, and combinations of these. He also discusses the possibility of mounting some of these engines on ships, as Marcellus had done at Syracuse.

Most of this adds but little to what Vitruvius and Philon had already said. However, Athenaios does propose one new idea: a three-wheeled belfry in which the front wheel of the tricycle is mounted on a post that can be rotated by a steering rope. So far so good; but then Athenaios winds this rope around a drum on the main axle, so that, in nearing a hostile wall, the belfry shall follow a zigzag course as the front wheel of the tricycle turns round and round, to disturb the foe's aim.

Practically speaking, the idea is absurd. Still, this is the first known attempt at a steering wheel.

From the following century, +I, comes the work of the ablest technical writer of antiquity: Heron of Alexandria. His date was long uncertain. This uncertainty was deepened by the fact that, in his work on catapults, this man gave his name as "Heron Ktesibiou."

Now, *Ktesibiou* means "of Ktesibios." Following normal Greek usage, the whole name would mean "Heron the son of Ktesibios." But the famous Ktesibios, the clock-and-organ man, lived in early –III, and Heron was certainly later than that. In fact he describes an eclipse of the moon, and the only eclipse that fits his description was that of +62.

That leaves unexplained the term "of Ktesibios." Perhaps Heron used the term in the unusual sense of being a follower of Ktesibios or an heir to his tradition. Perhaps his father was another Ktesibios. Perhaps this father was descended from the original Ktesibios. Maybe there was a whole family in which, in accordance with the usual Greek and Persian custom, the eldest sons were named for their fathers' fathers, so that they were called alternately Ktesibios and Heron for several generations. Perhaps the family held some hereditary post of Chief Mechanic in the Museum. . . . But all this is guesswork; we may never know.

Heron's engineering works comprised *Mechanics, Pneumatics, Siegecraft, Automaton-making,* and *The Surveyor's Transit.* He also wrote a geometry book, *Measurement,* and an optical work, *Mirrors.* There are fragments of another mathematical work, of a book on water clocks, and of a technical glossary. Some of these works have disappeared in the original Greek and exist only in Arabic or Latin translations.

Heron's *Mechanics* is mainly concerned with mechanical advantage. He describes the various means for obtaining this advantage—the windlass, the lever, the compound pulley, the wedge, the worm, and the gear train—and shows how they can be combined. Being a skilled mathematician he shows how, with such a device, the force applied and the distance moved vary inversely according to mechanical advantage. In fact, Heron comes close to discovering the modern technical concept of "work" (= force × distance). He also describes cranes, methods of raising large building stones by means of tongs and keys, and screw presses of the kinds mentioned by his contemporary Plinius.

Heron's *Siegecraft* (*Belopoiika*) tells about several designs for catapults. One is the crossbow; the others are conventional dart and stone throwers. Heron gives details of parts like the shackles that held the tension skeins in place, which we might not otherwise know about. Explaining that, with a very large stone thrower, the slide is too heavy to be pulled forward to the loading position by hand after each shot, he provides a second windlass to make this easy.

His *Surveyor's Transit* (*Dioptra*) describes a more sophisticated instrument than the Roman groma. On the pedestal is mounted a detachable sighting table, with worm gears for adjusting the table about the vertical and horizontal axes. This part could be removed and replaced by a water level, having at each end a little glass cylinder in which the water rises to the level of an eye slit in a bronze plate. Heron also provided leveling posts to sight on, with plumb bobs for setting them vertical and sighting disks hoisted up and down by pulleys. These posts looked much like modern stadia rods.

Heron's surveying system does not seem to have come into use. It was probably too far ahead of its time. There were not enough mechanics with the skill needed to make so advanced an instrument as the dioptra.

Heron's best-known work is his *Pneumatics*. He begins with a discussion of theory, which he probably took from Straton. Heron not only understands the compressibility of air, but he has also weighed fuel before and after burning and has come close to understanding the nature of combustion.

In the main part of the work, Heron describes many gadgets like those set forth by Philon and Ktesibios and tells of others besides. Even when the main idea of the device is old, Heron likes to arrange things in his own original way. Many of his devices are worked by siphons and air tubes. A pitcher pours only when the finger is removed from an air hole; another pitcher mixes wine and water in the desired proportions. There are also displays of automata: Herakles slaying a dragon, bronze birds that twitter, and bronze beasts that drink on command.

Some of Heron's inventions seem to have been for the benefit of the priesthoods of Alexandria, to enable them to awe their worshipers. In one, the fire of a burnt offering causes air to expand in a tank, forcing water out into a bucket, whose descent causes temple doors to open. In another, opening a temple door makes a trumpet toot.

A combination of vessels and siphons apparently turns water into wine, as an older contemporary of Heron is alleged to have done at Cana in Galilee. The biblical incident is probably based upon tales of Heron's magical device, just as the story of Joseph and the seven lean years is derived from the Egyptian legend of King Joṣer, his minister Imḥotep, and the god Khnum; and the story of Moses' parting the Red Sea goes back to the old Egyptian tale of King Ṣeneferu, his wizard Jajamankh, and the lost bangle.

Another example of Heron's sacred sleight-of-hand is the original coin-in-a-slot device, combined with a holy-water dispenser. It is baldly labeled: "Sacrificial Vessel which flows only when Money is introduced." Heron's *Mirrors* gives directions for making spooks appear by means of mirrors. Anybody as useful as Heron was to the crafty priesthoods of Alexandria, with their purifying bronze wheels, their ever-burning lamps with asbestos wicks, and their spirit-reflecting mirrors, must have done right well by himself. He must also have contributed to the great wave of supernaturalism that finally killed Roman science.

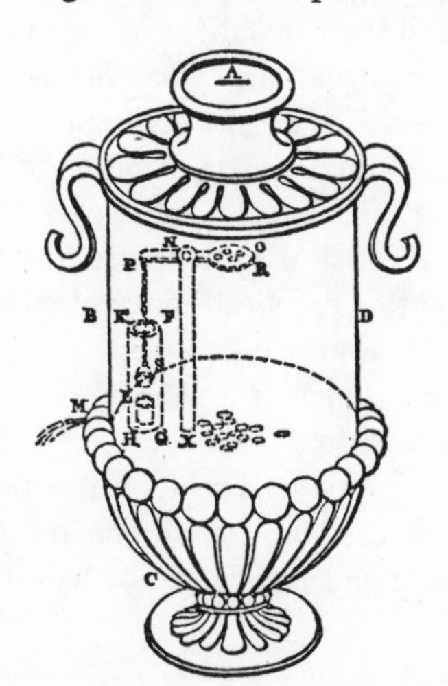

Fig. 15. Coin-in-the-slot holy-water dispenser, designed by Heron of Alexandria (*from Woodcroft's edition of Heron's* Pneumatics).

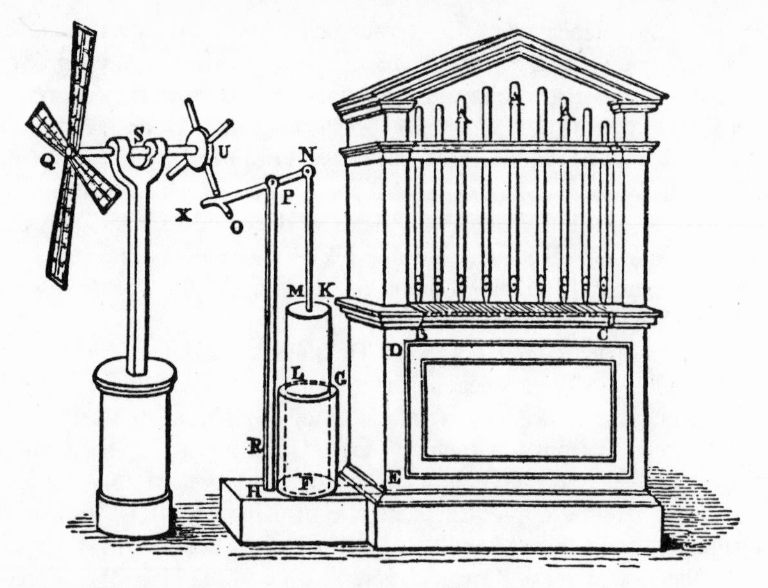

Fig. 16. Heron's wind-powered pipe organ (*Woodcroft*).

Heron also shows more useful devices. He understands the check valve and the float valve, such as is used in a modern water closet. He describes a hydraulic pipe organ like that of Ktesibios.

He also describes another, smaller organ powered by a little windmill. This is the first mention of a windmill of any sort. Heron calls his gadget an *anemourion* or "wind vane," and the illustrations in the manuscripts, copied and recopied from the originals, show that he had in mind something like a modern toy pinwheel. Perhaps such pinwheels were used as toys in Heron's time. Heron's mechanism, connecting the windmill with the organ, is unconvincing, and he probably never built such a machine. A striking oversight of classical engineering was the failure to develop the windmill from a toy to a practical source of power.

Heron also describes a force pump with a nozzle that can be turned up, down, and sideways for fighting fire. Apparently Alexandria had some sort of hand-operated fire engine not unlike those used in Europe and America in early +XVIII. Such an engine was merely a wooden tank or tub, with handles at the corners for carrying and a pump on top. At the fire, the tub was filled by a bucket brigade, while two or four men worked the pump handles.

In fighting fires, the ancients used buckets and large syringes. When

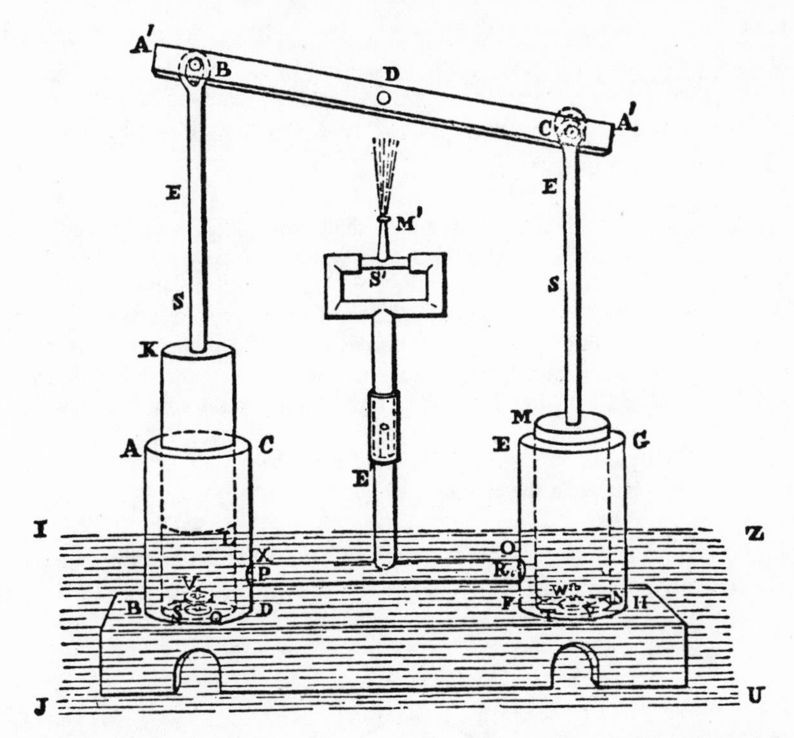

Fig. 17. Heron's two-cylinder pump, with a dirigible nozzle, for fire fighting (*Woodcroft*).

there was no river, fountain, or other water supply close at hand, they probably did what firemen in London did in early +XIX: dug a hole in the street and waited for it to fill with ground water.

Marcus Licinius Crassus, the Roman banker and triumvir (–I), had a private fire brigade of 500 lusty slaves. When a house caught fire, Crassus' brave fire laddies chased away the laddies of rival companies while Crassus bargained with the owner for the property, "so that the greatest part of Rome, at one time or another, came into his hands."[20]

Later, Augustus scotched this extreme form of private enterprise by organizing a corps of *vigiles* or "watchmen," who served both as firemen and as policemen. The vigiles were recruited from freedmen, who might, after faithful service, transfer to the army and thus become citizens. Some other cities had a similar system. Elsewhere, fire fighting was assigned to soldiers or to members of the builders' guild, neither so effective as trained vigiles. At best, ancient fire-fighting methods were so far below our standards that any great city could expect a conflagration at least once each century.

Of all of Heron's inventions, the one most pregnant with future possibilities was his steam engine. Heron modestly calls it "A ball rotated by steam" and explains:

> Place a cauldron over a fire: a ball shall revolve on a pivot. A fire is lighted under a cauldron, AB, containing water, and covered at the mouth by the lid CD: with this the bent tube EFG communicates, the extremity of the tube being fitted into a hollow ball, HK. Opposite to the extremity place a pivot, LM, resting on the lid CD; and let the ball contain two bent pipes, communicating with it at the opposite extremities of a diameter, and bent in opposite directions, the bends being at right angles and across the lines FG, LM. As the cauldron gets hot it will be found that steam, entering the ball through EFG, passes out through the bent tubes towards the lid, and causes the ball to revolve, as in the case of the dancing figures.[21]

Heron's steam engine—or "aeolipile,"[22] as it is sometimes called—worked on the reaction principle of the rotary lawn sprinkler. Heron

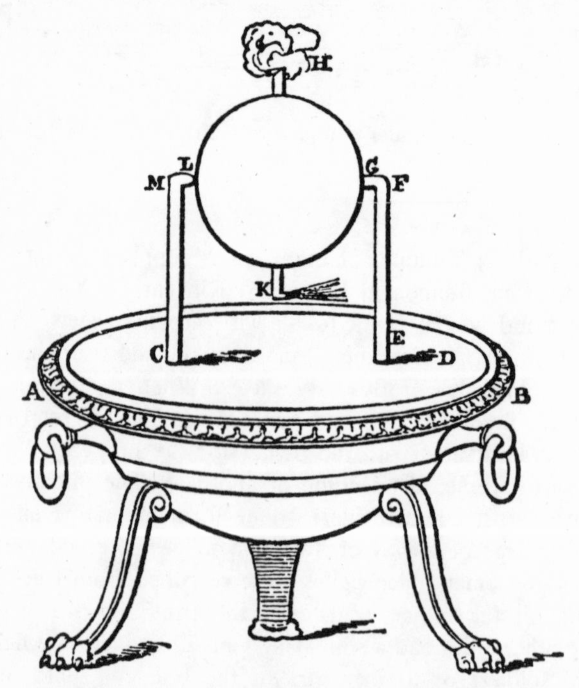

Fig. 18. Heron's steam engine or aeolipile (*Woodcroft*).

never claimed that it was more than a toy. In another design, he puts this principle to use by causing heated air, issuing from the ends of a swastika-shaped pipe assembly, to whirl a disk, attached to the pipes, on which statuettes are mounted. Hence the reference to "dancing figures" in the above quotation. Elsewhere, as we have noted, he used heated air to work a piston to open temple doors.

Heron had available nearly all the elements needed to make a practical working steam engine of either the reciprocating or the turbine type. Yet none of the three devices he described was such a steam engine; that is, none could have worked a pump or turned a millstone.

Still, if some emperor had hired Heron to devote a few years of intensive research to the steam-engine problem . . . But it is useless to expect people to attack problems that they do not know exist. As it was, Heron did more creative work than a hundred ordinary men do in the course of their lives.

Although Heron's book was not widely read in later centuries, it was never altogether forgotten. A curious echo of it appeared in the medieval European legends of Virgil the Magician. In these tales, the gentle poet of Mantua, Publius Vergilius Maro, became the wizard Virgil, who founded Naples on eggs, owned a brazen head that prophesied, built a whirling castle, and kept a demon captive in a glass. He also maintained an army of robots to trounce his foes and otherwise do his will. These yarns are derived from the little figurines of Heron's puppet shows, expanded to life size, made mobile, and run by magic instead of by machinery.

Later, the engineers of the Renaissance studied Heron with lively interest. In +XVII, when Europeans began to harness the power of steam, they remembered Heron's achievements. In the 1670s, a Flemish Catholic missionary in China, Ferdinand Verbiest, made a two-foot model steam cart driven by an engine copied from Heron's aeolipile. Verbiest was one of a group of Jesuits who for several decades operated the Imperial Chinese observatory of Peking. His steam car ran, at least a few inches.

At the same time the Marquis of Worcester, who had also read about Heron's whirligig, was experimenting with his "water-commanding engine." This was a steam-powered vacuum pump for pumping out coal mines. Although Worcester never realized the profits he sought from his invention, he got far enough so that his successors Savery and Newcomen built up a thriving business installing similar pumps. It was in trying to improve a Newcomen engine that James Watt made his portentous discovery.

Heron's was the last of the great classical treatises on the mechanical arts. In early +IV an eminent mathematician, Pappos of Alexandria, devoted the eighth and last book of his mathematical treatise to mechanical problems. But Pappos merely described the standard means of gaining mechanical advantage—the windlass, lever, compound pulley, wedge, screw, and gear train—and various ways of combining them. He added nothing to what Philon and Heron had already said.

The fall of the West Roman Empire is usually dated +476. In that year, one of the Empire's many German generals, Odovacar, deposed the boy-emperor Romulus and declared himself king of Italy, only to be defeated and treacherously slain soon after by Theodoric the Goth. The various German kings, who already occupied large parts of Gaul, Spain, and North Africa, sometimes acknowledged the supremacy of the other emperor in Constantinople; but within a few years all effective Roman rule in these lands faded away. During the following century, the eastern emperors recovered parts of Italy, Spain, and North Africa. This recovery did not last, and soon all of North Africa, and Syria as well, were wrested from the Byzantines by Muḥammad's Arabs.

Formerly, when Europeans assumed that Europe was the only civilized part of the world, this revolution was described as the downfall of civilization. The "Dark Ages" followed, and civilization did not revive until about +XI or +XII.

This is of course nonsense. Even at its greatest extent, the Roman Empire comprised but a fraction of the world's civilized lands, and the Chinese Empire under the Tang dynasty (+VII) was much larger and equally civilized. What really happened was that civilization withdrew from its westernmost outposts, where it had never been too securely established anyway, but it continued as usual in the East.

The shrunken East Roman Empire—the richer and more populous half of the Empire—proved a tougher nut for barbarian hordes to crack than its western sister. In fact, it held out against constant assault for a thousand years, until Constantinople fell to Sultan Muḥammad II in 1453. Even then, the last vestige of Byzantine power was not extirpated until Trebizond on the Black Sea, which had its own "Roman Emperor," surrendered to the Turks in 1461.

The total time from the legendary founding of Rome in −753 to the fall of Trebizond is 2,214 years. Considering that this is a much longer time than the life span of any other government in history, Rome can hardly be said to have failed, even if it finally fell. (Egypt and China existed longer as recognizable nations, but not as governments, because their history is broken into segments by division within or subjection from without.) Like the dinosaurs, who dominated the earth for 150,-000,000 years, the Roman rulers must be counted a success even though they became extinct at last.

Yet, while civilization continued in the eastern Mediterranean lands and in Iraq, Iran, India, and China as if nothing had happened, the fall of the West Roman Empire was no small event. Over a vast area, Roman roads, aqueducts, and harbors fell into ruin, because there was no central government strong, rich, and enlightened enough to maintain them. Literacy, which had been fairly high under Rome, declined almost to the vanishing point. Science became superstition of the most idiotic sort, and engineering declined to rule-of-thumb craftsmanship.

Some of the German kings who ruled the conquered lands made a pretense of preserving civilization. Thus when Theodoric the Great, king of the East Goths, ruled Italy under the nominal overlordship of the Emperor (early +VI) he told his minister, the philosopher Boethius, to make a water clock and a sundial as a royal gift to Gundibald, king of the Burgundians. For a while, some people in these lands were even better off than they had been, because they were freed from the smothering web of Imperial regulations. But then the Avar, Viking, and Magyar raids (+VII to +X) reduced western Europe to a state as barbarous as before the rise of Rome.

To appreciate how far this decline went, you have to read some of the surviving Dark Age literature. Although the Romans had their intellectual limitations, they seem brilliant by comparison.

For instance, a leading man of late +VI was Pope Gregory I, called "the Great." This man burned the library of the Palatine Apollo, lest its secular literature distract the faithful from the contemplation of heaven. He also wrote a book, a collection of saints' lives called *Dialogues,* which displays an infantile credulity that would shame a tribal shaman.

Thus Gregory tells how the pious youth Honoratus, by calling on Christ and making the sign of the cross, halted a boulder that was rolling down the mountain upon his monastery. When the monk Placidus fell into a lake, the monk Maurus rescued him by walking on the water and hauling Placidus out by his hair. A nun ate a lettuce in a monastery garden without crossing herself and consequently swallowed a demon, who possessed her until an abbot exorcized it. The demon reasonably complained:

"What have I done? I was sitting there on the lettuce, and she came and ate me!"[23]

Later, the much more intelligent Pope Sylvester II, or Gerbert (+X), studied in Spain in his youth, collected classical writings, built a water clock and an organ, and wrote a treatise on the abacus. For these not very alarming deeds he was reputed a magician. It was whispered that he had studied astrology, necromancy, and other occult arts at Muslim Córdoba, stolen a wizard's book and daughter, and built a golden head, which answered questions.

Historians have proposed as many different reasons for the downfall of Rome as have been suggested for the extinction of the dinosaurs, and they have found it just as impossible exactly to determine any one true cause. Some of their suggestions are: the demilitarization of Romans in the interior of the Empire when soldiers were more and more recruited either from people living on the frontiers or barbarians living beyond them; the loss of money from the Empire, drained off to India and China to pay for imported luxuries; the loss of manpower in the western provinces as a result of the great plagues of +165 and +250, the civil wars from +235 to +284, and the barbarian invasions thereafter; the growth of local separatism in the provinces as these acquired the techniques and amenities of Roman Italy and so no longer felt that they needed Rome; the restrictive laws of Diocletian and his successors, fixing prices and wages and compelling every man to labor at his father's trade forever.

At least, modern historians agree that Rome did not fall because the Romans reveled in luxury or indulged in orgies. Although there have been, in all times and places, some people who liked a little orgy now and then, the great period of Roman laxity had occurred in –I and +I. The Romans of +IV and +V were different. Many had been Christianized. In any case, as nearly as we can tell, their standards of morals and manners had become a good deal stricter; in fact, these standards could probably be compared with our own.

Some take the view that there was no one cause of Rome's downfall. Instead, the Roman government, like every other, was confronted by a series of problems. For a long time it succeeded in solving them well enough to carry on. At last, however, as much by luck as anything else, a number of these problems piled up all at once at a time when the Western Empire lacked strong leadership. The wonder is not that Rome fell but that it managed to keep going so long.

The principal problem, of course, was the barbarians. Before +400 their incursions had been only raids, destructive but not fatal. However, in +406 the Vandals, Suevi, and Alans burst into Gaul and headed for

Spain at a time when the emperors of East and West were too busy fighting each other to defend the frontiers. A few years later the Franks, Burgundians, and others came in, settled, and refused to leave. Although willing at first to acknowledge the Emperor's rule, they proved too numerous to absorb and too strong to oust; so it was only a matter of time before they took over the rule of the lands they occupied.

Another factor in Rome's fall was that, while the Romans were almost standing still in science and engineering, the barbarians were advancing. They were learning the Roman arts of peace and war, the former by trading contacts and the latter by mercenary service in the armies of Rome. The arts of peace enabled their lands to support denser populations, while the arts of war made them as formidable, man for man, as the Romans.

The tremendously long land frontiers of the Empire had been no great danger when those beyond it had been only a thinly spread population of primitives. But, when the barbarians waxed in numbers, knowledge, and power, these long frontiers became indefensible.

The barbarians across the frontiers had long shown that they could not only learn from the Mediterranean peoples but also devise inventions of their own. Before the Romans conquered them, the Celts of Gaul and central Europe invented soap and the barrel. They were experts at making enameled ware and built advanced four-wheeled carriages, with graceful thin-spoked wheels. It was probably they who had the idea of making a metal tire all in one piece, heating it, and shrinking it on the wheel.

A Celtic native, or perhaps a Roman overlord in a Celtic land, devised the first harvesting machine. On large farms in Gaul, harvesting was done with a kind of header–that is, a device to cut the heads only from the wheat stalks. It looked something like a large modern garden cart, with an arrangement of fingers and slots along the leading edge to catch the heads of the wheat. A beast of burden pushed it from behind.

In Republican Roman times, also, the nomads who roamed the steppes of southern Russia and Turkestan made a number of inventions that revolutionized man's use of the horse.

For one thing, they built a more substantial saddle than was used around the Mediterranean. Classical riders rode either bareback or with a thin pad strapped to the horse's back. The Scythian saddle, on the other hand, was padded, with a definite pommel and cantle. These made it easier to stay on the horse when it suddenly started or stopped, instead of falling off behind or before on one's head like the White Knight. The Romans adopted this saddle in late Imperial times.

Now we come to an invention of even more moment: the stirrup. As the stirrup is closely connected with the rise of the feudal system, it was one of the world's most influential inventions, and its origin has been much disputed. According to one theory, about −1000 the Sarmatians, who roamed the plains north of the Caucasus, invented the stirrup. This is doubtful; but what the Sarmatians or their kinsmen the Scythians did develop was the "tree saddle" with raised pommel and cantle, so that the horseman was less likely to be pushed backwards off his horse's rump. Others credit the Indians, who, riding barefoot, hooked their big toes into the girth or into a pair of rings attached to the girth.

down gave rise to the legend of the Amazons.

Before this invention, all riding had been without stirrups. As Xenophon pointed out in encouraging his Ten Thousand: "The foot soldier can strike harder and with truer aim than the horseman, who is precariously poised on his steed, and as much afraid of falling off as he is afraid of the enemy."[24] Although himself an expert rider who wrote on horsemanship, Xenophon scoffed: "Did anyone ever die in battle from the bite or the kick of a horse?"

Lacking stirrups, classical cavalry was used to scout, to fight other cavalry, and to cut up a foe who was already broken and fleeing. But, against a well-trained block of spearmen, it could do little, save in the hands of an exceptional general like Alexander or Surena, the Parthian who destroyed Crassus' army at Carrhae. During the Hannibalic War, the battle of Lake Ticinus began as a cavalry skirmish but ended as an infantry fight because so many of the riders had fallen off.

Riders long sought a means to assure themselves a more secure seat than that afforded by a simple knee-grip. Xenophon advised the rider to seize the horse's mane while galloping uphill or jumping–a method that would bring rousing sneers from modern riding instructors. Insecure riders in the Near East were sometimes provided with a footboard hung from one side of the saddle, on which they rested their feet while sitting sidewise on their mounts.

The Romans, too, sought greater security for the rider. They fitted their saddles with a projection on each side, just forward of the rider's knee, which could be grabbed in an emergency. When Apollodoros of Damascus built Trajan's Forum, the decorations included a number of marble reliefs showing horses and riders. Later, Constantine removed these reliefs and installed them in the Arch of Constantine, which still stands a block from the Colosseum. Several horses in these reliefs have such knobs on their harness.

Furthermore, a bronze statuette of Marcus Aurelius on horseback, found recently at Velia in southern Italy, has an object shaped like a capital D, with the straight part vertical and the curved part aft, where the horse's breast strap joins the forward edge of the saddle pad. So the simple knob of +100 had, by late +II, evolved into a handgrip or hand stirrup.

The invention of stirrups changed all this. Use of single leather loop to help the rider to mount may have long been known to the peoples of the steppes, though we do not know when and where the invention originated. About a century ago, a Russian scholar found, in a grave barrow at Chertomlýk in the Ukraine, a parcel-gilt silver vase 2 feet, 4 inches high, which is the acknowledged masterpiece of Graeco-Scythian art. The date is uncertain but is usually put in −IV. A strainer in the neck indicates that this jug was for kumys (fermented mare's milk) which the jolly Scythians were wont to drink from cups made of their enemies' skulls.

In addition to birds, gryphons, and other objects, the reliefs decorating the Chertomlýk vase include scenes from Scythian life. The steps in breaking a horse to the saddle are pictured. One of the scenes shows the horse saddled, with something like a strap hanging from the saddle. Some consider this object to be such a proto-stirrup, though others deny it; and in any case the relief is too small to be sure.

Before stirrups, a man who wanted to mount a horse was obliged either to vault on to the animal's back, to step on a mounting block, or to get somebody to give him a leg-up by clasping his hands together. Hence, if not athletic enough to vault, a man unhorsed on the steppes was in a fix.

The soft-leather loop was probably the first answer to this problem. However, such a loop would not be much help in actual riding, because the rider could not kick his foot free is his mount began to fall, or if he began to fall off and had to free his foot quickly.

Archaeological discoveries in the last quarter-century have narrowed the invention of the stirrup down to China (or Korea) in late +V. The invention was brought to Europe in +VI by the Avars, a people of Turkish affinities from the steppes of central Asia.[25] They were the second wave of central Asian nomads, the Huns being the first, to make life interesting for Europeans. They fought for and against the Byzantines and settled in the plain of Hungary, which somewhat resembled their homeland. Under an able +VI khagan, Bayan, they ruled a large central-European empire, largely of Slavic peoples. They raided far into Germany and Italy until finally crushed by Charlemagne and assimilated.

Astride a saddle with pommel and cantle, feet braced in stirrups, a mounted fighter was much more formidable. Experiments by a Briton, Philip Barker, suggest that the tree saddle and stirrups made little difference to the horse archer. But the Avars also brought in the more powerful reflex bow in which, when unstrung, the ends curve around forward. The warrior could charge home with a lance, while the tree saddle kept him seated. Stirrups were of most value in close combat. Without them, a horseman who swung a sword or an ax at a foe was likely to swing himself clear off his horse.

However, a Greek military writer of early +VI implies in his description of cavalry tactics that stirrups had been in use for some time. The first definite mention of stirrups in the West is in the treatise *On Stratagems* published about +600 under the name of the emperor Flavius Tiberius Mauricius, or Maurice. Moreover, it is said that stirrups are mentioned in China as early as +477.

As a result of these changes in horse trappings, armored cavalry became the mainstay of all western armies, civilized and barbarian, in late Roman and Byzantine times. Infantry remained unimportant for nearly a thousand years, until the Flemish burghers at Courtrai, afoot, slaughtered the French king's chivalry under Robert of Artois in 1302. The Age of Chivalry was thus, as its name reveals, the age of the horseman. The continental European words for "knight"–*chevalier, caballero, Ritter*–all mean simply "horseman." The feudal system was, as much as anything, a method of government worked out by trial and error to enable the countries of Europe to support a class of professional warriors who spent all their time practicing the complicated art of fighting on horseback–and, what is more, to support an especially large, voracious, and costly breed of war horses for these knights to ride.

The Belgian, Shire, and Percheron breeds are descendants of medieval destriers. While big enough to carry a man in 70 to 100 pounds of armor, they had to be used with care not to exhaust them. In 1396, King Sigismund of Hungary crusaded against the Turks in the Balkans, with volunteers from many nations. At Nicopolis, inflamed by visions of personal glory, French knights insisted on a headlong charge into Sultan Bayezid's well-trained army. They broke through the first lines but kept on until their horses were blown. Thereupon fresh Turkish forces surrounded and cut them to pieces.

Another improvement was the iron horseshoe. The Mediterranean peoples had a kind of horse sandal, made of thin iron, straw, or the like, which could be tied to the foot of a horse or a mule in an emergency. The heavy iron shoe, permanently nailed in place, enabled the horse to run safely over hard or stony ground without cracking or splitting its hooves. Opinions differ as to the time and place of this invention; but at some Roman sites in Britain, iron horseshoes with a distinctive scalloped outer margin, apparently Roman, have been found.

Then there was the matter of draft harness. An ancient horse could pull only a fraction of the load that a modern horse of the same kind can pull, because of the harness it wore.

The first animal used for heavy pulling was the ox. Now, the ox holds its head low, while its back rises to a hump over its shoulders. In a wild state, the ox depends upon its horns for defense and therefore stands with its head down, ready to use them.

The horse, having no horns, stands with its head up, so that it can see its foes from afar and flee. When men began to use the ox for pulling, they found that they could easily yoke a pair of oxen to a plow or a wagon. The yoke, carved to fit over the necks of the oxen in front of their humps, was held in place by bows: thin wooden collars that went under the animals' necks.

Later, when men began to harness horses, they tried to modify the ox yoke to fit the horse. They put a strap around the horse's neck and hitched it to a yoke. But, since the horse holds its head up, the load bore upon the creature's throat instead of its shoulders and back. When the horse pulled hard, the strap choked it and brought it to a stop.

The first improvement was a Chinese harness with a breast strap, held low down by the other straps. The nomads of the steppes brought this harness (sometimes called the postillion harness) to the West in late Roman times. Then, during the Dark Ages, an even better scheme was developed: the horse collar, carved to fit the base of the horse's neck and padded so as not to hurt the animal's skin. Horses are shown plowing with collars in the Bayeux Tapestry, which tells in cartoon form the story of William of Normandy's conquest of England in 1066.

These improvements in harness quadrupled the load a horse could pull and thus increased the power available to man. For the first time it became usual to plow with horses instead of with the strong but sluggish ox.

Note that these advances were all of barbarian or Far Eastern origin. Some thoughtful Romans warned that the barbarians were showing greater ingenuity than the people of the Empire. Vegetius the military writer (about +400) said: "The barbarous nations, even at this day, think this [military] art only worth attention, believing it includes everything else." An anonymous writer remarked a few decades earlier that: "although the barbarian peoples derive no power from eloquence and no illustrious rank from office, yet they are by no means considered strangers to mechanical inventiveness, where nature comes to their assistance."[26]

This nameless writer, usually called *Anonymus,* is an interesting character. Addressing the emperors Valentinian I and his brother Valens (about +370) he urges that they mechanize the Roman army to make up for the manpower shortage and enable the Romans once again to face the invading barbarians on better than equal terms.

To this end, besides various financial reforms, Anonymus urges the emperors to reduce his military inventions to practice. These include: a mobile, rapid-fire catapult; a catapult with an iron bow, cocked by the power of treadwheels; a spiked battle car to be pushed against the foe; a buckler to be nailed to the battle car; a javelin with leaden weighting and a lateral spike to hamper the foe when it sticks in the ground in front of him; another weighted javelin; an unmanned scythe-wheeled chariot drawn by two horses bearing armored lancers, with means for raising the scythes to the vertical; a one-horse version of the same; another version, with two horses urged on by automatic whips; a thick shirt or haqueton to be worn under armor; a raft held up by inflated skins; and a warship driven by three pairs of paddle wheels, each pair of wheels powered by a pair of oxen walking capstan spokes around on deck.

Plainly, Anonymus was a very ingenious man but one without practical engineering experience. As with the siege engines of Biton and Athenaios, many of his proposals would not work in practice. Still, his heart was in the right place. We can, for lack of contradiction, hail him as the inventor of the paddle wheel as a means of propulsion. Even though his ox-ship would have been too slow to be practical, the idea, which he may have gotten from watching a Vitruvian water mill, is there.

However, the proposals of Anonymus and the warnings of Vegetius alike fell on deaf ears. A more typical Roman opinion was voiced by our old friend Frontinus, the water commissioner. In a book on military stratagems, Frontinus said that he would ignore "all considerations of works and engines of war, the invention of which has long since reached its limit, and for the improvement of which I see no further hope in the applied arts."[27]

In other words, all possible inventions, at least in the military field, had already been made. Had Frontinus been able to look ahead a couple of thousand years, he would have had quite a surprise.

The opposed ideas of Frontinus and Anonymus bring us to another Roman "fall," distinct from the political fall of the West Roman Empire. This is the fall of classical science. Why, under the later Empire, did science and engineering stand still, letting the barbarians catch up in power?

One of the most striking intellectual movements of the Principate was the movement away from science. Alexandrian science was certainly less vigorous in the time of Augustus than it had been under the first three Ptolemies. But it could still produce first-raters like Heron and his contemporary, the astronomer Claudius Ptolemaeus.

After this pair of first-century giants, however, the list thins out. There are a few good mathematicians–Diophantos, Pappos, Theon–but no more of the stature of Hipparchos or Archimedes. Outside of mathematics, there is practically nothing.

Historians have put forward almost as many reasons for this halt in scientific work, under Rome, as they have for the fall of the Empire. They blame the prevalence of slavery, the rigor of Roman rule, the anti-intellectual Roman tradition, and the rise of Christianity. But slavery declined under the Empire. The early emperors were on the whole no more tyrannical than the Ptolemies had been. Upper-class Romans of the Empire were strongly Hellenized. Lastly, Christianity had only local influence through its first three centuries.

A more likely cause of Roman technical stagnation was the fact that the natural resources needed for the next major industrial advances–coal and iron, and to some extent wood and water power–were rather scarce around the Mediterranean, though abundant in northern Europe.

Another argument concerns the interdependence of pure science and engineering. Science, it is said, went as far as it could without instruments and mechanical aids, such as the wheeled clock, the microscope, and the telescope. As slavery fostered a traditional contempt for manual tinkering, classical scientists had no urge to invent these gadgets themselves. So they got stuck and could go no further until these inventions were made, over a thousand years later. Even if they had conceived these inventions, the craftsmanship of classical times was not skilled enough to make such devices workable.

Although there is probably some truth in these contentions, they are not the whole story. We have seen that Heron had most of the ingredients for a successful steam engine. Likewise the techniques that could make elaborate water clocks and geared astronomical computers of the Antikythera type could also have produced a wheeled, weight-driven clock. All that was really needed was for somebody to invent an escapement.

As for optical instruments, burning glasses had been known since –V, and magnifiers were also known though made little use of. Plinius noted that: "a glass globe full of water, if placed opposite the sun, becomes hot enough to set garments on fire." Furthermore, Seneca observed that "Letters, however small and dim, are comparatively large and distinct when seen through a glass globe filled with water."[28]

No scientist, Seneca thought that the wetness of water had something to do with this magnification. Further, he shows that men of his time understood the principles of curved mirrors. He tells of a Roman named Hostius Quadra, whose whim it was to indulge his lusts, natural and otherwise, in a room lined with curved mirrors, so that he could behold his own acts grotesquely magnified.

So optical instruments could have been invented in Roman times if anybody had taken the trouble. So could other useful devices like the printing press, the canal lock, and the full-sized windmill. But they were not. A patent law protecting inventors might have helped, but even that might not have overcome the two great anti-scientific and anti-technical forces, which took possession of the Roman Empire and which we shall soon discuss.

Nor would it be right to say that everybody agreed with Frontinus that all discoveries had already been made. Seneca, a much more influential writer than the water commissioner, declared:

> The day will yet come when the progress of research through long ages will reveal to sight the mysteries of nature that are now wholly concealed. A single lifetime, though it were wholly devoted to the study of the sky, does not suffice for the investigation of problems of such complexity [as those of astronomy]. . . . The day will yet come when posterity will be amazed that we remained ignorant of things that will to them seem so plain.[29]

Even if Seneca was a bit of a windbag, here he really grasps the idea of scientific progress. But alas! when he looks about at Nero's world, what sees he? That ". . . far from advance being made toward the discovery of what the older generations left insufficiently investigated, many of their discoveries are being lost."[30]

There was no active discouragement of science under Rome, at least before Christianity came into power. But science has to compete for attention, belief, respect, interest, and financial support with politics, commerce, art, literature, religion, entertainment, sport, war, and all the other areas of human activity. If any one of these interests greatly expands its appeal to the people, it does so at the expense of the others.

Now, the great intellectual movement of the classical world under the Principate was towards supernaturalism; that is, religion, mysticism, and magic. There was nothing new about these. All went back to primitive times. Mystery cults had flourished in Greece from –500 on and probably earlier if the record were complete.

But, under Rome, those who exploited the human love of the marvelous and fear of the unknown made striking advances in methods, just as metal workers learned to make brass and glaziers to make windowpanes. They found that it added to their following and advanced their own power, glory, and wealth to flourish a body of sacred writings wherewith to confound the heathen; to promise lavish rewards and punishments after death, in order to right the injustices of earthly life; to set up a tightly knit, far-flung, conspiratorial organization; to expound a verbose and seemingly logical body of spiritual doctrine; to impose a fixed code of morals and tabus–some reasonable and some purely arbitrary–on their followers; and, most of all, to incite a fanatical hatred of rival groups and a grim determination to win the world to one's own faith.

All these procedures were inventions, just as much as Heron's toy steam engine. With these new techniques, the priests, prophets, and magicians could more effectively compete for public attention and support. Credulity they redefined as "faith," and fanaticism as "zeal," while respect for the laws of cause and effect was condemned as "blind materialism."

The success of a cult depended upon its fidelity to these principles. This success, needless to say, had nothing to do with the objective truth of its doctrines, any more than the success of a modern advertising campaign has to do with the virtues of the cigarette or detergent being sold.

As a result, the world witnessed a great "return to religion." The new cults grew swiftly. The old Greek and Roman polytheisms, with no central organization, no theology but a mass of childish and inconsistent myths, and no particular doctrine of future life, crumbled before the tide. The worship of Kybelê, Mithra, Isis, Yahveh, Serapis, Abraxas, and Christ waxed mighty. The magical cults of the Neopythagoreans, the Neoplatonists, and the Gnostics throve. The pseudo-science of astrology flourished, and hedge-wizards like Simon Magus swarmed.

The methods listed above had already, before the Christian era, appeared to some extent in Judaism and Zoroastrianism. So it is not surprising that the strongest of the new religions should prove to be Christianity and Mithraism, heretical offshoots of Judaism and Zoroastrianism respectively.

Of all the mass religions, Christianity made the most effective use of these principles. Possessed of the tightest organization, the most bewildering logic, the most impressive sacred literature, and the most fanatical spirit of any, it captured the Principate while Christians were still a small minority in the Empire. Then, armed with the terrifying doctrines of exclusive salvation, eternal damnation, and the imminent end of the world, and backed by the Emperor's executioners, it soon swept its rivals from the board.

This, however, happened long after classical science had withered in the sirocco of a resurgent supernaturalism, of which Christianity at first was but a small part. It withered because few paid any attention to science any more. Who wanted to spend years in pursuit of some obscure natural law, or pay somebody else to do so, when by joining one of the new cults you got drama, passion, mystery, a feeling of superiority to all non-members, and a promise of eternal bliss in heaven?

Supernaturalism could offer these things, since nobody came back from the dead to complain that he had been swindled. Science could not, so science lost out. People who might have become scientists became mystics or theologians instead; it seemed more worth while.

This is not to say that supernaturalism does no good whatever. Strabon of Amasia, the great geographer of the time of Augustus, knew what he was talking about when he remarked: "The great mass of women and common people cannot be induced by mere force of reason to devote themselves to piety, virtue, and honesty. Superstition must therefore be employed, and even this is insufficient without the aid of the marvelous and the terrible."[31] But supernaturalism is not, on the whole, friendly to science and technology, and there is no use pretending that it is.

The collapse of classical science affected engineering. Engineering, however, never ground to so complete a stop as pure science. Although its pace slowed, it forged ahead somehow. Rulers still wanted grandiose buildings and other public works. There remained a demand for millwrights and other engineering craftsmen, even in parts of the West Roman Empire that had come under barbarian rule.

Thus, by +350, when the West was distracted by civil war and barbarian raids, the city of Antioch had installed the world's first system of public street lighting. In late +V, when in Italy Odovacar was contemptuously tossing little Romulus off the Imperial throne, an overshot water mill was set up in the marketplace of Athens. And around +500, some unknown craftsman built, at Gaza on the frontier between Egypt and Palestine, the most elaborate water clock that had yet been seen.

We know about this clock from a battered manuscript in the Vatican, attributed to Procopius of Caesarea, the historian of Justinian's wars, together with some details added by al-Jazari in early +XIII. According to Procopius' *Description of a Clock,* the mechanism was housed in an open marble building about 8 feet wide and 19 feet high. The actual clock stood at the rear, while a marble barrier topped by iron spikes ran around the sides and front to keep the vulgus out. Marble pillars at the front corners upheld the roof.

Above the clock, near the roof, a Gorgon's head glowered down. Below this decoration stood a horizontal row of twelve small doors. At night these doors were opened, and time was told by a lamp, which the clockwork moved slowly from right to left behind these doors.

Lower down was a second row of twelve doors, and below these a number of large statues, including one of Herakles in the center, holding a gong in one hand and his club in the other.

During the day, a tiny statue of Helios, moving along the molding below the doors, told the time. At each hour, as the Helios passed the appropriate door, the Gorgon's eyes rolled, the big statue of Herakles beat the gong, the door next to the Helios opened, another statue of Herakles popped out, an eagle above the door bent forward and dropped a crown of victory on the hero's head, this little Herakles bowed and slid back into his compartment, and the door closed. Each of these twelve "jacks," as such timekeeping statues are called, illustrated one of Herakles' twelve labors.

Instead of striking the numbers of the hours from one to twelve, however, this clock sounded them from one to six, beginning over again at noon. When the day was done, a statue of the god Dionysos blew a trumpet. Evidently there was still some engineering talent at large in the world, to construct so imposing a machine. Remains of a similar clock can still be seen in the Bu'ananiyya Mosque in Fez, Morocco.

Withal, engineering did slow down all over the Roman Empire after +I. Advances during the next few centuries were few and modest. Only a few new inventions such as the onager, the truss, and the pendentive can be credited to the late Roman Imperial period, and their dates are not closely known.

Another factor in the collapse of science and the paralysis of engineering under Rome was the "reforms" of Diocletian. By a series of laws, beginning in +301, this emperor fixed prices and wages and bound every man in the Empire (with few exceptions) to his father's trade.

Diocletian did this to stabilize economic conditions in the Empire, which had just gone through a half-century of destructive civil war. In this fifty-year period, there had been twenty-seven emperors and at least twice as many would-be usurpers.

These wars had made a shambles of the Empire. For, despite all the centurions of the type of "Fetch-another" Lucilius, soldiers of the Empire were a disorderly lot, much given to mutiny and to killing emperors who did not please them. In times of civil war, Roman soldiers got completely out of hand and marched through the countryside, robbing, burning, and massacring like the barbarians against whom they were supposed to protect the people, while the emperors and generals who commanded them did not dare to try to stop their ravages for fear of being murdered.

Since the laws of economics had not yet been discovered, Diocletian could not understand that his restrictive legislation, however justified it may have seemed in an emergency, in the long run only made matters worse. He and his successors succeeded in combining the faults of socialism and capitalism without the virtues of either.

In particular, the laws confining people to their ancestral occupations gave the Roman Empire a system of universal serfdom like the caste system of India. These laws probably affected the masses less than they would today because, in pre-industrial nations, most sons follow their fathers' trades anyway. Many people even like a caste system, because it assigns them a secure and definite place in the social order and largely relieves them of the need for competing and thinking. As Bertrand Russell once said: "Most people would die sooner than think–in fact, they do so."[32] Nevertheless, the ever stricter ordinances issued by Diocletian's successors show that there were still many people who wanted to quit their occupations, simply because, under the Diocletianic system, they could no longer make a living at them.

How did this Imperial strait-jacket affect science and invention? It helped mightily to strangle them. Such restrictions are fatal to technical progress, as is shown by the long technical stagnation of ancient Egypt and of more recent India under similar dispensations. What baker's son will fool around with an idea for a bicycle when he is compelled to make bread all his life willy-nilly? And when all occupations are hereditary, how could anyone even imagine an occupation, like that of a maker of bicycles, if it did not already exist?

With the fall of Rome, the ancient world, as usually defined, came to an end. Engineering was not dead; only stunned by supernaturalism and hog-tied by caste legislation. Several centuries had to pass before the beginning of modern times in the technical field, and many things were to happen. Let us look them over.

THE ORIENTAL ENGINEERS

EIGHT

With the decline of civilization and engineering in Europe after +500, the scene of our story shifts eastward. Here, in the Balkans and Asia Minor, the Byzantine Empire—that eastern rump of the Roman Empire—tenaciously clung to life for many centuries more. Beyond Byzantium, the ancient kingdom of Iran, now ruled by the vigorous Sassanid dynasty, still flourished. Beyond Iran lay India and China, whose technologies were already old.

During the thousand years that followed the fall of the West Roman Empire, the Byzantine Empire continued certain trends begun in the later years of the united Roman Empire, while at the same time it was evolving into a medieval Greek kingdom. Although Greek replaced Latin as the official language, and although in +XI the Normans wrested from the Empire its last Italian possessions, the Byzantines still called themselves *Romaioi*. Their ruler still termed himself the Roman Emperor and his domain the Roman Empire, and he was insulted when Crusaders called him "the King of the Greeks." We use the term "Byzantine" as a convenience, but the Byzantines themselves never did.

One of the late-Roman trends that carried over into the Byzantine Empire was towards an extremely centralized, bureaucratic, despotic government, minutely regulating the lives of its subjects. Russia was to inherit the Byzantine governmental tradition, with results of worldwide moment for our own times.

In contrast to the early Roman Emperors, who affected to be merely the "first citizens" of the Empire, the Byzantine Emperors lived, like the ancient watershed emperors, in godlike seclusion, surrounded by a huge apparatus of servants and agents. Diocletian began this practice in his implacable search for means of stabilizing the Empire, and his successors exaggerated it. When the emperors appeared in public, their appearance was made the occasion for ceremony of stupefying formality, with a parade of precious metals and jewels, gorgeous costumes, pompous titles, and gestures of servility. The Byzantines thought that this ostentatious display of wealth would overawe the simple barbarians who came to pay their respects. In fact, it merely whetted their appetites.

Another trend was toward extreme religiosity. All the keen Greek intellect that had gone into art, science, and philosophy was now turned into channels of subtle theological dispute. Next to chariot racing, this became the leading obsession of the Byzantine masses. As early as +IV, the bishop Gregorius of Nyssa complained that at Constantinople:

> People swarm everywhere, talking of incomprehensible matters, in hovels, streets and squares, marketplaces, and crossroads. When I ask how many oboloi I have to pay, they answer with hairsplitting arguments about the born and the unborn. If I inquire the price of bread, I am told that the Father is greater than the Son. I call a servant to tell me whether my bath is ready; he rejoins that the Son was created out of nothing.[1]

When church councils were called to settle these matters, partisans of one theory or another concerning the Trinity proved their points by setting upon their opponents with clubs and beating them senseless.

Later, when the Crusaders arrived in Byzantium in late +XI to fight beside the Byzantines and regain the Holy Land for the Cross, it was a case of dislike at first sight. The Crusaders found the Byzantine Greeks pompous, subtle, sly, bitter, and cynical; the Greeks for their part found the "Franks" (so called because most Crusaders were Frenchmen) dirty, noisy, rude, and violent. Each party found that it liked the Saracens, the common foe, much better than it liked its nominal ally.

In such a world it is not surprising that science and engineering did not advance far. Although the Byzantine Empire performed functions valuable to civilization, by preserving part of the classical Greek literature and rolling back wave after wave of comparatively barbarous Huns, Bulgars, Arabs, Avars, Cumans, Russians, and Turks, the culture of the Byzantines would not seem attractive to most intelligent people today.

In the Byzantine Empire, as in the Latin West, intellectual activity became the monopoly of churchmen. The Mesopotamian monk Severos Sebokht (+VII) brought to the notice of the West the Indian system of numbers, which we now call "Arabic" numerals. History also flourished; although today only professional scholars read the works of the excellent Byzantine historians.

A few of these churchmen were interested in science; some even did scientific work of their own. About +500, for instance, Ioannes Philiponus pondered earnestly on physics and even, it would seem, performed an experiment or two. He cleared up some of Aristotle's mistakes about motion and grasped the idea that a fluid offers resistance to the motion of a solid body through it, instead of helping it along as Aristotle had thought. Thus he approached the modern concept of friction. He performed, or at least recorded, the experiment that Galileo is supposed, on doubtful authority, to have done more than a thousand years later:

> For if you let fall from the same height two weights of which one is many times as heavy as the other, you will see that the ratio of the times required for the motion does not depend upon the ratio of the weights [as Aristotle had asserted], but that the difference in time is a very small one.[2]

The attitude of most of the Church Fathers towards science, however, was one of indifference or hostility. Eusebius, the noted historian of the early Christian Church, said of scientists: "It is not through ignorance of the things admired by them, but through contempt of their useless labor, that we think little of these matters, turning our souls to better things" such as preparing one's soul for the imminent end of the world. Basil of Caesarea declared it "a matter of no interest to us whether the earth is a sphere or a cylinder or a disk, or concave in the middle like a fan." Lactantius called the study of astronomy "bad and senseless." Like many other churchmen, he combated the pagan Greek notion that the earth was round and argued on scriptural grounds that it must be flat.[3]

Justinian, who reigned from +525 to +567 and temporarily recovered parts of the Western Empire from the Germans, might be called either the last Roman or the first Byzantine Emperor. He, of course, never imagined himself to be anything other than *the* Roman Emperor. The transition from classical Rome to medieval Byzantium was spread out over so many centuries that Justinian could not know that he marked an end or a beginning.

Like Augustus and Hadrian, Justinian was a great builder. But, where the earlier emperors built temples, baths, and circuses, Justinian's constructions ran more to churches and fortifications. In his church building, Justinian employed Anthemios of Tralles and Isidoros of Miletos, among the ablest architects of their time. Anthemios also built machines, including some sort of steam or hot-air engine, which shook his house when he ran it. To get rid of unwanted callers, Anthemios had only to start his engine, and the visitors, thinking that the earth was quaking, fled.

A Christian church was always less of a god-house and more of a meeting hall than the pagan temples had been. Therefore, when Christianity became legal, Christians sought a type of building more suitable for churches than the temples of Zeus and Minerva and the rest.

The most efficient plan for a church building was found to be that of the old Mediterranean *basilica* or town hall, where officials conducted their business, judges held court, and guilds and other associations met. This was a building of simple, rectangular, boxlike form, usually with a row of columns down each side of the interior, dividing the hall into three aisles and upholding a raised central section of roof. The gap between this high monitor roof and the lower roofs along the sides was filled by clerestory windows.

One end of the early church ended in a half-dome, beneath which the sacred functions were performed on the raised area called the apse. Later, a cross-hall called the transept was placed in front of the apse, giving the whole building the plan of a cross.

Many early churches used columns and other marbles taken from pagan temples. As time went on, the Byzantines developed ornate church-building styles, using a great deal of mosaic, which even ran in corkscrew bands up the columns. The rows of arches that were often supported by these columns rested on pieces of stone called impost blocks, tapering downward, circular below (where they rested on the tops of the columns) and rectangular above (where they supported the arches).

Most characteristic was the use that the Byzantine architects made of the dome supported by pendentives. The church that most strikingly carried out Byzantine ideas of sacred architecture was Justinian's Santa Sophia in Constantinople.

Near the Imperial Palace and the Hippodrome, which stood at the end of the main peninsula on which Constantinople was built, there had been a Church of the Christians, more popularly known as the Church of Santa Sophia, or Divine Wisdom. During the Nike rebellion

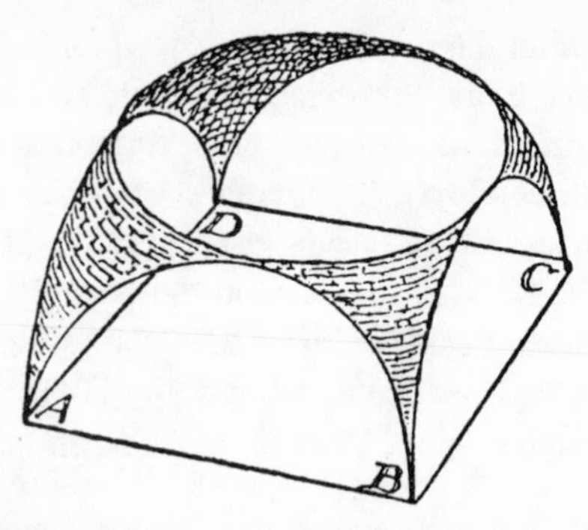

Fig. 19. The pendentive dome (*from T. G. Jackson:* Byzantine and Romanesque Architecture, *1913*).

of +532 (so called from the war-cry *nikê,* "conquer!" used by the rioters) this church was burned by the mob.

To replace the old church, Anthemios and Isidoros built an enormous new Santa Sophia of stone and brick. The central part covered an area 120 feet square. At each corner of this square rose a huge 100-foot stone pier. In building these piers, sheets of lead were laid between the courses of stone to avoid concentration of the load, which otherwise might have cracked the stones.

These piers in turn supported four arches, each 60 feet high. The arches upheld a pendentive dome; that is, a dome cut out to make four pendentives joined at the corners. You can get an idea of its shape by setting a half-orange flat side down, making four vertical cuts around the edge so that the half-orange is square in plan and shows a semicircle of peel on each side, and then slicing the top off level with the tops of these semicircles.

Thus the pendentive dome formed a transition between the square below and the circle above. On this circle the architects mounted a single low dome. This was not a complete hemisphere, but a smaller fraction of a sphere, so that the silhouette was flattened at the top.

On the east and west sides of this square, half-domes supported by semicircular walls buttressed the structure and provided room for the chancel and nave. The inside of the church was partitioned by arcades mounted with impost blocks on rows of columns plundered from temples at Athens, Ephesos, and Ba'albakk. Additional buttressing and half-domes strengthened the structure from without. Most of the ornate decoration was inside, in the form of gilding, mosaics, and colored marbles; the outside was left bare.

In +558 an earthquake damaged Santa Sophia. When Isidoros the Younger, grandson of the first Isidoros, repaired the damage, he dismantled the original dome and built a new and taller one.

For many centuries Santa Sophia, the most sumptuous building in the world, dazzled visitors to Constantinople. In +987 the mighty Prince Vladimir of Russia decided to abandon the paganism of his forefathers, with its hearty human sacrifices, for one of the religions of the neighboring lands. Accordingly he sent envoys abroad to learn how the followers of the Roman, Greek, Muslim, and Jewish faiths conducted their affairs.

The envoys reported that the Muslim Bulgars seemed a sad and dirty lot, so their religion could not be much good. The Roman Catholic churches of the Germans struck them as ugly. But, in Constantinople, the representatives of Orthodoxy welcomed them to Santa Sophia and turned on the full splendor of the Greek ritual. The dazzled envoys told Vladimir: "We no longer knew whether we were in heaven or on earth."[4] So, thanks to the charm of Santa Sophia, the terrible Slav embraced the Greek Orthodox faith, sent his 800 concubines packing, and spent the rest of his life in good works.

When Muḥammad II captured Constantinople in 1453, he paused in awe at the entrance of Santa Sophia. Inside he found a Turk chopping up the floor with an ax and asked:

"Why are you doing that?"

"For the Faith," replied the soldier.

The Sultan hit the man. "You've got enough by pillaging and enslaving the city!" he cried. "The buildings are mine!"[5]

Muḥammad had the mosaics whitewashed to conform to the Muslim tabu on images and turned the church into a mosque, adding minarets at the corners. The modern Turkish Republic has made the building into a museum and, by taking the whitewash off the mosaics, has at least conjured up a ghost of its ancient splendor.

Justinian was also active in fortifying the cities and frontiers of the Empire and furnishing the cities with water supplies.

For instance, the city of Daras, on the Persian frontier, straddled a small tributary of the Khabur. Although the stream flowed through the city, enemies were kept out by an iron grate at each end. Once, however, a flood swelled the stream until it washed away part of the city. To keep this from happening again, the architect Chryses of Alexandria built a flood-control dam:

At a place about forty feet removed from the outer fortifications of the city, between the two cliffs between which the river runs, he constructed a barrier of proper thickness and height. The ends of this he so mortised into each of the two cliffs, that the water of the river could not possibly get by at that point, even if it should come down very violently. . . . The barrier was not built in a straight line, but was bent in the shape of a crescent, so that the curve, by lying against the current of the river, might be able to offer still more resistance to the stream. And he made sluice gates in the dam, in both its lower and its upper parts, so that when the river suddenly rose in flood, should this happen, it would be forced to collect there and not go on with its full stream, but discharging through the openings only a small volume of the excess accumulation, would always have to abate its force little by little, and the city wall would never suffer damage.[6]

Thus Chryses invented the horizontal arch dam, like an arch lying on its side. After Chryses, however, this invention seems to have been forgotten—despite Procopius' clear description—until it was revived in modern times. Besides the dam, Chryses built an aqueduct leading from the dam to the city, with the usual Roman water-distributing system.

When a similar flood of the Skirtos devastated Edessa, Justinian's engineers dug an overflow channel about the city, even though they had to raze a hill to do so. They also added to the harbor works of Constantinople, not only on the shores of the main city, but also in the suburbs across the Bosporus in Asia Minor. Justinian weaned his wilder Armenian and Caucasian subjects from brigandage by building roads through their mountains and encouraging them to take part in honest trade.

In their fortifications, the Byzantines continued the Roman system of crenelated walls and towers. The huge triple wall built by Theodosius II stretched across the base of the peninsula on which Constantinople stood, while a sea wall surrounded the city along its waterfronts. Later Byzantine Emperors extended and strengthened the city's walls, which might have remained impregnable for aye, had not cannon been invented.

In military technique the Byzantines, like their Roman predecessors, relied upon armored cavalry. When he became Justinian's commander-in-chief, Belisarius found himself in command of a motley collection of imperial guards, feudal retainers, local levies, and barbarian mercenaries—mostly Huns, Germans, and Arabs. He reformed the army and adopted a particular type of horseman, well armored and equipped not only with lance and sword for close work but also with a powerful bow, as standard.

For several centuries these highly trained *kataphraktoi* or knights were the world's best soldiers. Used in carefully thought-out ways by adroit and crafty Byzantine generals, they again and again defeated superior numbers of Byzantium's swarming foes. Most of these troopers were farm boys from the provinces of Asia Minor. After the Empire lost these provinces to the Turks, following the disastrous battle of Manzikert (1071), the Empire was forced to rely on mercenaries, and its doom was sealed.

The only significant new military invention by the Byzantines was the so-called Greek fire. Incendiary mixtures go back at least to –IV, when Aineias the Tactician wrote a book *On the Defense of Fortified Positions*. Said he:

> And fire itself which is to be powerful and quite inextinguishable is to be prepared as follows. Pitch, sulfur, tow, granulated frankincense, and pine sawdust in sacks you should ignite if you wish to set any of the enemy's works on fire.[7]

Later it was found that petroleum, which seeps out of the ground in Iraq and elsewhere, made an ideal base for incendiary mixtures because it could be squirted from syringes of the sort then used in fighting fires. Other substances were added to it, such as sulfur, olive oil, rosin, bitumen, salt, and quicklime.

Some of these additives may have helped–sulfur at least made a fine stench–but others did not, although it was thought that they did. Salt, for instance, may have been added because the sodium in it gave the flame a bright orange color. The ancients, supposing that a brighter flame was necessarily a hotter flame, mistakenly believed that salt made the fire burn more fiercely. Such mixtures were put in thin wooden casks and thrown from catapults at hostile ships and at wooden siege engines and defense works.

In +673, the architect Kallinikos fled ahead of the Arab invaders from Helipolis-Ba'albakk to Constantinople. There he revealed to the emperor Constantine IV an improved formula for a liquid incendiary. This could not only be squirted at the foe but could also be used with great effect at sea, because it caught fire when it touched the water and floated flaming on the waves.

So Byzantine galleys were armed with a flame-throwing apparatus in the bow, consisting of a tank of this mixture, a pump, and a nozzle. With the help of this compound the Byzantines broke the Arab sieges of +674–76 and +716–18 and also beat off the Russian attacks of +941 and 1043. The incendiary liquid wrought immense havoc; of 800 Arab ships that attacked Constantinople in +716, only a handful returned home.

By careful security precautions, the Byzantine Emperors succeeded in keeping the secret of this substance, called "wet fire" or "wild fire," so dark that it never did become generally known. When asked about it, they blandly replied that an angel had revealed the formula to the first Constantine.

We can, therefore, only guess the nature of the mixture. According to one disputed theory, wet fire was petroleum with an admixture of calcium phosphide, which can be made from lime, bones, and urine. Perhaps Kallinikos stumbled across this substance in the course of alchemical experiments.

One treatise on military engineering has come down from Byzantine times. This is a *Book of Machines of War* by a writer who calls himself "Heron the Mechanic" or "Heron the Second." He is generally known as "Heron of Byzantium." Whether his name really was Heron, or whether he used that name to capitalize on the repute of the famous Heron of Alexandria (whom he calls "Heron the Mathematician") there is no way of telling.

His treatise, however, contains very little that is new. The author frankly states that his purpose is to combine the works of the classical poliorcetists and to express their teachings in up-to-date language, for Greek had changed as much between Hellenistic and Byzantine times as English has changed since Shakespeare wrote.

Heron's main source is Philon of Byzantium. Like Philon, Heron tells of mantlets, jugs used as tank traps, the ram, the bore, the flying bridge, the sambuca, and catapults.

One of Heron's less practical ideas, taken from Philon, is a pneumatic scaling ladder–that is, a ladder whose uprights are a pair of inflatable sausage-shaped leather bags. Another is a flying bridge decorated with a pair of gaping model heads of lions or dragons to frighten the foe. The section on catapults is short and perfunctory, as if the author did not understand these mechanisms very well.

In fact, Heron adheres so closely to his classical models that there would be little reason to date the treatise so late, save that the author refers to a Bulgar invasion in +X and to Greek fire. To keep the defenders of a wall from overthrowing scaling ladders, he advises that a flying bridge be maneuvered up to the wall and that on this bridge a man stand and "attack the faces of the enemy with fire by means of a portable flame-throwing weapon."[8]

Despite the keenness of the Byzantine mind in many directions, technology was not its strong point. Craftsmen in Constantinople continued to make mechanical toys like those of Philon of Byzantium and Heron of Alexandria many centuries before, but they failed to develop these gadgets into machines of practical utility.

In the desperate 1440s, when the Turks were closing in, the patriarch Bessarion, one of the Empire's most learned scholars, wrote from Italy to Constantine Palaeologus, lord of one of the surviving fragments of the Empire. Bessarion urged Constantine to send young men to Italy to study Western techniques, which had far outstripped those of Byzantium. He spoke of Western improvements in textiles, glass, arms, ships, metallurgy, and the use of water power to save labor in sawing and in pumping bellows. But then it was too late.

East of the Euphrates, the warlike Iranians ruled their vast, thinly peopled land of parching deserts and towering crags from the expulsion of the Macedonians in –II to the conquest by the Arabs in +VII. During the first half of this period the realm was called the Parthian Empire, because it began in the province of Parthia, or the Arsacid Empire after its founder Arshaka. In +III, a Persian kinglet overthrew the Parthians and set up a Persian dynasty called the Sassanid after Sassan, an ancestor of the founder.

The government under these dynasties was in some ways a continuation of the Achaemenid Empire. In others, it resembled the feudal regimes that later arose in Europe, with grades of nobility and a class of knights who fought as armored horsemen. Although Iran was far inferior to the Roman Empire in population and resources, the valor of its knightly class enabled Iranian rulers for seven centuries to meet Romans and Byzantines on equal terms. As a result, the Romans and Byzantines kept half their army on the eastern frontier, next to the only adjacent state that was comparable to their own in culture and power.

Like the Byzantines, the Iranians were not very technically minded. But the kings kept up the roads and the irrigation works, especially the great canal network of Babylonia. From time to time they either hired architects from the Roman world or employed those they captured in war with the Romans to put up buildings, bridges, and other structures. A great irrigation dam at Shuster, still used in modern times, was built by captives from Antioch taken by Shapur I in his invasion of +260. But the Persians did not go in for temple building, as the only regular Zoroastrian sacred structure was a pair of fire altars out of doors.

The Iranians built many bridges, however, of which some survive from the time of Shapur I (+III). Eight such bridges, of which the foundations at least probably go back to Sassanid times, still stand in Luristan. Another bridge near Susa, of which traces have survived down to modern times, was described by the Iranian encyclopedist Zakariyya al-Qazwîni (+XIII):

> Idady is a country lying between Ispahan and Khuzistan, subject to earthquakes. The greatest wonder there is the bridge Harah Zad, built by Ardeshir's mother over a dry stream-bed. This runs water only when a heavy rain falls; then its water rises like a sea, at which time it extends over a breadth of 1,000 cubits and reaches a depth of 150 feet. The structure, furthermore, narrows as it rises. The gap between the foot of the banks and the bridge, which rises about 40 cubits from its base, is filled with a poured-out mass of iron slag mixed with lead. The abutments on which the bridge rests, which project from the highest peaks above the bridge, are composed of this substance. Thus the entire interval between the stream banks and the bridge is filled by a mixture of clinkers and lead. The bridge has but a single arch, a marvel of construction on account of its internal reinforcement. However, the bridge was destroyed after the time of Mazdai and long remained in this condition to the detriment of travel; during which time also the neighboring peoples took away a large part of the lead which reinforced the structure. At last the repair of the work was again undertaken by 'Abû 'Abdallah Muḥammad ibn-Aḥmad al-Awmu, who was the vizier of al-Ḥasan ibn-Buyas. He hired workers and surveyors and directed all their means and effort to this project. Most of the workers had to be lowered to the highest part of the bridge by a windlass in baskets and nets to rebuild the bridge, whose completion took many years. Most of the workers were brought from Ispahan and Idady to this job; their daily wage, moreover, was over 350,000 pieces of gold, spent on things connected with this bridge, which still endures to our time and is a wonder to see.[9]

Building up the abutments with a mixture of clinkers and lead shows that the Sassanid Persians followed the extravagant Roman system of using molten lead as we should use concrete. Qazwîni's tale also shows what happened to all such installations sooner or later.

The Sassanid kings of +IV also built a great palace at Ctesiphon. This was their capital on the Tigris, northeast of deserted Babylon and downstream from the village of Baghdad. Part of this palace still stands, including most of the vaulted dining hall, "the widest single-span vault of unreinforced brickwork in the world."[10]

This vault is 77 feet wide at the base and 112 feet high. The hall has small entrances at the sides and at one end, while the other end was apparently left open or was closed only by a huge curtain, giving the effect of a colossal tent made of brick.

The open end was flanked by a pair of façades connected by an arch. One façade collapsed in a flood in 1909; the other, reinforced, still stands. Although these façades have been described as "a masterpiece of bad taste, a surprising example of the unimaginative application of the bare principles of copy-book Roman architecture,"[11] the palace is still an imposing piece of work for a non-technical people, and the soundness of its structure is shown by the length of time that parts of it have endured.

East of Iran lies India, a peninsula jutting out from the continent of Asia, having just about the same area and population as Europe.

Although civilization in India is almost as old as in Iraq, much less is known of its ancient history. There are several reasons for this state of affairs. For one thing, after the Aryan conquest, writing did not again come into wide use for many centuries. For another, the usual writing material was a paper made from palm fronds, which did not last long in that country of climatic extremes and abundant insect life.

Finally, the Indians showed but little interest in history. Indian philosophy, with its doctrines of endless cycles and the unreality of the material world, led Indian interests away from mundane events. Therefore most of what we know of Indian history before the Muslim conquest is based on the findings of archeologists and the writings of a few foreigners—Greeks, Chinese, and Muslims—who visited the land.

Back in the days of Khammurabi of Babylon and Amenemhat III of Egypt—about −2000—a civilization flourished in western India, in the valley of the Indus River. This civilization is best known from the ruins of two of its chief cities, Harappâ and Mohenjo Daro ("Place of the Dead"). The culture is sometimes called "Harappan." These names are all modern, as we have no idea what the people who lived in these cities called them. For, although the Harappans had writing, its decipherment has only begun.

The Harappan cities were made of brick and laid out on a standard grid pattern with a large temple area, a central bathing pool, and an elaborate drainage system. To some archeologists the layout suggests the authoritarian rule of a priestly caste. The work, though carefully planned and neatly finished, displayed little artistry or decoration. The walls tapered upwards with a pronounced batter, as in Egypt. There were corbelled arches but no true arches.

In the middle of the second millennium B.C., the Aryans, as the eastern branches of the Indo-European-speaking nomads called themselves, overran northwestern India. These barbarians extended their rule across the north of India and down the Ganges Valley to the Sea of Bengal, but in the south their conquests were halted. In the areas they conquered the cultural level declined, as it did in western Europe after the fall of Rome. People went back to simpler buildings of mud brick, wood, and bamboo, of which no trace remains today.

The Aryans either imposed on the darker but more civilized folk they conquered, or (more probably) took over from them, a peculiar form of social organization. This system is marked by division of society into castes. Each caste member is confined to certain occupations, all strictly hereditary. Men are forbidden to marry or even to eat with persons outside their respective castes.[12]

The Aryans naturally gave themselves the highest ranks in the system. The priests developed the doctrine that, if a man was born into a low caste, it was a punishment for a sin in a prior incarnation. If, in this life, the sinner behaved himself, and if he respected and obeyed those in the higher castes, he might be promoted in his next incarnation. This is perhaps the most diabolically clever method of making the downtrodden resigned to their lot that an ingenious species has yet thought up. From the religions of the Aryan invaders and their Harappan subjects evolved the modern religion of Brahmanism or Hinduism, with its bewildering variety of gods, cults, tabus, and philosophies.

The reason for going into these social matters is that the Indian social system affected India's technology. It made the country extremely conservative. It purchased order and stability at the cost of progress and adaptability. Indian workmanship shows the qualities to be expected when the workman is born into his trade and has no thought or hope of leaving it: extreme technical skill and finish coupled with an almost complete lack of progress from age to age.

An added factor in blocking progress in India was a Hindu religious tabu against sea voyages. This tabu discouraged Indians from traveling abroad and learning what the rest of the world was doing.

Despite India's long technical and scientific torpor, some modern Indians like to assert that India had airplanes or nuclear power ten or fifteen thousand years ago. The basis for these ridiculous claims is the references in ancient Indian myths, legends, and epics to their heroes' magical powers and devices. One could argue with equal effect, from the lines in the *Odyssey:*

> 'Neath it the deep black water is swallowed by mighty Charybdis.
> Thrice in the day doth she swallow it down and thrice she rejects it.[13]

that the ancient Greeks had electric washing machines.

When the great Darius conquered the Panjâb around –515, Persian building methods were introduced into India. However, the Indians long continued to prefer wood to stone for building. Even city walls were of wood, except for Râjagriha, the capital of the powerful kingdom of Magadha on the lower Ganges, which had a wall of cyclopean stonework.

Later, Seleukos I (a successor of Alexander) sent Megasthenes (–III) as ambassador to India. An adventurer named Chandragupta Maurya had conquered the Panjâb and Magadha and had built himself a new capital at Pâṭaliputra (modern Patnâ) on the Ganges. Megasthenes found the self-made emperor reigning in a fine palace with vast hypostyle halls in the Persian manner–but all of wood. Wooden pillars upheld the roofs, and round these pillars twined golden vines whereon perched silver birds.

While Darius was conquering the western corner of India, a man named Siddhârtha Gautama, born an aristocrat in a small north-central Indian state, was walking about India in search of wisdom. When he became convinced that he had found wisdom, he spent the rest of his life preaching it. Therefore his disciples called him Buddha, "the enlightened one." Although he accepted the Indian belief in reincarnation, he ignored the swarming legions of Indian gods and put all his stress on right conduct and unselfishness.

In its original form, the austere creed of Buddha was less a religion than an ethical philosophy, comparable to the Stoicism of the later classical world and the Confucianism of China. Subsequently, Buddhist theologians fancied it up with a hierarchy of gods, demons, Buddhas, ex-Buddhas, and apprentice Buddhas.

Our interest in Buddhism, however, arises from the changes it brought about in Indian building. The new creed spread over India. A later Maurya king, Ashoka[14] (–III), conquered most of India and made Buddhism his state religion.

Before the Mauryas, most Indian religious structures had been unpretentious little wooden shrines. The Buddhists, however, developed new types of religious structure. One was a monastery, with scores of cells ranked about a compound or placed in tiers up a hillside.

Another was a shrine called a *stûpa,* housing the relic of a Buddhist saint. The main feature of the stupa was a large domed structure over the actual relic. There were also four symbolic gateways on the four sides, a railed walk around the base of the dome, and in front of the main gateway a pillar bearing the Wheel of Life.

At first the domes of stupas were made of brick, but under Ashoka they began to be made of stone. Some later stupas had the dome squared off like the crown of a Victorian derby; some were stepped pyramids; but the standard form was the simple hemispherical dome. These domes, which littered the land like roc's eggs, were topped with stone finials of many shapes.

After Ceylon (whose early kings built one of the most elaborate irrigation systems ever attempted in ancient times) went Buddhist, the stupas built there reached enormous size. In some the dome was over 300 feet in diameter and the whole monument over 250 feet high, so that in bulk they compared with middle-sized Egyptian pyramids.

When the Ceylonese king Duthagamini (–II) began such a stupa of brick and rubble, he laid the first brick himself while a band played, a host of dancing girls performed, and 40,000 men stood at attention. This king's construction resembled modern reinforced concrete. The interior was built up of layers of stones, clay, and networks of iron. To pack down the mixture, the king had his elephants, shod with leather boots, march back and forth over the rubble.

The third distinctive Buddhist structure was the rock temple. This was a temple dug into the side of a rocky hill, as the temple of Rameses II was dug at Abu Simbel. Such temples might have some construction in stone and wood outside, forming an ornamental entrance, but the main part was a series of artificial caves. Such structures were practical only in hilly or mountainous regions. It is thought that the work required to excavate such a temple was actually less than that needed to erect in the open, by the primitive hauling and hoisting means that the Indians had, an equivalent structure of stone.

The main part of such a cave temple was the *chaitya* or hall of worship. This had an ornamental arched entrance, opening into an elongated hall with a row of pillars down each side, a roof shaped like a ribbed barrel vault, and a small stupa at the far end.

We must, however, distinguish these "arches" and "vaults" from those of the Mediterranean and Euphratean worlds. Until the Muslim conquests, India knew not the true or voussoir arch, dome, and vault. The wooden buildings of India used structures shaped like arches, domes, and vaults, but all of wood. The wooden arch of India differs greatly from the voussoir arch in its engineering principles, possibilities, and limitations.

In the village houses from which these wooden arches evolved, the shapes were attained by means of semicircular hoops of bamboo, thatched with reeds. In temples and palaces, the same shapes were used. But here the hoops became curved wooden beams, built up of shorter pieces mitred and dovetailed together with the immemorial skill of the Indian carpenter.

When the Buddhists began digging their temples into hillsides, they simply imitated, in stone, the shapes employed in wooden buildings. Hence these pseudo-arches and pseudo-vaults in solid stone. In fact, the masons copied the details of wooden construction with such meticulous care that we know quite a lot about the wooden buildings of Buddhist India, even though not one joist or lath from these buildings has come down from ancient times.

Hinduism revived in –II. For several centuries Hindus and Buddhists struggled for control with almost the zeal and cruelty of the monotheistic faiths of the West. In +III a Hindu empire, that of the Guptas, arose. Thenceforth Buddhism declined in India and virtually disappeared by +VII, although it had spread far and wide over the other lands of eastern Asia.

When Hinduism rallied, Hindus took to building cave temples also and continued to do so down to +X. The last Hindu work of this type was a set of huge temples at Ellora in western India, standing in the open but carved out of the solid rock of a hill, which was cut away around them. These monolithic temples had a curious parallel: the monolithic Christian churches, chiseled from outcrops of solid rock in +XII in Abyssinia.

Resurrected Hinduism not only copied the Buddhist cave temples but also built large stone temples in its own style. Such a temple might be composed of a series of separate buildings, or these buildings might be run together into one. A typical temple consisted of four main parts: a porch or entranceway, a hall of assembly, a vestibule, and a sanctuary.[15] The sanctuary comprised the dark, cell-like shrine room, housing the main ikon, and towering over it a spire, the *shikhara*.[16] There might also be other auxiliary structures.

The shikhara, sometimes rising over 200 feet, was square or octagonal in plan, with rounded corners and an intricately ribbed pattern. On top rested a large stone disk shaped like a flattened pumpkin or a round cushion, so that the whole shikhara has the look of a huge, ornate salt shaker with a screw cap.

The phallic symbolism of these towers is well known. The temples of aberrant sects, especially the Temple of the Sun at Konârak in the East and others at Khajurâho near the center of the country, are notorious for their erotic carvings.

In the South, where the Aryans never established lasting rule, temple architecture followed other paths. The salient feature of the South Indian or Dravidian temple compound was the *gopuram* or monumental gateway. This was a tall, tapering structure, like a huge wedge standing upright. Four such portals were placed on the four sides of the temple inclosure.

A gopuram usually has a foundation of stone and a superstructure of brick and plaster, the whole covered with statuary. Being averse to blank spaces, Hindu architecture lavishly covers them with statuary and reliefs.

The Indians used post-and-lintel construction and also corbelled arches and domes. They built these domes by filling the space beneath the structure with sand and shoveling the sand out afterwards. Like the classical Greeks, they made but little use of mortar, preferring to hold the stones of their temples together with iron dowels. Large stones, like the capstone of a shikhara, were raised on high by hauling them up earthen ramps.

Though technologically backward in some ways, the ancient Indians were skilled in iron work. One of their monuments is the Iron Pillar of Delhi, a 24-foot shaft set up by a king[17] about +415 to bear the statue of a *garuḍa* or man-bird, the steed of Vishnu. Although the garuḍa has disappeared, the pillar still stands. Its remarkable resistance to rust is laid to the extreme purity of the iron and to the fact that most of the year the climate is too dry for rust to get a start. The pillar is thought to have been made by welding together a series of iron disks and smoothing down the outer surface by hammering and filing. About the same time, the Chinese Buddhist monk Fa Hsyen, traveling in India, discovered that the Indians made suspension bridges held up by iron chains.

Ancient India was one of the few lands where the secret of good steel was known. In Roman times, two of the world's main iron centers were Austria and India, whence steel of high quality was exported all over the civilized world. Then and later, ingots of Indian steel were taken to Damascus, where Syrian smiths made them into the famous

Damascene sword blades. Elsewhere iron was made and used locally in small amounts.

In the northwest of India, where Indian civilization arose, few architectural relics of ancient times have survived; for, in this region, man and nature have combined to destroy them. From +VI on, the land was devastated by a series of invasions by Huns and Turks, who expressed their hatred of Buddhism and Hinduism by trying to destroy all the monuments and temples of these religions. Of the northwestern temples that the invaders overlooked, most were toppled by a great earthquake in 1819.

Little by little the Turks extended their control over the whole Indian peninsula; for the Indians, while they fought bravely enough, had with their usual conservatism not changed their methods of warfare in a thousand years. Muslim rule became complete with the foundation of the Mughal Empire by Babur of Kabul in +XVI. The thoughtful, hard-drinking, tubercular Babur defeated the vast but ill-organized Indian armies with that terrible new weapon, the cannon. The Turkish conquest halted the development of native Indian technology and ends our concern with that land.

The conquests of Babur formed the closing act of a movement that began with Muḥammad ibn-Abdallah, a sometime traveling salesman, caravan leader, and shopkeeper of Makkah or Mecca. Early in +VII, Muḥammad led a religious revolution in Arabia. By the time of his death in +632, Muḥammad ruled all of Arabia proper. A little over a century later, the domains of Islam (literally "peace") stretched from Spain to Turkestan.

At first this whole vast empire was under the rule of the Khalîfah ("successor") or Caliph. In time the central power weakened. Parts of the realm broke away under separate Muslim dynasties. Other parts became independent in fact even though they paid lip-service to the Khalîfah. Finally the Turks seized the secular power in most of the Caliphate, leaving the Khalîfah with spiritual authority only. For a time, however, the Khalîfah's writ ran from Lisbon to Samarqand.

The first four Khalîfahs, friends and relatives of Muḥammad, ruled from Arabia. After battle, murder, and sudden death, the Caliphate came into the hands of Mu'awiyyah (+661) of the Meccan clan of Umayya. Mu'awiyyah made Damascus his capital.

For ninety years the Umayyads ruled from Syria, using the existing Byzantine bureaucracy. Their rule was on the whole tolerant. Islam they regarded as a racist Arab movement. Therefore, they tried to convert all Arabs–Christian, Jewish, and pagan–to Islam, but they made it hard for non-Arabs to become Muslims. They wanted a Muslim Arab aristocracy to rule a non-Muslim, non-Arab, taxpaying populace.

In conquered Iran, however, this policy was for various reasons reversed. Here everyone was forced to become Muslim. Zoroastrianism was persecuted out of existence. The religion of Darius and Sassan died out save among refugees who took it to India, where they became the Parsees.

Thereafter Muslim policy vacillated between tolerance and forced conversions. Where wholesale conversions were encouraged or compelled, the original Arab aristocracy was soon diluted out of existence. Muslims, though speaking Arabic, worshiping Allâh, and conforming to many Arabian customs, were superficially Arabized Syrians, Iraqis, Egyptians, and so forth. Of the "big four" of Arabic science–ar-Râzi, al-Birûni, ibn-Sîna, and ibn-al-Haytham[18]–all but the last were Iranians, and ibn-al-Haytham was born on the borders of Iran at Basra. They were "Arabs" only in the sense in which Heron of Alexandria and Apollodoros of Damascus were "Romans."

In +VIII, dissident Muslim elements overthrew the Umayyad Caliphate and set up another line, of the Arabian clan of Abbas. In +762 the second Khalîfah of the Abbassid line, the energetic and boundlessly treacherous al-Manṣûr, built a new capital on the southwest side of the Tigris, at the site of the ancient Babylonian town of Baghdad. Laid out by the Jewish astronomer Mashallah and the Persian engineer al-Nawbakht, Manṣûr's Baghdad was (like Plato's dream-city of Atlantis) a perfectly circular city, 3,000 yards across. A moat and a dyke surrounded it. Inside the moat rose two circular walls close together. In the center of the circular city a small circular wall inclosed the royal precinct and the governmental buildings.

By +X, however, the wandering river had destroyed part of Manṣûr's circular walls, and the bulk of the city had shifted to the northeast bank. Today Baghdad sprawls for miles along both banks of the Tigris. Nothing is left of Manṣûr's round town, and few structures from the later Abassid period survive. In 1258 Hulagu Khan's Mongols burned the whole city and massacred 800,000 inhabitants, so that Baghdad did not again become important until modern times.

Under the Umayyad and Abbassid caliphs occurred an Arabic renaissance of science. This is not so strange as it might seem. For one thing, the original Arabian conquerors were more civilized at the start than the Germans who conquered the West Roman Empire. They were quicker than the Germans at mastering the arts of those they conquered, and they never killed off whole populations as did the Mongols. Although many of the Arabs of the conquering armies were desert nomads, Arabia also had cities, writing, agriculture, and well-developed trade and commerce. Muḥammad himself had been an urban merchant.

Arabian engineering, as shown by the great dam of Ma'rib and the multistory houses of the Ḥadhramawt, was far from primitive. The greatest of these tall buildings was put up in +I by an Arab king[19] to protect his townspeople against the nomads. This was the castle of Ghumdân in Yaman. Later writers said it had twenty stories, each ten cubits high, of granite, porphyry, and marble. Although the size and splendor of this Arabian skyscraper have probably been much exaggerated, there is little doubt that Ghumdân was an imposing structure.

The mere setting up of a large and reasonably well-run empire, with a single official language and without internal barriers, brings about material prosperity for a while. The abolition of national boundaries stimulates trade, which causes wealth to accumulate, which provides the means for scientific research and engineering works. Making travel easier in itself stimulates technical progress because, as I have already explained, it increases the size of the intercommunicating, interconnected population, so that a discovery made in one place quickly spreads to all.

It is only later that the shortcomings of such a government begin to catch up with its virtues. In the case of the Caliphate, one shortcoming was the primitive political naïveté of the Arabs. Politics in Islam never got beyond a simple, cutthroat scramble for unlimited power and privilege.

There is no trace in early Islamic history of the evolution of constitutional government. Government was despotism tempered by inefficiency. It was as absolute as under the ancient watershed empires, without even the stability that the king's divinity gave to those realms. Hence the Caliphate, like the third-century Roman Empire, was kept in constant turmoil by palace plots and murders, revolts, and civil wars; and it proved easy for the Turks to subvert and the Mongols to destroy. Things have not utterly changed in modern Muslim nations.

Arabic science owed something to an earlier movement that took place in Mesopotamia and southwestern Iran. In +VI, Justinian closed the pagan philosophical schools of Athens. He and other Byzantine emperors also persecuted dissident Christian sects, the Nestorians and Monophysites. Hence many Byzantine scholars fled to Iran, where the Sassanid king Khusrau helped them to set up a medical school at

Jundishapur. Here, writings in Greek and Sanskrit were translated into Syriac and Persian. For two centuries Jundishapur was the world's scientific capital.

After Baghdad was founded, it outshone Jundishapur as an intellectual center. For a century and a half, until about +900, the Khalîfahs employed scholars, mostly Christians and Jews, to translate the wisdom of the unbelievers into Arabic.

As these translations took effect, original Arabic science developed. In science, the Arabs were weaker in theory than the classical Greeks but stronger in practice. Where the Greeks had been mainly interested in knowing things, the Arabs were interested in accomplishing things. Their main advances were in medicine, mathematics, chemistry, astronomy, and geography.

During the great age of the Caliphate, Arabic engineering, while not negligible, was not so outstanding as that of the Roman and Hellenistic periods. Although the Khalîfahs kept up the roads and canals, they did not create such spectacular structures as the Pharos of Alexandria or make such striking advances as Sennacherib's aqueduct or the Appian Way. In traveling, the Arabs were satisfied, like their forebears, to ride on the backs of animals. Hence there was no general paving of roads and no revival of the wealth of wheeled traffic that had rumbled along the roads of Rome. And when the last ancient bridge across the Tigris collapsed in +X, it was not replaced.

In fortification, the Arabs continued the system of battlemented walls and towers that they had inherited from the Romans and Byzantines. They preferred the square tower, which if more vulnerable was easier to build, to the more effective round tower.

The Arabs made one advance in fortification (+VIII) which Europeans later copied. This was machicolation: the mounting of a parapet, not on the main wall, but on a row of stone corbels or brackets extending out several feet beyond the main wall. Between each pair of brackets was an opening in the pavement, filled with a trapdoor. When the trapdoor was taken up, the defenders could shoot arrows, drop stones, and pour boiling oil, molten lead, or red-hot sand straight down on the heads of unwanted visitors without exposing themselves to return fire.

At Blarney Castle in Ireland, visitors today are lowered down headfirst through one of the machicolations to kiss the Blarney Stone, which forms part of the bottom course of stones of the parapet. The idea of machicolation probably arose from the stone latrines, projecting out from the side of the outer wall, with which castles were often furnished from Byzantine times on.

The Arabs also revived the ancient practice of making the entrance to a fortress in the form of a dog leg. He who would enter had to make a right-angled turn or two and could not, from the outer gateway, see or shoot into the inner courtyard.

The building of houses continued in Islam much as it had before. The inside-out courtyard house was everywhere common.

The one distinctively Muslim type of building was the mosque. The first mosques were plain little boxlike brick buildings, but the Arabs soon began to build more splendid houses of worship. They borrowed wholesale from the architecture of the lands they conquered, using the Byzantine dome and the Syrian pointed arch. They often made the voussoirs of their arches alternately of black and white stone, as in the Alhambra of Granada and Suleyman's Mosque in Damascus.

The inside of a mosque has always been rather plain. The only permanent features, aside from the structure itself, are a niche in one wall indicating the direction of Mecca, and the pulpit. Muḥammad's severe tabu on graven images caused Muslims to avoid all pictorial art other than conventionalized foliage, especially in mosques. It also incited the Arabs, in the first fine iconoclastic frenzy of conquest, to destroy a multitude of works of art that had come down from former ages. The Arabs themselves used geometrical patterns and verses from the Qur'ân for decoration instead. But the Iranians, who had a well-developed pictorial art at the time of the conquest, refused to take this prohibition too seriously, even after their conversion.

Another distinctive Muslim structure is the minaret, a tall slender tower from which the *mu'adhdhin* calls the faithful to prayer. Minarets in Cairo are now fitted with electronic public-address systems, and the call to prayer is made by a recording. The shape of the original minarets may have been inspired by that of the Pharos, and they may in turn have inspired the medieval Italian bell towers.

The most remarkable minaret was that built by the Khalîfah al-Mu'taṣim (+IX). Mu'taṣim raised a corps of Turkish soldiery who, though excellent fighters, proved a trial in other ways. They galloped full-tilt through the streets of Baghdad and rode down the Baghdadis, who therefore hated them and lynched them whenever they caught them alone. So Mu'taṣim built a new capital at Samarra. This included a conical minaret with a spiral path to the top. This tower, looking like an ancient ziggurat, still stands.

Mu'taṣim's son Mutawakkil tried to bring water to the new city by means of a canal. But his engineers miscalculated, and the water failed to flow. So, after Mutawakkil's death, the capital was moved back to Baghdad.

The Umayyad and Abbassid caliphs were great canal builders. The former extended the canal system by which Damascus receives and distributes the water of the Barada River. The Abbassid caliphs did likewise on a grand scale with the canals of Mesopotamia. They cleared out the old Sassanid canals and dug new ones. They redug the great Nahrwân canal from Baghdad to Kut-al-'Amâra, which dates back to Babylonian times.

Arabic engineers also made water clocks like those of their predecessors. In the more elaborate Islamic clocks, when the hour passed, balls fell into bronze vessels, trumpeters tooted, drummers drummed, cymbalists clashed their instruments, and lights appeared and vanished. For the astrologically inclined, indicators showed in what signs of the zodiac the sun and moon currently dwelt. The jack work was usually operated by strings, pulleys, and weights attached to a float that rose or sank.

Gearing, long known but rarely used because of the difficulty of making well-fitting gears, now became common. The learned Iranian al-Birûni (+XI) described a geared astrolabe by which the phases of the moon could be determined for any desired date by turning the main shaft. A similar device, made in +XIII by another Iranian engineer and looking something like a large brass watch, is preserved in the Museum of the History of Science at Oxford. And in 1232, Sultan Saladin[20] of Egypt, who gallantly led the Saracen host against the warriors of the Third Crusade, sent the Emperor Frederick II a gift of an orrery showing the movements of the sun, the moon, and the planets.

Al-Jazari, a leading mechanician and clockmaker of +XIII, introduced a water wheel as part of his striking mechanism. Water, poured into the buckets of the wheel at striking times, made the wheel turn so as to trip lugs and move the jacks.

The idea of perpetual motion arose in Islamic times in connection with water wheels and water clocks. Arabic and Indian experimenters mounted closed tubes, partly filled with water or mercury, slantwise around the rim of a wheel. They hoped that the liquid would run to the outer ends of the tubes on one side of the wheel and back to the inner ends on the other, so that the side on which the liquid was farthest from the center would overbalance the other side and provide free power.

So began another of man's many snark hunts. For centuries, men tinkered with wheels, levers, pulleys, gears, pumps, and magnets, trying to get a perpetual-motion machine that would work. Of course none ever did. The poor experimenters thought it was just a matter of calculation and workmanship; they did not realize that they were up against a law of nature.

This idea traveled to Europe along with the rest of Arabic science. The notebook of the medieval French craftsman Villard de Honnecourt (+XIII) has a sketch of a perpetual-motion machine, consisting of a wheel with a circle of mallets pivoted round the rim. Not until +XIX, with the discovery of the law of conservation of energy, did the hopelessness of this quest transpire.

Arabic millwrights actively developed water power. They floated barges on the Tigris on which they mounted mills of various kinds, driven by undershot wheels.

The first practical use of the windmill also took place in the Caliphate. In the vast dry plains of Seistan, in northeastern Iran, a tremendous gale blows from the same direction for months at a time. Here, a thousand years ago, the dwellers in this forbidding land set up batteries of windmills to grind their grain. Like Iraq, Seistan was a populous land until the Mongols destroyed the irrigation system. Timur the Tatar devastated the country in +XIV in revenge for a wound he received there, and after that the land returned to desert and swamp.

These windmills, some of which are still in use, are quite different from those we are used to. Instead of a horizontal shaft like that of an electric fan, the Persian windmill has a vertical shaft like that of a revolving door. To make the wind blow on the vanes of one side only, the Seistanis put a wall of brick around half the structure, leaving an open space for the wind to blow through the other half. This scheme works only so long as the wind blows consistently in one direction.

Knowledge of the vertical-shaft windmill traveled to China, where the Chinese improved upon it. In the Chinese version, the shaft is also vertical. A bamboo framework carries a circular array of slatted sails, like those of a junk, so pivoted that they catch the wind from any direction.

Arabic science fell into decline in +XII and +XIII, and Arabic engineering declined along with it. The most obvious reason for the stoppage of effective Muslim scientific work was the Turkish and Mongol invasions.

The Turks, newly Islamized Mongoloid shepherds from the steppes of Central Asia, filtered into the Islamic world from +IX on. First used as mercenary soldiers, they soon seized power in the Caliphate, much as the Germans seized the rule of the West Roman Empire. By 1100, although a Khalîfah still reigned in Baghdad, the real power in Islam was wielded by the Turkish sultans.

These were no longer Turks in a strict ethnic sense. As a result of the Turks' boundless appetite for the women of the conquered peoples, their original Mongoloid racial type disappeared by dilution in a few generations. Suleyman the Magnificent (+XVI), the most famous of all their sultans, was a perfect example of the Armenoid type, which has predominated in Anatolia ever since Hittite days.

The Turks excelled in the martial virtues. Even their foes the Crusaders found them brave, chivalrous, and sometimes even honest. But they had the defects of their virtues. They had little imagination and little interest in technical and scientific matters. These were a people who destroyed much but created little. They looked upon those they subdued much as they looked upon their sheep and camels. They built, or commanded their subjects to build, some fine mosques and palaces, but they invented nothing.

The only technical field in which the Turks showed a progressive spirit was that of warfare. In this art, for a couple of centuries, they led the world. In +XV, when European armies still relied upon armored knights, clattering about on plow horses, the Turkish armies included companies of musketeers, clad in uniforms, marching in step, and firing by volleys. In 1453 Muḥammad II, the Othmanli sultan, took Constantinople with the world's most powerful artillery, including the biggest cannon of the age: "A measure of twelve palms is assigned to the bore; and the stone bullet weighed above six hundred pounds."[21] Although these measurements are not consistent and therefore cannot be accurate, it was still quite a gun. The Turks' supremacy in war enabled them to swallow up the Balkan peninsula and twice besiege Vienna, until the Europeans caught up with them in the technique of organized bloodshed.

The effect of the Turkish invasions on the civilization of Islam was compounded by the Crusades and by the outburst in +XIII of the Turks' kinsmen, the Mongols. Both Christian and Mongol invasions, especially the latter, were highly destructive. If the Turks looked upon those they conquered as domestic animals to be exploited, the Mongols looked upon them as nuisances to be extirpated.

The Mongols erased the strongest Muslim power of the time, the kingdom of Khuwaresm in Iran and Turkestan. Northern Iran, which had given Islam some of its greatest scientists, was almost depopulated. Jengis Khan ordered his soldiers, when they had captured a Khuwaresmian city, to cut off and stack the heads of all the corpses, lest any human being should escape by shamming death. Baghdad was razed, the books of the great libraries of that city thrown into the Tigris, and Iraq turned into a waste.

In +XIV another of the same breed, Timur the Tatar, repeated the performance, dotting Asia with pyramids of human heads. After taking Ispahan he made one of 70,000.

As a result of these devastations, Islam underwent another change: a return to religion. Many caliphs had been indifferent Muslims–skeptics, materialists, winebibbers. In view of the disasters that had fallen upon Islam, the time was ripe for a return to the true faith.

In +XII and +XIII lived two of the greatest philosophers of the age. The first was an Iranian, al-Ghazzâli; the second, a Neapolitan, Tomaso d'Aquino or Saint Thomas Aquinas.

The pious and learned Saint Thomas (1225–74) spent much of his life arguing, at enormous length and in tiny illegible handwriting, that there was no conflict between science and religion; that all truth was one, and that therefore Aristotle's logic must fit the Christian faith. In fact, Saint Thomas promoted Aristotle to a kind of pre-Christian saint.

The pious and learned Ghazzâli (1058–1111) also studied the science and philosophy of the Greeks but came to different conclusions. After mature and searching consideration, he decided that these studies were harmful, because they shook men's faith in God and undermined religion; "they lead to loss of belief in the origin of the world and in the creator."[22]

Europe followed Saint Thomas, while Islam followed Ghazzâli. For example, in 1150 the Khalîfah of the moment proved his piety by burning the books of a philosophical library of Baghdad. As a result of these diverging trends, science and technology flowered in Europe so richly and advanced so swiftly that the rest of the world is still breathlessly trying to catch up. On the other hand, science in Islam withered away.

The real irony is that Ghazzâli was right and Saint Thomas wrong. Science *does* shake men's faith in God and undermine religion. It has been doing so for many years and shows every sign of continuing to do so. As to how it will all end, and whether this is a good thing or a bad thing, only our remote descendants–if any–will be able to say.

At the sunrise end of the Old World lies China, bounded by the China Sea, the jungles of southeast Asia, the high mountainous plateau of Tibet, the Gobi Desert, and the once-forested Manchurian plains. The jungle, plateau, and desert cut off China from the rest of the Main Civilized Belt—except for a slender tentacle of territory where travel was possible.

This strip of traveled land begins in China's rugged but fertile northwest, in the province of Kansu. Thence it runs westward, between the Tibetan plateau to the south and the Mongolian desert to the north. Precariously skirting the feet of the Tibetan mountains, it creeps west to the oval Tarim Basin, larger than Texas, where the Tarim River, like a sluggish snake, winds across a howling waste to lose itself in a cluster of reedy lakes.

Some travelers skirted the basin to the north, along the bases of the Tien Shan Mountains, while others went south along the Kunluns. After these routes rejoined at Kashgar, the travelers crept over the passes of the Pamirs. Winding between the Afghan peaks and the Turkmen desert, they straggled down into Iran, whence the way lay clear to the south and west. It was a rugged route but, once men had the camel, not impassable save when the tides of war washed over it or it shriveled in the flames of brigandage.

East of the Tarim Basin, from early times, dwelt a vast population of Tibetan mountaineers, Mongolian nomads, and Chinese peasants. All were of the same stocky, flat-faced, yellow-skinned Mongoloid race. All, probably, had once come from Siberia, for the Mongoloid racial type is adapted to cold, just as the Negroid type is to heat.

Although connected by the slender umbilical cord of the Tarim Basin route with the other nations of the Main Civilized Belt, the East Asiatics were also isolated by the difficulty of the route. They were even more isolated than the people of India were by the mountains and deserts west of India. Hence it is no surprise that Chinese civilization arose later than the civilizations of the Near and Middle East.

Once it had arisen, however, there was continuous traffic over the Tarim Basin route, from China to India, Iran, and the West. But, because of the length and rigor of this journey, the route itself acted as a filter. Things passed along it more easily than thoughts.

Traders carried goods back and forth, but seldom did any single trader make the entire journey. Rather, each had a beat of a few hundred miles. Back and forth he plied this beat, picking up goods at each end and reselling them at the other, but never getting more than a few days' ride from his home. Thus an artifact might be passed from hand to hand from Anyang to Babylon, but knowledge had a much harder time in making the trip.

To judge from archeological remains, the first stirrings of Chinese civilization took place along this trade route. In Kansu a prehistoric

Chinese culture, comparable to that of early Sumer, blossomed around –2000. When trustworthy Chinese history begins, however, the scene has shifted eastward a thousand miles, to northern Honan. Here, around –1500, the Shang emperors of the first historical dynasty reigned from Anyang.

Before trustworthy Chinese history comes a lot of pseudo-history. This is a mass of legend in which a few historical facts may be embedded, as in the tales of Moses, Achilles, and King Arthur. We cannot, however, strain the mixture to separate the facts from the fiction.

Thus the Chinese historians tell of the emperor Fuhi, who invented marriage, the calendar, and the domestication of animals; of Shênnung, who invented agriculture; of Hwangdi,[23] who built roads and invented the oxcart, while his empress founded the silk industry and his ministers invented well-digging, clothes, arithmetic, and other useful things. We read that in the reign of Yau,[24] a great flood caused the waters of the Hwang-ho–the Yellow River–to mingle with those of the Yangtze, making a great inland sea. The eminent engineer Yü spent many years bringing it under control and later became emperor in his turn.

Actually, these people are no more real than were Osiris, Prometheus, and Noah. But, with the rise of the Shang, real history comes to light. Many or all of the Shang emperors may have been living mortal men. China then had a well-developed bronze-age civilization, with artistic bronze vessels and picture writing on bones and strips of bamboo.

These early emperors did not rule anything like the area of modern China. They ruled only what we should call the north-central part, within a few hundred miles of Anyang. Their realm was the watershed of the Hwang-ho.

Nor was their power so absolute as it later became. The early emperors seem to have been more like high priests, representing the people in dealings with the gods, than secular rulers. Most political power lay in the hands of local kings and lords. Nevertheless, when an emperor was buried, hundreds of people were sacrificed and buried with him.

In –1122 the Jou[25] dynasty succeeded that of Shang and continued until –220. During the last three centuries of this dynasty, however, the central power became so weak and the lesser rulers so strong and quarrelsome that the time is called the Age of Contending States.

In late Jou times, trade over the Tarim Basin route became brisk. Traders learned that westerners would pay its weight in gold for Chinese silk. Silk was already manufactured on the Aegean isle of Kôs from the cocoons of the local, wild silk moth, and Aristotle knew whence came this filmy textile. But, when the superior Chinese silk appeared in the

Mediterranean in Hellenistic times, Greek silkworm culture languished.

It was supposed that Chinese silk had some origin different from that of Mediterranean silk. Perhaps, thought the Greeks, it was a plant fiber. The question was not settled until +VI, when a pair of Iranian monks smuggled silkworm eggs from China to Constantinople in their walking sticks.

The silk trade, which made the Tarim Basin route into the "Silk Route," stimulated traffic in ideas as well as in material things. Buddhism traveled the route from western India to China. Chinese iron was brought to the West. Later, the inventions of paper, printing, and gunpowder traversed the same route to westward.

Several themes run through Chinese history. One is an alternation between periods of strong central power on one hand and of feudal disorder on the other. Nevertheless, even in times of invasion and anarchy, there was usually an emperor, or somebody who claimed to be such. This continued from Shang times right down to the "Republic" (which turned out to be just another clutch of Contending States) of 1912.

Hence there is a continuity and unity to the history of China over 3,500 years that is not found in any other civilization for any such period. There is no such sharp distinction in China between "ancient" and "modern" as we are used to in European history.

Another Chinese theme is periodical invasion and conquest by the nomads of Mongolia. These, though racially much like the Chinese, spoke languages of another group, the Ural-Altaic. They had adopted the horse-riding, stock-herding, nomadic way of life from their Aryan neighbors, such as the Sakas, to the west. As one or another tribe–Huns, Avars, Turks, Uighurs, Tatars, Mongols, Uzbeks–rose to brief supremacy, most of the nomads for a time went by the name of the dominant tribe. But the distinctions among them were never hard and fast.

Possessing the extreme hardihood, valor, and practicality that nomadic life demands, these people invaded China whenever that country was weak and made themselves lords of large parts of it. In time they were either driven out or absorbed.

Another Chinese theme is the internal political conflict between a progressive or radical faction, interested in foreign explorations and adventures and in scientific and engineering enterprises, and a conservative faction wishing to restore the ceremonies and alleged virtues of former times.

Despite the many revolutions, invasions, and times of anarchy that have befallen it, Chinese civilization has remained remarkably true to the features it developed in early days. These culture traits include its emphasis on courtesy and ceremony, its callous indifference to human life, its deep-rooted xenophobia, and its skeptical, rational, practical bent. The sheer size of the land and of its population, its remoteness from other great powers, the difficulty of mastering its written language, and its centralized bureaucratic government have given it powers of resistance to foreign influence and of absorption of foreign invaders unmatched elsewhere in the world.

The materialistic Chinese practicality, which contrasts sharply with Indian mysticism, governs the history of Chinese science and engineering. Chinese science, like that of the Arabs, has always been weak in theory and strong in practice. Before the European scientific revolution, China was one of the world's main sources of ingenious inventions. For instance, Jugo Lyang,[26] a general and inventor of the time of the Three Kingdoms (+III), is said to have invented that valuable if humble vehicle, the wheelbarrow. However, because of the semi-isolation of China, these inventions have sometimes taken many centuries to reach the West; it took the wheelbarrow a thousand years.

Although China has sometimes led the world in technology, she has usually lagged in pure science. One reason is that the two leading Chinese schools of philosophy have been anti-scientific.

The most important school was Confucianism. For thousands of years this was the official ideology, which aspirants to the well-organized civil service–another Chinese invention–had to master. Confucianism stressed ethics, morals, filial piety, and etiquette. Like Sokrates, its founder Kung Fu-dz[27] (–VI) or Confucius scorned science in favor of relationships among people.

The next most important school, Dauism or Taoism, affected to admire nature. But the Dauists admired nature in a romantic, mystical way. The founder Lau-dz,[28] an older contemporary of Kung Fu-dz, taught that man should adapt himself to the universe. Lau-dz never thought of adapting the universe to man.

Another handicap to science in China was the nature of the language. This tongue is very odd indeed. The classical or literary form of the language is made up of comparatively few sounds, and these may be combined in only a limited number of ways. Only 412 different syllables are possible.

Moreover, another rule of the language was: One syllable per word. This meant an absurdly small vocabulary. The use of different tones to distinguish words otherwise identical in sound enlarges the list of possible vocables to 1,280, but this is still a ridiculously small number for a civilized tongue.

As a result, any one syllable may have scores of meanings. To distinguish these meanings, the Chinese use a system of compounding. It is somewhat as if we had only the one word "cat" for all members of the cat family and had to distinguish the lion, tiger, cheetah, and pussycat as king-cat, stripe-cat, dog-cat, and house-cat. All languages do this to some extent, but none to the degree that Chinese does. In the spoken language such compounds are tending to become permanent, forming polysyllabic words; but not so in the literary form. The language is therefore ill-adapted to scientific thought, which needs a large vocabulary capable of minute distinctions.

Turning to Chinese engineering, let us first consider the art of building in China. Since China has always been a large and varied country, with good internal communications and a rich variety of building materials, Chinese builders were never limited to one or two materials, in the way that the Mesopotamians were confined to brick.

Hence the Chinese early adopted a composite style of architecture, with stone foundations, wooden superstructures, tile roofs, and brick here and there. Even when a house has walls of stone or brick, the weight of its roof is still carried, not by the walls, but by wooden columns.

The Chinese also knew the barrel vault from early times. But, despite their skillful use of wood, they never discovered the truss. Instead, builders held up their roofs by a complex system of posts, lintels, and brackets.

The larger Chinese private houses have for thousands of years been built on the inner-courtyard pattern of the Near East. In pretentious projects like palaces, buildings of modest size are scattered about an enormous complex of terraces, courtyards, and monumental stairways.

A distinctive feature of Chinese building is the use of a curved roof, presenting a concave surface to the sky. In the Han dynasty, which flourished at the time of Christ, Chinese roofs were straight as they are in most places. Some time later, they became curved. The roof corners are often given an additional upward sweep, like the toe of a Turkish slipper. The use of several roofs, one above the other, is characteristic of East Asian building.

Regional styles evolved in China through the centuries: comparatively simple in the North, floridly ornate in the South. The peculiar richness of Chinese building styles comes from the lavish use of lacquer, gilding, and colored tiles. A distinctive type of Chinese building is the *ta* or pagoda, which evolved from the Buddhist stupa. This is a memorial tower, usually attached to a temple. It has a number of stories, each with a roof running around it.

The Chinese also built bridges of the common primitive types: floating, beam, and suspension. Later they developed more sophisticated bridges. The exact history of Chinese bridgebuilding, however, is hard to follow, because the existing old bridges have been repaired and replaced so often that we cannot tell much about their original forms.

However, the Chinese did develop some original methods of bridgebuilding. They hung suspension bridges from cables made of bamboo fiber, a strong and durable substance which they also used for ship's cordage and towropes. Bamboo is a two-phase material, made of cellulose fibers imbedded in the woody substance lignin. The Chinese buried bamboo stems until the lignin rotted away and then dug up the fibers to weave into ropes and cables. After the Buddhist monk Hsüan Dzang[29] returned from India (+VIII) the Chinese began building suspension bridges upheld by iron chains, like those of India.

The Chinese also built curious wooden cantilever bridges, in which beams projecting out from stone piers act as brackets to support the deck. And they built many stone arched bridges, with marble parapets and wooden roofs. They anticipated the European invention of the segmental-arch bridge, as in Li Chun's +VI bridge on the river Xiao, which still stands.

After China had long been convulsed by wars among the Contending States, the ambitious king of Tsin, Tsin Jêng,[30] conquered all the other Contending States, chopped off the head of anybody who showed a desire to go on contending, and about –220 took the name of Tsin Shï Hwang-di.[31] We may translate this with tongue only slightly in cheek as "Mr. Tsin, the First Divine Autocrat."

The Chinese soon found that they had exchanged King Log for King Stork with a vengeance. Tsin, China's Rameses II, gave that nation its first strong, centralized, autocratic rule. He dotted the land with 60-ton statues of himself, standardized weights and measures, and slew his grandparents when they opposed him. He conquered China south of the Yangtze, where dwelt a number of comparatively backward peoples. He sent a fleet to look for a rumored fairyland, the Isles of the Genii; the explorers found Japan and settled there, glad to escape Tsin's energetic tyranny.

Naturally, these drastic measures aroused mixed feelings. Conservatives bitterly opposed them. To thrash the matter out, Tsin called a meeting of his chief officials and leading scholars. After one of the scholars had criticized Tsin's innovations as severely as he dared, the chancellor Li-sz[32] spoke:

> Beware these idling scholars! Bred on the past with senseless veneration of everything that is old, they cannot appreciate anything fresh. If you issue an edict, they criticize its language; if you order a new project, they declare it is unprecedented. Their one test is, has it been done before? They go about sowing unrest and sedition among your subjects. Their influence must be broken if the empire is to prosper. It is founded on books; destroy then the books. Their occupation will be gone, and none can arise to succeed their generation of them. Some books of course there are which are of value. Preserve all that relate to medicine, husbandry, and divination; preserve also the records of this illustrious reign. Let all else be destroyed, break with the past . . . with natural science, religion, medicine, and law be content, and let the mere literary classics cease to curse the land![33]

Tsin so ordered. While the ruthless ruler could not make a clean sweep of all the literature in the empire, he came close to it. Because paper had not yet been invented, books were still written on strips of wood and bamboo. And because these strips were awkward and bulky, like Ashurbanipal's clay tablets, books were far rarer than they later became. When horrified scholars protested Tsin's book-burning, he had 500 of these reactionaries buried alive.

But Tsin Shï Hwang-di also created a more lasting monument than his repute as a book burner. It was, in terms of sheer size, the greatest single engineering work of antiquity. Alone of the works of man, it could be seen from the moon. It is the Great Wall of China, a greater wonder than any of the Seven Wonders of the Greek writers.

When he came to supreme power, Tsin disbanded the many little feudal armies of China. To cope with raids by the Mongolian nomads, known at this time as Huns, he commanded the building of a wall clear across the northern boundary of China, from the sea to the far northwest.

And so it was done. Under the direction of General Mung Tien, hundreds of thousands of workers—conscripted artisans, soldiers, criminals, and paupers—were rounded up. Their first step was to lay out farms along the course of the wall and get crops started so that the army of builders should have a secure supply of food.

The construction of the wall varied in different sections, according to the local materials available. Parts of the old walls built by kings of the Contending States against each other and against the Huns were incorporated where possible. In most places, the wall was about 30 feet high and 25 feet thick at the base, tapering to 15 feet at the crest. A paved roadway ran along the top. On the northern side of this roadway rose a 6-foot crenelated parapet; a 3-foot parapet protected the southern side.

In most sections, the core of the wall was rammed earth or rubble, while the sheathing was of harder material–ashlar masonry, rough field-stone, or brick and mortar. The roadway was paved with slabs of stone or brick. In the sections sheathed in squared stone, the masonry is of excellent workmanship. Naturally! Tsin ruled that a workman who left a crack between stones wide enough to stick a nail into should be instantly beheaded.

At intervals of about 250 yards, watch towers, averaging 35 feet square and 45 feet high, straddled the roadway on top of the wall. Gate towers, to control traffic through the wall, were built at longer intervals. The road atop the wall was a practical route for marching soldiers, though the steep grades and the narrow arched openings in the towers made it impractical for the ponderous Chinese chariots.

Additional watch towers and beacon towers were set up along the route, separate from the wall itself. There were altogether 25,000 towers straddling the wall and 15,000 detached towers. North of the wall, a zone 200 yards wide was fortified by a triple ditch with obstacles on the embankments. Where practical, these ditches were filled with water.

Like a colossal caterpillar, the wall stretched over mountain and vale for a crow's-flight distance of 1,400 miles, so that if its eastern end were in Philadelphia the western would be in Kansas. Counting all the bends, branches, and loops, it totals about 2,550 miles, farther than from New York to Reno. In the eastern parts, north and west of Peking, the wall forked and rejoined to make a loop, and there was a long southward extension. Near the western end, a double loop and branch guarded the Kansu-Shensi region not only against the Huns to the north but also against the Tibetans to the southwest.

The general form of the wall was much like that of the standard Chinese city wall. The design of the wall represented no special engineering advance, except for the scale on which it was built and the way it functioned. Tsin, you see, had a theory about his wall.

The wall was not high enough or thick enough to halt invaders all by itself. Nor was it possible to station enough soldiers along its entire length to stop all nomadic raids. The Huns could always rush a section of wall, drive the local defenders to cover by massed archery, and get over the wall with scaling ladders.

But, while a man can climb a ladder, a horse cannot. Therefore the raiders had to leave their horses on the northern side of the wall and go clumping about the Chinese countryside in their heavy boots. Without their horses they lost most of their fearsomeness and were easily attacked by the local garrisons, especially when they were trying to get back over the wall with their booty.

A large Hunnish army, by bringing up battering rams to knock down a section of wall or by besieging one of the gate towers, could get its horses through. But these tasks took time, during which larger Chinese armies could assemble near the threatened point.

Tsin's theory proved successful. The Huns, repulsed from China, surged westward, driving other nomadic peoples before them. Among the fugitives were the Goths. These Germanic folk, originally from Sweden, had set up a short-lived Russian empire, being the dominant power on the steppes until elbowed out by the Huns. In their retreat, the Goths forced their way into the Roman Empire, to whose fall Tsin thus indirectly contributed.

Of course, if the Chinese were weak enough and the nomads strong enough, or if the nomads were led by a military genius like Jengis Khan, the wall did little good. The Jurchen Tatars proved this in +XII by conquering the northern half of China and setting up the Gin[34] dynasty. A century later the Mongols repeated the feat, conquering not only the Gin but also the native Chinese dynasty of Sung in the south as well.

In the ages after the wall was built, weak emperors and fragmented empires let the wall decay, while strong emperors repaired it. Therefore it is hard to say how old the various parts of the existing work are. Long stretches in the West are in ruinous shape, since the peasants have taken away the sheathing of stone and brick and the remaining earthen core has eroded down to a mere ridge.

In the East, however, much of the wall is still in good condition. Although the battlements crumble here and there, weeds grow up between the stones, and little gray lizards scuttle over the masonry, the wall would still be an effective defense against sheepskin-clad, fur-capped archers on shaggy ponies.

Tsin Shï Hwang-di did not succeed in founding a lasting dynasty. When he died in −210, his son Tsin Êr-shï[35] promptly sliced the heads off his twelve brothers, ten sisters, and all other relatives who might conceivably covet the throne. But his ability did not match his ruthlessness. Revolts broke out. Aspirants fought the armies of Tsin and each other. Eight years later, a general named Lyou Bang[36] ascended the throne as the first emperor of the Han dynasty.

Meanwhile the First Divine Autocrat had been interred under the sacred Mount Tai, where, says folklore, his body floats on a pond of mercury to foil tomb robbers. Later Chinese historians portrayed him as a villain, which he was. But so are most autocrats; they have to be. In any case, Tsin ranks as one of the greatest builders of history; and his wall was not only a grandiose idea but, in its time, an outstanding success.

For thousands of years, China has been a land of canals. Chinese officials took much the same pride in the canals they dug that Roman patricians felt in the roads and aqueducts they built.

Most of these canals were of modest size, suitable for irrigation but navigable only to a limited extent. Although Chinese engineers led canals down slopes through a series of sluice gates, they never discovered the true canal lock. Today, as a result of 3,000 years of diligent digging, there are about 200,000 miles of canals in China, the greater part in the lower basins of the Hwang-ho and Yangtze rivers.

The foremost canal of China is the Grand Canal, the Yün-ho, which runs 1,200 miles north and south through the eastern part of the land, from Tientsin to Hangchow. No one emperor built it; it is the result of engineering labors over more than a thousand years.

The Yün-ho was built piecemeal. Some parts probably go back to Jou times, before –III. The main work of linking the parts was done by two emperors of the Swei dynasty, who reigned around +600.[37] A book of later times tells us that 5,430,000 workers were conscripted for the work and guarded by 50,000 policemen. Shirkers were beheaded whenever caught. Two million perished in the course of construction.

These numbers have probably been inflated like those of Xerxes' army, but the picture of masses of men toiling with simple tools under ferocious discipline is probably true enough. In any case, the Swei emperors aroused so much opposition by their vast public works that in +617 the dynasty fell and was replaced by the Tang.

Subsequently, the great Kublai Khan not only renovated the Grand Canal but also built a branch to his capital of Khanbaliq (Marco Polo's Cambalu) near Peking. The finished canal was 40 paces wide. Roads ran along its banks, which were planted with willows and elms. Polo noted that it was "navigated by so many vessels that the number might seem incredible. . . . It is indeed surprising to observe the multitude and size of the vessels that are constantly passing and repassing, laden with merchandise of the greatest value."[38]

Knowledge of iron reached China from the West about –700, though for a time it was used only for tools and implements. Bronze remained the standard metal for weapons.

Then, however, Chinese iron-working technique swiftly improved. By –IV the Chinese had discovered how to make cast iron. In Imperial Roman times China, like India, became an exporter of high-grade iron to the West. From +X to +XV the Chinese made pagodas entirely of cast iron, of which two or three still exist. In +X they cast an iron statue of a lion 20 feet high, which still stands at Tsang-jou. The Chinese led the more westerly nations in the development of iron-working machinery, such as a forge bellows in the form of a double-acting piston, which appeared some time in the first millennium of the Christian era.

The most portentous Chinese inventions, however, had to do with writing. In the latter years of +I the eunuch Tsai Lun[39] became a privy councilor of the emperor Ho-di and later inspector of public works. An official history written some centuries later explained:

> In ancient times writing was generally on bamboo or on pieces of silk, which were then called *ji*.[40] But silk being expensive and bamboo heavy, these two materials were not convenient. Then Tsai Lun thought of using tree bark, hemp, rags, and fish nets. In +105 he made a report to the emperor on the process of paper making, and received high praise for his ability. From this time paper has been in use everywhere and is called the "paper of Marquis Tsai."[41]

For years the mortar in which Tsai Lun was supposed to have mashed his rags and fish nets was kept in an imperial museum. In actual fact, the invention of paper–a thin felt of fibrous materials–may have been a more gradual process than the story indicates. Several men in succession probably contributed to the process, as has been the case with most major inventions.

Subsequently Tsai got involved in intrigues among the emperor's womenfolk. This was an occupational hazard with imperial eunuchs. When commanded to appear before judges for examination, Tsai "went home, took a bath, combed his hair, put on his best robes, and drank poison."[42]

The new invention appeared when Tsin's decrees against the classics had just been rescinded. It turned out that while much had been lost in the great burning, much also had been saved. Some books had been hidden away, while some old scholars had memorized parts of those that had been utterly destroyed. So the classics, with some gaps, were reconstituted.

The new invention, paper, spread slowly. It had reached Turkestan in +VIII, when the armies of Islam entered the land from the other side. Two Turkish tribes feuded; the Arabs took the side of one, the Chinese of the other. The forces of the Prophet defeated those of the Tang emperor at a mighty battle at the Talas River (+751) and drove the Chinese out of Turkestan.

The Arabs, however, captured some Chinese paper makers and brought them to Samarqand to practice their trade. By +793 the art had reached Baghdad; by +900, Cairo; by 1100, Morocco; and by 1150, Spain. During +XIII and +XIV paper spread over France, Italy, and Germany, ousting papyrus altogether and relegating parchment to official documents.

The invention of printing, which followed that of paper and was dependent upon it, is not so simple and straightforward a tale. Printing in the modern sense grew out of a number of different processes, and we cannot tell which of these processes contributed the most.

For example, people had been stamping seals on documents for thousands of years. The seal might be made of stone, brick, wood, or metal; it might make an impression in clay or wax. Babylonian kings had their names stamped on the bricks of their buildings. In China, about +V, a new method of impressing seals came into use. In this method, inked characters were impressed on paper by means of a wooden seal, as we do today with a rubber stamp.

Another process that may have contributed to printing was the block printing of patterns on textiles. This method was known in the Mediterranean in Roman times. If it did not originate there, it probably spread thither from India.

The supernatural also entered into the evolution of printing. The Chinese placed great faith in written charms and spells to avert misfortune. To make the charms more effective, the Chinese made up seals bearing the charms, which they impressed in great quantities, first on clay and then on slips of paper. It is possible though not certain that playing cards, another early application of printing, may have evolved from these charm prints.

When this practice of printing charms reached Japan, the Empress Suiko and her co-ruler, the regent Shotoku (+VIII) had one million charms printed. Each charm was put into a tiny wooden pagoda, and the pagodas with their charms were distributed among the temples of Japan to guard the kingdom from maleficent influences.

All these experiments with stamps and printed charms resulted in Chinese block printing. The oldest known printed book was found by Aurel Stein, the intrepid Hungaro-British explorer, in 1900. Stein learned of a sealed manuscript chamber in a Buddhist cave temple at Dun-hwang, in the northwest panhandle of Kansu. The chamber contained over 15,000 manuscripts (now distributed among British, French, and Chinese museums) including this printed version of a Buddhist scripture.

The printed book consists of seven large sheets (one with a woodcut picture) pasted together to form a roll or scroll. It ends with the statement (the date being translated into our calendar): "Printed on May 11, 868, by Wang Jye,[43] for free general distribution, in order in deep reverence to perpetuate the memory of his parents."

We have already seen that all ancient books on papyrus, parchment, or paper took the form of the roll. In +I the codex—a book made of separate sheets, glued or sewn together at one side to make a tome like a modern book—appeared in the Mediterranean and gradually took the place of the scroll. For one thing, it was less likely to be torn if the reader dropped it. Furthermore, turning from one part of the book to another was much easier with a codex than with a roll, of which the reader had to roll up one end while he unrolled the other.

The codex must have traveled eastward as paper went westward, for another book at the Dun-hwang temple was a codex, printed in +949. It was a codex of a transitional type. The text was written in the usual Chinese columns on one side of a long strip of paper. This strip was then folded zigzag, like an accordion. The folds at one side were pasted together, so that each page of the book was a double sheet of paper printed on the outer side.

During the time of disorder that followed the fall of the Tang in +907, printing books from wooden blocks became a well-established Chinese practice. Governments encouraged it, not to enable the masses to own books—an idea that never crossed the officials' minds and would have alarmed them if it had—but to make sure that the texts of the Confucian classics and other important works were reproduced with absolute accuracy.

We should not speak of a Chinese printing press, because no press was involved. The Chinese block printer holds a pair of brushes at opposite ends of the same handle. With one brush he inks the wooden block, on which the characters of a page have been carved in relief. With his other hand he lays the paper on the block and runs the dry brush over the paper to insure full contact. Then he peels off the paper. An expert can thus print 2,000 or more sheets a day.

A by-product of Chinese block printing was paper money. After some preliminary experiments, paper currency was put on a regular basis by the local government of the kingdom of Shu in +X. When the Sung dynasty reunited the country, this practice continued. It worked fine as long as the central government was strong, wealthy, and prudent enough to keep its notes at par.

When hard times and Tatar invasions around 1100 put a strain on the national finances, however, the inevitable happened. Finding it much easier to spend printed money than to extract it from the people by taxes, the government allowed the number of notes in circulation to rise steeply. Hence the value of this paper sank out of sight. China underwent all the hardships of runaway inflation until the Mongols conquered the country (1234–60). The Mongol or Yüan dynasty issued its own paper currency, which likewise held its value until the dynasty got into trouble.

Paper money reached Iran in +XIII and led to the usual inflation. This hazard has persisted ever since in every country that has adopted this convenient but treacherous invention.

A Chinese, Bi Shêng[44] (+XI), invented movable type: that is, type in which each character forms a separate piece, which pieces can be assembled to print a page and then broken up and redistributed until they are needed again. Bi used clay types glued to an iron plate. Later, types of tin and wood were found to be better. In +XIV the Koreans developed the art of printing from movable cast-bronze type.

However, movable type proved less important to printing in the Far East than it did in Europe. The reason is the nature of the Chinese written language. As in early Egypt and Sumeria, this script began as a series of pictures, one picture per word. In China, however, these pictures never evolved into a phonetic alphabet. Although conventionalized and compounded to make up the familiar Chinese characters, they have remained in the state of logographs (word-signs) or ideographs (idea-signs), a different sign being needed for each word.

Knowledge of printing probably did not come through the Arabic world, which transmitted so much else. The Muslims never took to printing, even when they knew about it. The only things they ever printed in quantity down to modern times were charms and sacred texts believed to act as charms, as the Chinese had done. Being in the state of mindless conservatism that followed the Arabic age of science, they rejected with horror the idea of printing the holy Qur'ân. It had always been handwritten, they said, and therefore it would be impious to reproduce it otherwise.

What was brought from China to Europe was the idea of block printing. The idea of movable type probably did not make the journey, because the first European printing was block printing. In early +XV, enterprising Europeans began to do what their opposite numbers in China and Islam had done. In response to a demand for cheap sacred pictures and texts believed to act as magical charms, they printed and sold single sheets. The existing screw press, then used on clothes and grapes, was adapted to this new use.

About the middle of the century, several Europeans were experimenting with movable type, of wood and of metal. One was a Dutchman, Laurens Janszoon Coster.[45] Another was Johann Gutenberg of Mainz. Although Gutenberg was in partnership at various times with several people in Mainz, these partnerships usually broke up in lawsuits, because Gutenberg seems to have been a litigious man.

With the backing of his partner Fust, Gutenberg in 1456 printed a tremendous new edition of the Bible. This was as big an advance over previously printed books as Fulton's steamboat was over its predecessors.[46] Little is known of Gutenberg's later life, except that the archbishop of Mainz pensioned him during his last years.

Because of lack of records and Gutenberg's own secrecy, we do not know exactly what he invented and how the credit for the different steps in the invention should be apportioned. The only direct account of these events appeared in the *Chronicle of Cologne* for 1499, half a century after they happened. Although the *Chronicle* calls Gutenberg the "inventor of printing," he certainly was not. Some unknown Chinese had beaten him to it back in +IX.

Some think that Gutenberg invented movable type, independently of Bi Shêng. But some credit this invention to Coster, while others deny it to both. Some think that Gutenberg's invention was the adjustable type mold; others, that he was the first to adapt the screw press to its new task.

In any case, printing, like the steam engine, had no single inventor. It was developed by a long series of inventors, known and unknown, over many centuries. There is no question, however, that it is one of the most important of all inventions in its contribution to civilization.

The Chinese possessed geared machinery from an early age—some think as far back as –IV. They readily took to the devices developed in the West in the Hellenistic Age and in +I began using water wheels and norias. Chinese engineers preferred the horizontal water wheel to the vertical. The earliest known application of the water wheel in China was to work a bellows in the furnace of a forge.

Water clocks of the Hellenistic type reached China as early as +VI. Chinese mechanicians elaborated them. Since the wheels of these complex clocks required more power to drive them than could easily be furnished by a float rising or falling in a vessel, the Chinese engineers drove their clocks by means of water wheels.

The biggest Chinese step in clockwork was the escapement. This mechanism makes it possible closely to regulate the speed of a clock and to drive it with a comparatively small power source. The first known clock with an escapement was built about +724 by Lyang Lingdzan.[47] This apparatus included a celestial sphere that turned with the heavens, a model sun and moon that went around the sphere as the real ones seem to do about the earth, and jacks that struck bells and beat drums to mark the passage of time.

The bell of Lyang's clock marked the Chinese "hour" or *shï,* which is twice the length of one of ours. The drum sounded a shorter period, the *ko*. This is $\frac{1}{100}$ of a solar day, or 14 minutes and 24 seconds on our time scale. Like more westerly peoples, the Chinese originally divided day and night into intervals, which stretched and shrank with the seasons. Later, about +II, the Chinese adopted a system of equal, permanent periods that stayed the same regardless of the wanderings of dawn and sunset. This change made clock-making easier.

In Lyang's clock, "Water, flowing [into scoops], turned a wheel automatically, rotating it one complete revolution in one day and night." The machinery of the clock included "wheels and shafts, hooks, pins and interlocking rods, stopping devices and locks checking mutually."[48]

"Pins and interlocking rods" describes the escapement, which was needed to make the wheel revolve so slowly. The escapement was presumably a simple system of tripping lugs that held the water wheel against rotation until one scoop had been filled and then allowed it to move only far enough to bring the next scoop into the filling position. The water flowed from a system of vessels like that of the traditional water clock of the Ktesibian type.

Lyang's clock kept better time than anything seen before, although it would no doubt have seemed impossibly inaccurate by our standards. The delighted Emperor[49] composed a poem, which was inscribed on the ecliptic ring around the celestial sphere:

The moon in her waxing and waning is never at fault
Her twenty-eight stewards escort her and never go straying,
Here at last is a trustworthy mirror on earth
To show us the skies never-hastening and never-delaying.[50]

After Lyang's time, corrosion of the parts of bronze and iron put the clock out of action, and it was retired to a museum. Later mechanicians built grander clocks. In +976, Jang Sz-hsün[51] built a clock that occupied a pagodalike tower over 30 feet high. This had nineteen jacks, which not only rang bells and beat drums but also popped out of little doors holding signs to show the time. Other parts showed the movements of the heavens, the sun, the moon, and the planets. To keep his clock from being stopped by the freezing of the water in winter, Jang rebuilt it to use mercury instead of water as the working fluid.

The grandest of these imperial water clocks was built by Su Sung in 1090. Su Sung's memorial to the emperor Shên Dzung[52] describes his clock, with diagrams, so that if anybody wished to do so he could reconstruct the clock with fair accuracy today.

At this time the Sung dynasty ruled most of China, though a nomad tribe, the Kitans,[53] had conquered some of the northern provinces. Su Sung had a long career in the imperial bureaucracy. His eventual list of titles included: Official of the Second Titular Rank, President of the Ministry of Personnel, Imperial Tutor to the Crown Prince, Grand Protector of the Army, and Kai-gwo Marquis of Wu-gung.

When Su was sent on a mission to the Kitan court to congratulate the khan on having passed the winter solstice, he found that he had arrived a day too early. The Sung astronomers had erred by a quarter-hour in calculating the exact time of the solstice. Su saved his sovran's face by blandly lecturing on the difficulty of exactly calculating such events.

But, when Su returned to the Sung capital of Kaifêng,[54] he urged the Emperor to let him build a clock accurate enough to avoid such contretemps. Receiving approval, Su, like any competent engineer, built a couple of wooden pilot models, one small and one full-sized, to get the bugs out of the design before tackling the final clock.

The finished machine occupied a tower at least 35 feet high, counting the penthouse on top. Water, flowing through a series of vessels, filled the thirty-six scoops of a water wheel, one after the other. An escapement allowed the wheel to rotate, one scoop-interval at a time. The wheel revolved once in nine hours, while the water fell from the scoops into a basin below the wheel.

The wheel turned a wooden shaft in iron bearings. This shaft, by means of a crown gear, turned a long vertical shaft that worked all the rest of the machinery, to which it was connected by gearing. The machinery included an armillary sphere (a set of graduated intersecting rings corresponding to the horizon, the ecliptic, and the meridian) in the penthouse on top. There was also a celestial sphere, with pearls for stars, in the top story, and five large horizontal wheels bearing jacks.

Some of these jacks carried signs to indicate the *shï*, the *ko*, sunrise and sunset, and the watches of the night. Others marked these events by ringing bells and beating gongs and drums. Altogether Su's clock must have been an impressive spectacle, what with the continual splashing, the clatter of the escapement, the creak of the shafts in their bearings, and the frequent outbursts of drums, bells, and gongs.

One shortcoming of the clock was that it was not so placed that a natural source of water could power it. Hence it had to be "wound up" from time to time. This was accomplished by hand-turned water wheels, which raised the water from the basin, into which it cascaded from the scoops of the main water wheel, to the reservoir above this wheel.

In early +XII another Tatar people, the Jurchens, whose kings reigned under the dynastic name of Gin, conquered the Kitan lands and some Sung provinces as well. In 1126 they captured Kaifêng and carried away, to their own capital of Peking, Su's clock together with mechanics to run it. The captive horologists built a new tower and succeeded in making the clock run, after they had adjusted its astronomical parts for the change of latitude.

After a few years, though, the parts wore out, the clock stopped, and lightning wrecked the upper part of the tower. The Gin emperors left the remains of the clock behind when they fled before the Mongols in the 1260s, and it disappeared.

Meanwhile the Sung emperors wanted another imperial clock. But Su Sung was dead, and nobody could be found who knew the subject well enough to build such a complicated mechanism. Similar clocks were, however, built under the Mongol or Yüan dynasty.

Although he was an incapable ruler, the last Yüan emperor[55] made a hobby of mechanical engineering and took part in the construction of tail-wagging dragons and other automata. But when the Ming overthrew the Yüan in 1368, all the clocks, mechanical dragons, and other machines built for the Mongol emperors were scrapped as "useless extravagances."

Thereafter Chinese clockmaking languished. The Chinese went back to simpler clocks, powered by water or falling sand. In late +XVI, Jesuit missionaries arrived and began importing European clocks or building clocks according to European designs to please the emperors.

Meantime, however, knowledge of the escapement mechanism, pioneered by Lyang Lingdzan and Su Sung, reached Europe in early +XIV. As with printing, it is not known who brought this knowledge, or when, or in how much detail. Europe at this time relied on water clocks of the traditional Ktesibian type, as well as on sundials, hourglasses, and time candles.

But European mechanics were already tinkering with improved time-keeping mechanisms. One such improvement was to drive the clock by means of a falling weight suspended from a chain wrapped around a drum. This is an obvious development from the Ktesibian water clock, with its counterweight attached by a string and pulley to the rising float. To slow the descent of the weight they tried brakes of various kinds, such as a little windmill driven through gearing, or a small drum, like the classical tympanum, charged with mercury that leaked through a pinhole from one compartment into the next.

The Chinese escapement was just what the clockmakers needed, and in the early 1300s they put it to use. In 1364, Giovanni di Dondi, of an Italian clock-making family, published a description of a weight-powered, escapement-regulated clock which, except for improvements in detail, is essentially a modern clock. Dondi became famous, and astronomers came from foreign lands to look at his marvelous clock. Galileo later substituted a pendulum for Dondi's crown-shaped balance wheel, but in watches and small clocks we still use Dondi's device.

The main difference between our clocks and Dondi's masterpiece, aside from improvements in detail and workmanship, is that clocks have been drastically simplified since Dondi's time. Clockmakers gradually gave up trying to show not only the time but also the movements of all the heavenly bodies. Many grandfather clocks, however, still show the phases of the moon.

A few years after 1500, Peter Henlein of Nuremberg invented the spring-driven watch, so called because it was originally used by watchmen. Henlein's "Nuremberg egg" was about the size of a large modern alarm clock, had a single hand, and hung from a chain around the neck. Early watches gave their owners much difficulty; as Maximilian I of Bavaria (+XVIII) used to say: "If you want troubles, buy a watch."[56]

When the Mongol army arrived before Kaifêng–once the capital of the Sung but now that of the Gin–in 1232, the armies of the Gin checked the invincible Mongols for a while by secret weapons. One called "heaven-shaking thunder" was an iron bomb lowered by a chain from the city's walls to explode among the foe. The other, an early rocket called an "arrow of flying fire," whistled among the Mongols with much noise and smoke and stampeded their ponies.

Charcoal and sulfur had long been known as ingredients for incendiary mixtures. As early as 1044 the Chinese learned that saltpeter, added to such a mixture, made it fizz even more alarmingly. We do not know who first learned that if you grind charcoal, sulfur, and saltpeter up very fine, mix them very thoroughly in the proportion of 1:1:3.5 or 1:1:4, and pack the mixture into a closed container, it will, when ignited, explode with a delightful bang.

But somebody did so between 1044 and 1232. It has been suggested that experimenters, believing that salt made a fire hotter because it made it brighter, tried various salts until they stumbled on potassium nitrate or saltpeter. The mixture was applied both to fireworks and to the primitive military devices used at Kaifêng. An old Chinese custom was to throw bamboo stems into open fires so that the bamboo should burst with a loud bang and scare away evil spirits. Firecrackers made of the new powder proved an even better demon repellent.

The rocket probably evolved in a simple way from an incendiary arrow. If one wanted to make a fire arrow burn fiercely for several seconds, using the new powder, one would have to pack the powder in a long thin tube to keep it from going off all at once. It would also be necessary to let the flame and smoke escape from one end of the tube. But, if the tube were open at the front end, the reaction of the discharge would be in the direction opposite to the flight of the arrow and would make the missile tumble wildly. If the tube were open to the rear, on the other hand, the explosion would help the arrow on its way.

In fact, it was probably soon discovered that with a discharge to the rear the arrow did not even have to be shot from a bow. The forward pressure of the explosion, inside the tube, would move the device fast enough.

Although these new weapons failed to save Kaifêng, the lesson was not lost. It soon reached Japan. A Japanese woodcut of 1292 shows a brave archer loosing his bow while hostile shafts whistle about him and a big black bomb explodes nearby. In late +XIII, several writings appeared describing gunpowder and its properties. Strictly speaking, it was not yet "gunpowder" because the gun had not been invented.

For instance, at some date in +XIII an otherwise unknown Marchus Graecus, "Mark the Greek," wrote *Liber Ignium,* or *The Book of Fire.* Marchus told how to make explosive powder by a mixture of "one pound of live sulfur, two of charcoal and six of salt-peter." This would give a weak explosion. Albertus Magnus gave the same formula as Marchus, while Albertus' contemporary Roger Bacon recommended "seven parts of saltpeter, five of young hazelnut wood and five of sulfur."[57] This would also produce a feeble bang.

Moreover, about 1280 the Syrian al-Ḥasan ar-Rammaḥ wrote *The Book of Fighting on Horseback and with War Engines.* Herein he told of the importance of saltpeter for incendiary compounds and gave careful directions for purifying it. He also told of rockets, which he called "Chinese arrows."

Thus knowledge of explosive powder was well established in Europe, Islam, and the Far East by 1300. As with printing and the clock escapement, we do not know just how this knowledge traveled from China westward: whether over the northern route through Russia, then under the Mongol yoke, or south through Islam.

The Chinese performed some further experiments in the military use of explosive powder. They put it into tubes of bamboo, making the first Roman candles. Early Roman candles had alternate packings of loose and compressed powder, so that as the powder burned down from the muzzle, the solid lumps were thrown out and burned as they flew. Solid particles such as stones could also be thrown with some force.

The Roman candle soon reached Europe. Soldiers found that:

> . . . it is useful to tie certain tubes of paper on wooden forms to the ends of the lances of the cavalry, or on the ends of the pikes of the infantrymen . . . Then, having put in a good fuse and turned the fire-emitting end toward the enemy and having tied it well to the lance or pike, at the proper time cause them to burn as you wish. From the thing made in this way you will see a fearful and very hot tongue of fire more than two cubits long, full of explosions and horror.[58]

While these Roman candles would not greatly harm troops who were used to them, they could frighten the foe's horses into bolting.

The Roman candle was as close as the Chinese came to the invention of the gun. Although the invention of the real gun is an obscure and disputed event, it probably took place in Germany. The Chronicle of the City of Ghent for 1313 states that "in this year the use of guns (*bussen*) was found for the first time by a monk in Germany."[59] Notwithstanding that some consider this document a forgery, the statement may be not far wrong, because from then on mention of guns becomes more and more common.

Thus a manuscript of 1326, Walter de Milemete's *De officiis regum,* shows a primitive gun called a *vasa* or vase. This is a bottle-shaped thing shooting massive darts. An Italian manuscript of the same year mentions guns. In the 1340s, Edward III of England and the cities of Aachen and Cambrai all paid bills for guns and powder.

A story about the origin of the gun appears in the anonymous *Fireworks-Book* of the 1410s. According to this, one Konstantin Anklitzen entered the monastery at Freiberg a century previously and took the religious name of Friar Berchtholdus, although he was better known as Berthold Schwartz; that is, Berthold the Black.

Friar Berthold, the tale goes on, dabbled in alchemy. He put sulfur, saltpeter, mercury, oil, and lead into a massive copper vessel, sealed it, and set it on the fire. The mixture exploded with a frightful roar, hurling the lid of the pot against the ceiling and convincing Berthold's terrified fellow-monks that the Devil had come for his own. Berthold, noting what his lid had done, went on to develop the gun.

Although it is a charming story, some historians of technology consider it fiction. Unfortunately the monastery of Freiberg and its records were destroyed in the Reformation, so there is no checking it at first hand.

Fig. 20. Hand-gunner of about 1400 firing his piece by means of a heated iron bar (*from Konrad Kyeser:* Bellifortis, *1405*).

Some of the earliest guns were of wood strengthened by iron hoops, or of copper and leather. Guns soon evolved into cannon and hand guns. The latter at first were small cannon lashed to poles, which the gunners held under their arms like lances at rest. Cannon evolved into long guns for direct fire and very short guns, called mortars from their shape, for high-angle fire. For a time, balls of iron or lead were used in hand guns and balls of stone in cannon.

Iron cannon balls soon replaced those of stone, because iron is much denser than stone and iron balls therefore carried more kinetic energy for a given bore. Now cannon had to be made stronger and of smaller bore, because if a cannon designed for stone balls were fired with an iron ball of the same size, the gun would burst. Biringuccio, writing in the 1530s, said of iron cannon balls:

> This is surely a very fine invention and a horrible one because of its very powerful effect. It is a new thing in warfare, because (as far as I know) iron balls shot from guns were never seen in Italy before those that King Charles of France brought here for the conquest of the Kingdom of Naples against King Ferrandino in the year 1495.[60]

The gun soon brought the feudal system tumbling down. This was not accomplished by shooting holes in the armor of knights. The early hand guns were not so effective as all that. They did not completely displace the crossbow until +XVI, and they did not put the armor makers out of business until +XVII, three centuries after guns came into use.

What the early gun did was to knock down the walls of the castles whence the local lordlings had domineered the countryside and defied their king. By shattering the feudal castle, just as it had the walls of Constantinople, the cannon prepared the way for the era of sun-kings ruling by divine right. The towering medieval fortresses with their draw-bridges and machicolations became mere relics. The new fortress, developed by Francesco di Giorgio Martini, an Italian engineer in the service of the Duke of Urbino (+XV), and others, was a low, sprawling, star-shaped structure of huge ditches and massive earthen embankments.

Meanwhile the hand gun in its turn improved until it outshone the cannon. As the flintlock musket, it became cheap enough for any citizen to own, simple enough for him to use, and deadly enough to enable him to face regulars. Then the stage was set for the fall of kings and the setting up of republics.

Another Chinese discovery was the magnetic needle, from which the mariner's compass evolved. Quietly and rather suddenly around 1200, the magnetic compass came into use all the way from the North Sea to the Yellow Sea.

Many nations, from the Finns to the Chinese, have been credited with the invention of this instrument. The story from the Chinese end is as follows: A history of +IV tells that the legendary emperor Hwangdi, along with all his other inventions, also created something called a *jï nan-gu* or *sz-nan-gu*. This may be translated as "south-pointing chariot."

Although Hwangdi is a legendary figure, there are several references to south-pointing chariots from historical times, from –XIII to +XII. While the earlier references to this device do not say just what it was, later allusions speak of it as a geared mechanism.

So far the south-pointing chariot does not sound at all like a magnetic compass and not much like a chariot, either. A modern theory holds that it was a cart on which was mounted a pointer–probably a statuette with outstretched arm–on a turntable. The turntable was connected with the wheels by a differential gear, made of wooden peg-toothed gear wheels. As the cart turned in one direction, the statue turned in the other at the same rate. Hence, if the statue were adjusted to point southward, it would continue to point to the south during a short journey, winding through city streets.

This would make an impressive magical device, like those which Heron invented for the Alexandrian priests. Probably the works were hidden and the gadget was used to awe visiting barbarians. However, the chariot could not be employed as a compass over longer distances, because to do so would have required more accurate manufacture and smoother roads than existed in ancient China. Derek Price has calculated that if the wheels differed by 0.1 per cent in diameter, the south-pointing statue would be pointing north by the time the cart had gone two miles.

The first definite allusions to the magnetic needle in China date from +VIII, when the Buddhist priest and astronomer Yi Hing is said to have observed magnetic declination; that is, the fact that the needle does not necessarily point to the true north and south. By +XI the Chinese were using the needle–but not for navigation. They used it, instead, for the branch of Chinese magic called geomancy. This is a system for locating houses, graves, and other constructions where the occult and spiritual influences will be most favorable.

The first Chinese reference to the mariner's compass comes from about 1100. A writer states that foreigners (Arabs and Persians) arriving at Canton navigated by a south-pointing needle. Of course the compass needle points south just as surely as it points north. But, since it was early associated in people's minds with the Pole Star—the South Pole of the heavens being invisible from Europe even if there were a star to mark it—Europeans fell into the habit of saying that the needle pointed northwards.

In Europe, the compass is first mentioned in a poem of 1190. A few years later Cardinal de Vitry, who had been to Palestine with the Fourth Crusade, wrote that the compass came from "India." One solution to the problem, then, is that the Chinese discovered the magnetic needle but used it only for magical purposes. Muslim navigators seized upon this invention for navigation, and that knowledge reached Europe by sea in the course of the mass movements of the Crusades.

An alternative, which some scholars prefer, is that knowledge of the compass—or at least of the magnetic needle—came first to Europe over the Silk Route. It was first adapted to nautical use in the Mediterranean. Several medieval historians mention a certain Flavio of Amalfi as the inventor of the compass, and this Flavio may have taken one or more of the steps required to change a simple magnetized needle, hung from a thread, into the more sophisticated instrument of later times, with its rotating card. Then the Muslims, the theory goes, carried this improved instrument back to China. There is no hard-and-fast proof as to which is right.

In any case, the Chinese soon borrowed back their magnetic needle in the form of the mariner's compass. The ships they navigated by this means were then the largest and most advanced sailing vessels afloat.

A junk of this type had a central, sternpost rudder at a time when westerners were still using quarter rudders or steering oars. The ship was driven by the Chinese lugsail, stiffened with slats or battens of bamboo. This was the most efficient kind of sail then in use.

A big ship had four permanent masts and two temporary ones that could be taken down in bad weather. The hold was divided into as many as thirteen watertight compartments, so that the ship could survive being holed. Crews numbered up to 600. In their day, these were the most seaworthy ships in the world.

In early +XV the Ming emperors sent fleets of these great ships, carrying 1,000 men apiece, to cruise the Indian Ocean and extort tribute from petty kings as far away as Ceylon. At least seven such cruises took place, most of them under command of the Grand Eunuch, Jêng Ho.[61]

The Chinese showed equal enterprise in developing fresh-water craft.

They built, about this time, boats propelled by paddle wheels turned by treadwheels.

In the coastal province of Chekiang they developed the quaintest method known to history, and still in use, of transporting salt. The salt is put in casks, which are put on small flatboats, which are tied together in trains of as many as six. Each train is towed by a water buffalo, which swims along the canal under the guidance of a driver on the first boat. The people of Chekiang have developed a special breed of buffalo for this damp duty.

Until the Ming overthrew the Yüan dynasty in 1368, China led the world in technology and engineering. Within the next half-century, however, China had lost this lead. Chinese technical enterprise slowed and languished as had that of the Mediterranean world under the Romans.

For one thing, the Ming dynasty represented a nationalistic, conservative, isolationist reaction against the rule of the hated Mongols. The Mongols were never accepted by the masses even after they had taken to Chinese speech and garb. As nationalists, the Mings and their supporters despised everything foreign and discouraged contacts between their own people and foreigners. The Silk Route was closed; the voyages to the Indian Ocean stopped.

In the case of the voyages, the officers of the army and the Confucian scholars of the bureaucracy ganged up on the clique of imperial eunuchs, who had promoted this and similar enterprises. Resentful of the power that these eunuchs had arrogated to themselves, the officers and scholars persuaded the emperors to confine the eunuchs to lesser duties. They contemned all foreign "barbarians" and anybody who dealt with them. The scholar-bureaucrats, who got their jobs by passing stiff examinations in the Confucian classics, were naturally stuffed with the anti-scientific views of Kung Fu-dz.

Another cause of Chinese technical stagnation, advanced by Joseph Needham, is that the overwhelming power of the centralized Chinese bureaucracy, brought into being by the need for large-scale hydraulic works, smothered the growth of the vigorous capitalism necessary for the promotion of the pure and applied sciences.[62]

Whatever the relative importance of these supposed causes, Chinese science and engineering certainly became torpid, just when Europe was beginning its scientific and technological revolution. Although the Ming emperors persuaded the Jesuit missionaries to cast cannon for them, the Jesuits and other foreigners were expelled in the early 1700s. Thereafter Chinese military technics remained so static that, when Britain in 1839–

42 attacked China to compel the Chinese government to let British merchants sell opium in China, the British easily scattered the spear-armed rabble the Chinese sent against them.

While China, Islam, and the Byzantine Empire stagnated and India remained sunken in mystical dreams of cycles and karma, medieval western Europe sprang into the lead in the useful arts of peace and war. By the time the older civilizations became aware of the threat from this vigorous new culture, it was too late to catch up, except at the cost of drastic and painful revolutions in their own civilizations.

THE EUROPEAN ENGINEERS

NINE

At first the decline of civilization in western Europe, after the fall of Rome, was gradual. We have the letters of two scholarly men, Sidonius and Cassiodorus, who lived while the barbarians were actually carving up the West Roman Empire into German kingdoms. These letters show little awareness that the writers were living at a time of great change, or that they deemed the upheavals around them unusual.

In fact, at the end of +VIII, civilization revived a little. King Karl of the Franks–Charlemagne to us–conquered most of Europe from the Oder to the Pyrenees. On Christmas, +800, the Pope crowned Charlemagne Roman Emperor, although this big, hearty, semi-literate German had even less in common with Augustus and Hadrian than did the other "Roman" Emperor in Constantinople. Nevertheless this shadowy title, assumed by a long line of German and Austrian kings, continued like an unlaid ghost to trouble the politics of Europe down to Napoleonic times.

Charlemagne, however, tried to live up to his imperial title. To revive the Roman road system, he appointed a road commission and commanded his feudal lords to co-operate with the commissioners. The commission repaired some old Roman roads and built others, but the work petered out after Charlemagne's death.

Charlemagne also built a huge wooden bridge across the Rhine at Mainz. It took ten years to build. Then it caught fire–whether by accident or arson–and burned up in three hours.

In addition, the new Emperor of Europe embarked on a grandiose plan to link the headwaters of the Rhine and Danube rivers by a canal. But his engineers were not up to their task. They were defeated by quicksands, heavy rains, and (according to the chronicles) by the hideous laughter of fiends at night, which terrified the workmen. The clergy, who wrote the chronicles, thought that God thus showed his disapproval of changing the face of nature. A Rhine-Danube canal was finally completed in +XIX, fiends or no fiends.

Charlemagne's empire broke up at his death, and the Vikings began raiding the lands of the former West Roman Empire. By +900 they had reduced western Europe to its lowest estate in a thousand years. At the same time, another wave of Mongolian nomads, the Magyars, ravaged Europe far and wide from their bases in Hungary before they settled down to adopt the rudiments of civilization.

During this age of chaos, Roman buildings were demolished to make crude fortifications of the stone. Communities tore up nearby Roman roads and broke down Roman bridges to make it harder for marauding armies to reach them. Most Europeans lived again in isolated villages, much as they had lived in the early days of the Agricultural Revolution, by farming and home manufacture.

Then, little by little, civilization began to recover. Mining revived in Muslim Spain and in Central Europe. Coinage, which had almost vanished, came back into use. Venice began her career as the leading shipping center of the Inner Sea. And brisk manufacturing activity sprang up in the cities of northern Italy.

Here too republicanism, which had seemed to perish from the earth with the triumph of Augustus, revived. And in +XIII some Italian city-states gained eternal honor by being the first governments in the world to abolish slavery.

Several importations and inventions helped to raise the dismal European standard of living. Cotton and sugar cane were for the first time planted on the shores of the Mediterranean. The horseshoe, already invented, came into general use. So did the new horse collar, which made it possible to plow with horses. Where oxen continued to be used as draft animals, improvements in their harness and the practice of shoeing them made their efforts more effective.

A new kind of plow appeared in northern Europe. This was a heavy wheeled plow with a colter or knife to slice open the turf in front of the share. The colter made it easier to plow the thickly grassed fields of the damp northern lands.

Another source of power, the water mill, became common everywhere.

But water power was not now limited to swift streams, because the tidal water mill appeared in +XI. Millwrights built basins in bays and estuaries. They let water run into these basins at high tide and out again at low, turning water wheels both ways.

During the Dark Ages, as the period from +VI to +X is called (though some modern historians dislike the term), engineering and architecture ceased to be recognized professions. The work of engineers and architects was carried on by craftsmen such as master masons. During the decline, construction in stone became rare. Wood and plaster were used instead, whence the familiar half-timbered medieval house.

Kings and popes, using clerics as architects, continued to put up churches in the Romanesque style: plain, massive stone buildings with small windows and many round arches. The curious "central" type of church flourished for several centuries. This was built in the shape of a circle or a polygon, with the altar and pulpit in the middle and the aisles radiating out from them like spokes. The emperor Constantine began the first of these churches, the octagonal *Domus Aurea,* at Antioch in +327.

The literature of the time consists mainly of dryly terse monkish annals, religious tracts, and highly fictional lives of saints. However, technology was not entirely mute. Some time in +X, a certain Eraclius wrote a book called *On the Arts of the Romans.*

About a century later, the German monk Theophilus wrote a longer book, which drew upon Eraclius and other sources and which was called *A Treatise on Various Arts* (*Schedula diversarum artium*). Theophilus' work deals with the arts that a cleric needed to know for decorating churches, making religious vessels and other accessories, and illuminating manuscripts. He tells how to make paint, glue, gold leaf, tin leaf, glass (clear and stained), metalworking tools, and wire. He explains how to temper a file, build an organ, cast a church bell, carve ivory, and cut glass. His first chapter is called: "On the Mixture of Colors for the Nude," which is a little startling for a medieval monk until we realize that Theophilus is merely giving directions for mixing the pigments for flesh tints in paintings.

Among Theophilus' formulas is an alchemical one that must have reached him from the Islamic world. It reads:

> There is also a gold called Spanish gold, which is composed from red copper, powder of basilisk, and human blood and acid. The Gentiles, whose skillfulness in this art is proverbial, make basilisks in this manner. They have, underground, a house walled with stones everywhere, above and below, with

two very small windows, so narrow that scarcely any light can appear through them; in this house they place two old cocks of twelve or fifteen years, and they give them plenty of food. When these have become fat, through the heat of their condition, they unite and lay eggs. Which being laid, the cocks are taken out and toads are placed in, which may hatch the eggs, and to which bread is given for food. The eggs being hatched, chickens issue out, like hens' chickens, to which after seven days grow the tails of serpents, and immediately, if there were not a stone pavement in the house, they would enter the earth. Guarding against which, their masters have round vessels of large size, perforated all over, the mouths of which are narrow, in which they place these chickens, and close the mouths with copper coverings and inter them in the ground, and they are nourished with the fine earth entering the holes for six months. After this they uncover them and apply a copious fire, until the animals inside are completely burnt. Which done, when they have become cold, they are taken out and carefully ground, adding to them a third part of the blood of a red man, which blood has been dried and ground. These two compositions are tempered with sharp acid in a clean vessel; they then take very thin sheets of the purest red copper, and annoint this composition over them on both sides, and place them in the fire. And when they have become glowing, they take them out and quench and wash them in the same confection; and they do this for a long time, until this composition issues through the copper, and it takes the color of gold. This gold is proper for all work.[1]

The mythical basilisk was a kind of pygmy dragon, which could turn people to stone by looking at them and therefore was hunted with mirrors in which it stared itself to death. Aside from this excursion into magic, however, Theophilus' work is sober and sensible.

Another factor in the revival of European civilization was the importation of literature from lands where letters had never fallen so low. Moreover, some Roman writings had been saved in the lands of the former Western Empire by Benedictine monks, who copied and recopied the manuscripts as a work of merit.

From +X on, other manuscripts trickled in from the Byzantine Empire and the Caliphate. Some were Greek, some Arabic translations from Greek, and some original Arabic works. As Greek and Arabic were almost unknown in western Europe, these works had to be translated into Latin, the international language of the time. The main center of the work of translation was Spain, and many of the translators were Jews.

At the beginning of +VIII, the Muslims conquered all of Iberia except a strip along the northern coast. Arab rule was in many ways more endurable than the Roman-Visigothic government whose place it took; but the Arabs, with their usual political fecklessness, quickly broke up into a lot of quarreling petty states. In the North a similar group of Christian states took form and began to win the Spanish peninsula back from the Muslims.

For most of two centuries, Iberia was a constellation of Muslim and Christian statelets, which fought incessantly. Not taking religion too seriously, a Christian and a Muslim lord often cheerfully formed an alliance to fight other Christian or Muslim lords.

Despite all this jovial murder and mayhem, Muslim Spain was at this time the most civilized part of western Europe. In this lively and liberal Christian-Muslim Spanish world, most of the translation of Greek and Arabic works into Latin took place.

At the end of +XII, however, a wave of Berbers from Morocco, newly converted to Islam and filled with the fanaticism of ignorance, overran the peninsula. Combining the political anarchism of the Arabs with a brutish hatred of thought and knowledge that was all their own, the Berbers soon ended the Golden Age of Muslim Spain, without being able to stop the advance of the Christian states.

Their fanaticism aroused a counter-fanaticism among the Christians. The Church became ever more powerful in Christian Iberia. It imposed such effective thought control that, even while Spain was rising to be the leading power of Europe (+XVI), it was also becoming the intellectual cipher that it has been ever since.

The terms "Middle Age" and "medieval" are not hard and fast in meaning. The age called by these names can be extended from the fall of the West Roman Empire around +500 to the beginning of the Age of Exploration around 1500. Some historians, however, prefer to consider the age a shorter period. Moderns invented these terms to denote the interval between classical times and their own.

Of course, medieval folk did not think of themselves as medieval, or as mere intermedia between the ancients and the future. They thought of themselves as modern, advanced, and enlightened people. The name "Age of Faith" better describes the medieval period, because during this millennium the conflicts between the major monotheistic religions, Christianity and Islam, and between the sects and schisms within these faiths, were a major fact of life.

The twelfth and thirteenth centuries are sometimes called the High Middle Ages, especially by people who idealize this stage of man's history. During these centuries, the things that we think of as particularly medieval reached their peak. This was the great age of knighthood, of castles and cathedrals, of the Crusades, of feudalism, of troubadourism, and of scholastic philosophy.

The High Middle Ages were a time of grim asceticism and riotous self-indulgence. Costume was wildly colorful, thanks for advances in the loom and in dyeing. The buttonhole, invented in +XIV, revolutionized dress as buttons replaced pins and laces. Gems glittered with new brilliance as a result of the recent discovery of faceting. There were frenzied outbreaks of religious hysteria, fanaticism, and cruelty, resulting in such monstrous mischiefs as the Children's Crusade and the massacres of heretics and Jews. Superstitious supernaturalism, typified by the twelve Holy Foreskins venerated as relics in various churches, ran wild. Literacy was still so scarce that in 1215, King John of England did *not* sign the Magna Carta, because he could not write. He affixed his seal instead.

For us, the High Middle Ages were the time when engineering regained most of the ground lost after the fall of Rome. In some ways, technology advanced well beyond its classical prototypes.

During this time, also, scholars thought and speculated about the behavior of matter. They pondered the nature of motion, force, and gravity. Later on, their reasonings became important to engineering when, in the early modern period, science and engineering came to depend on each other.

The work of the scholars did not, however, much affect the engineering practice of their own time. At that early date, engineers were anonymous craftsmen who worked by rule of thumb, using the skills handed down to them during their apprenticeship. They had little to do with scholars and scientific principles.

Medieval scholars included some very intelligent men and acute reasoners. They made some progress in straightening out physical concepts, although they worked under handicaps. For one thing, like the ancient Greeks, they made little use of experiment.

Furthermore, most medieval thinkers took Aristotle's doctrines in physics as basic truths. As these doctrines were nearly all wrong, the medieval scholars' advances in physics consisted for a long time of finding the errors in Aristotle and laboriously convincing their colleagues that "the master of those that know," as they called him, could make mistakes.

Finally, the scholars had difficulty in handling such concepts as force, mass, weight, distance, velocity, and acceleration because no hard-and-fast, clear-cut, agreed-upon use of words had yet been worked out. Hence, when a medieval scholar wrote of "quantity of motion," it is hard to tell whether he meant the distance an object moves, or its speed, or some combination of these.

A leading scientific scholar of the High Middle Ages was the English monk Roger Bacon (1214–92). Having studied at Oxford, Bacon spent most of his adult life as a lecturer at the University of Paris.

Although his voluminous writings summarized much of the scientific beliefs of his period and added something to human knowledge, Bacon has been overrated in modern times. For most of the last fifteen years of his life he was imprisoned by his fellow Franciscans for uttering "novelties," the nature of which is not known. Therefore, many have supposed that Bacon was a scientific martyr persecuted by the Church.

From the scanty records, however, it appears that Bacon was religiously more orthodox than those who condemned him. He may have gotten into trouble for any of several possible reasons: because he violated a Franciscan rule against writing books; because he was a quarrelsome man who berated his colleagues as ignorant asses; because his friend Pope Clement IV, on whose protection he counted, died; and lastly because he got involved in a power struggle within the Franciscan order.

Nevertheless, Bacon's works contain many interesting features. He wrote much of value on optics, although most of his ideas on this subject were taken from the Muslim scientist ibn-al-Haytham. He wrote the first known European account of gunpowder (at least, the first of whose date we can be fairly sure) and foresaw its use in warfare. He preached the importance of experiment when most clerical scholars avoided it as dangerously close to black magic.

On the other hand, much of his work we should call superstitious. He attributes the victories of the Mongols, then sweeping Asia from Honan to Hungary, to the Mongols' having better astrologers than their foes. And, because his writings did not begin to be published until +XVIII, half a millennium after Bacon's time, they had little influence on the development of science and technology.

In a famous passage, Bacon predicts the results of applying experimental science to the useful arts:

> Vessels can be made which row without men, so that they can sail onward like the greatest river or sea-going craft, steered by a single man; and their speed is greater than if they were filled with oarsmen. Likewise carriages can be built which are drawn by no animal but travel with incredible power, as we hear of the chariots armed with scythes of the ancients. Flying machines can be constructed, so that a man, sitting in the middle of the machine, guides

it by a skillful mechanism and traverses the air like a bird in flight. Moreover, instruments can be made which, though themselves small, suffice to raise or to press down the heaviest weight . . . Similar instruments can be constructed, such as Alexander the Great ordered, for walking on the water or for diving.[2]

Furthermore, Bacon's studies in optics convinced him that:

If a man looks at letters or other small objects through the medium of a crystal or of glass or of some other transparent body placed above the letters, and it is the smaller part of a sphere whose convexity is toward the eye, . . . he will see the letters much better and they will appear larger to him . . . Therefore this instrument is useful to the aged and those with weak eyes. For they can see a letter, no matter how small, sufficiently enlarged . . . The wonders of refracted vision are still greater; for it is easily shown by the rules stated that very large objects can be made to appear very small, and the reverse, and very distant objects will seem very close at hand, and conversely . . . Thus from an incredible distance we might read the smallest letters and number of grains of dust and sand . . . So also we might cause the sun, moon, and stars in appearance to descend here below . . .[3]

Now, Bacon never built a telescope or an airplane. Neither did he leave instructions that would enable others to build them. Therefore these statements by Bacon are imaginative speculations, not actual advances in science and engineering.

Nevertheless, Bacon's speculations have their place in the story of the growth of knowledge. An invention must be imagined before it can be built, even though he who imagines it may not be he who reduces it to practice. From Bacon's time on, more and more of such conceits found their way into writing. When these inventions had been imagined often enough, people became sufficiently used to the ideas so that some actually began tinkering. Thus, little by little, Bacon's imaginary devices became realities.

Bacon was but one of many medieval scholars who busied themselves with physical problems. His contemporary Jordanus de Nemore advanced the science of statics, the analysis of the forces in a solid structure bearing a load. He then attacked the problem of resolving a force into its components.

In the next century (+XIV) a group of English scholars at Merton College, Oxford University, straightened out some of the confusion in terms concerning motion, velocity, and acceleration. At the same time the French scholar Jean Buridan explained the behavior of moving bodies by a theory of "impetus" not far from the modern theory of inertia.

Others, too, speculated and invented. An English monk, Eilmer of Malmsbury (+XI) built a flying machine–a sort of glider attached to his arms and legs–and leaped from a height. He crashed and broke both legs but, though lame ever after, lived to old age. Perhaps he had heard of a previous attempt at flight by a physician, glass maker, musician, poet, and inventor of Muslim Córdoba, ibn-Firnâs (+IX), who also crashed, injuring his back.

Not much record remains of the thinking of the practical engineers of the Middle Ages–the cathedral builders, the catapult makers, and the millwrights. One of the few such records is the notebook of the craftsman Villard de Honnecourt, about 1230. It consists of thirty-three sheets of parchment covered with sketches and descriptions. Some of the sketches are too rough to give a clear idea of what Villard is driving at; in other cases his designs would not have worked. Still, we can see his ideas taking shape amid the general ferment of the time. Besides methods of cutting voussoirs for arches and driving piles for bridges, Villard shows a catapult, a water-powered sawmill, some automata of the Alexandrian type, and a perpetual-motion machine.

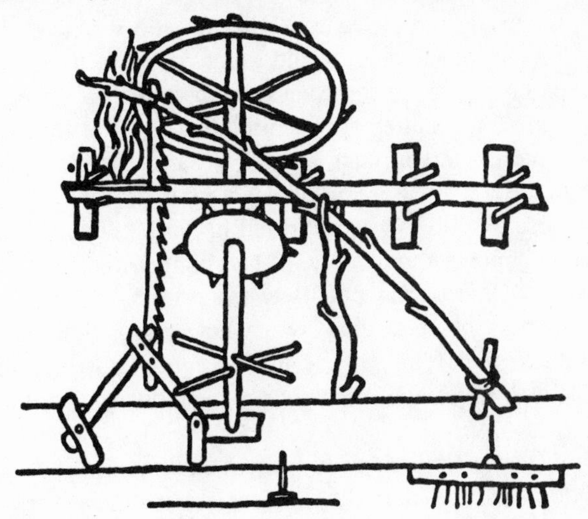

Fig. 21. Sketch for a water-powered sawmill, from the notebook of Villard de Honnecourt.

In connection with Chinese engineering, I have told you about the development of paper and the printing press. European printing, which matured in +XV, was preceded by another invention of equal moment. This was the eyeglass. Besides the other revolutionary changes that took place at the end of the Middle Ages, paper, printing, and eyeglasses together caused a revolution in reading.

All we know about the origin of eyeglasses is that in 1306, Friar Giordano of Pisa said in a sermon:

> It is not twenty years since there was found the art of making eyeglasses which make for good vision, one of the best arts and most necessary that the world has. So short a time it is since there was invented a new art that never existed. I have seen the man who first invented and created it, and I have talked to him.[4]

Evidently, eyeglasses were invented shortly after 1286. Perhaps the inventor–Friar Giordano unfortunately did not name him–was inspired by the optical writings of Roger Bacon, who died at about that time.

All the first spectacles had convex lenses to correct farsightedness. Most people, if not farsighted to begin with, become so with age as the lenses of their eyes lose their power of adjustment for short-range vision. Eyeglasses to correct for nearsightedness and astigmatism were not made until several centuries later.

Even so, the crude spectacles of +XIV and +XV extended the reading ability of many people who would otherwise have been cut off from this source of knowledge and pleasure in their forties. In earlier times most literate men, no matter how scintillating their intellects, were debarred from reading in their later years. They had either to be read to or to give up literature. Now men could go right on reading.

When the printing press was perfected in +XV, books became much cheaper and commoner than before, as had happened earlier when paper replaced papyrus and parchment. Between printing and eyeglasses, the amount of time spent in reading by a whole population was probably increased several fold. This fact by itself would almost account for the speed-up of intellectual and scientific progress during the last few centuries.

Of course, even after literacy had again become common in Europe, there were doubtless millions who, in spite of having learned their letters as children, read practically nothing as adults. Such behavior is not unknown among high school and college graduates of today.

Moreover, the reading revolution was not yet complete. Reading at night by candlelight was still an eyestraining task, which few cared to pursue. It took the invention in 1784 by Argand, a Swiss, of a really bright oil lamp to make nocturnal reading pleasant. Finally, it took nineteenth-century universal education and public libraries to bring about the modern state of affairs, when anybody can read about virtually anything that interests him.

Because most of the engineering of the High Middle Ages was done by unknown masons and other nameless craftsmen, we cannot say much about the medieval engineers themselves. But we can talk about their engineering. Let us see how the technology of the time developed, art by art.

Germany, with abundant supplies of coal and iron and its silver, copper, lead, and tin, led the revival of mining and metallurgy. The *mining* of coal developed in the Middle Ages, although coal had long been dug up and burned where it could be grubbed from the earth's surface. Even Theophrastos, Aristotle's successor, knew of coal as a fuel.

In ancient and medieval times, coal was used only for industrial purposes such as burning lime. It was not used for house heating because houses did not yet have chimneys. If anyone tried to heat a chimneyless house with coal, the suffocating smoke soon drove the dwellers forth.

After the fall of the West Roman Empire, there is no record for several centuries of the use of coal in Europe. Then medieval miners began digging up coal, not only from surface beds but also from shafts run into the ground as they did for other minerals. The use of coal greatly increased men's power over materials because coal furnished much more heat energy than other available fuels in proportion to its bulk and weight. This use waxed so swiftly that in 1307 the lime burners of London were forbidden to fire their kilns with coal, because the smoke bade fair to smother the city.

Rapid deepening of the shafts of coal mines aggravated the problem of drainage. In damp northern Europe, the shafts soon went below the water table, so that they filled up with ground water. By the end of the Middle Ages they had gone so deep that in some mines hundreds of men were kept at work passing buckets up ladders to keep the mine bailed out. This state of affairs led to the invention of horse- and water-powered mechanical pumps.

The smelting and refining of iron also advanced by great strides. Furnaces became larger, more permanent, and more carefully engineered. Early medieval smelters used the Corsican furnace, a semicircular funnel-shaped pit built against a stone wall. Then came the taller and more inclosed Catalan forge. Finally the completely inclosed Austrian *Stückofen* of +XV reached a height of 14 feet. The upper part evolved into a true chimney, which made possible a stronger air blast and higher temperatures.

At the same time other improvements appeared. One was the use of coal, at least in the early stages of smelting. The Chinese may have anticipated Europe in this use of coal. Coal could not be employed in the final stages of refining iron because the sulfur and phosphorus in the coal would damage the iron. Men had not yet learned how to turn coal into coke by roasting it, thus driving off the impurities and leaving almost pure carbon behind.

Another advance was the increasing use of water power. Water wheels worked stamping mills, in which ore was broken into small pieces in mortars. It worked bellows, providing a stronger air blast and higher temperatures than could be obtained with muscle-powered bellows. Finally, it lifted trip hammers to pound the finished iron into shape.

The higher temperatures attained in late medieval furnaces made it possible to melt iron, dissolve carbon in it, and cast it. The hotter molten iron is, the faster carbon dissolves into it. The higher carbon content, which makes iron hard and brittle, also lowers its melting point and makes it flow freely.

Ancient smiths had produced cast iron, which they considered spoiled and worthless, only occasionally, by accident. But in late medieval times it was found that, as the Chinese had already learned, cast iron was very useful for some purposes, such as stoves and cannon. Late medieval founders could turn out either wrought or cast iron at will, in masses of hundreds of pounds per charge. The wrought-iron bloom of the classical smith seldom weighed over 50 pounds.

Although the medieval ironmongers could now readily convert wrought iron to cast iron, the latter comprised only a small fraction of all the iron produced. As late as 1750, cast iron still amounted to but 5 per cent of all the iron made.

However, the larger quantities of iron produced in medieval furnaces and the new flexibility of methods of smelting and refining meant that iron became much commoner and cheaper than it ever had been. Little by little, iron began to be used in common tools and machines that had up to then been made entirely of wood.

In addition to the water-powered hammer, the Middle Ages saw the birth of other devices for molding metal: the rolling mill and the draw plate. The draw plate is an iron plate with holes in it for drawing wire. The wire is thinned by pulling it through a hole smaller than itself. The wire drawer of +XIV sat on a swing, which was pulled back and forth by a water wheel and crank. On each back swing, the craftsman seized the wire with tongs and drew it back with him, pulling it through the hole. Before the draw plate, all wire had to be beaten out with a hammer.

The early rolling mills were not used for rolling iron, because iron was too tough for them. Instead, they rolled out the H-shaped bars of lead that held in place all the many little pieces of stained glass in church windows. These inventions affected the production, not merely of iron, but also of all the common metals.

Novel chemical products appeared during the Middle Ages, opening up new possibilities for the manipulation of matter. Gunpowder I have already told you about. Alcohol, judging by its name, was probably an Arab discovery; but the first actual description of it comes from the Italian medical school at Salerno about 1100. Its first use was as a base for perfumes and cosmetics. In +XIII appeared the three strong acids: sulfuric, nitric, and hydrochloric; the first of these may have been discovered by the Arab alchemist, Jâbir ibn-Hayyân, as early as +VIII. Before this time, chemists had had no acids stronger than vinegar to work with.

In few respects was medieval Europe more backward, compared to the Roman Empire, than in the planning, building, and keeping up of cities. Down almost to modern times, most European cities were as innocent of sewer systems as some large Asiatic cities are today.

Most medieval European cities grew out of villages, which were sometimes built on the sites of ruined Roman cities. These towns displayed the typical village layout: a tangle of crooked alleys.

In the later Middle Ages and the Renaissance, these cities were faced with a difficulty much like that of modern cities confronted by the automobile age, for which they were never planned. The streets of medieval cities were adequate for pedestrians, with an occasional man on horse or muleback. But, when carts, wagons, and carriages increased, the old irregular layout proved inadequate.

So, from 1300 on, kings and councils issued a blizzard of decrees about parking, speeding, and making U-turns. In 1540, François I of France ordained:

> And, under the same penalties, We forbid wagoners and drivers, whether of carts, drays, wagons, or other vehicles, to turn in the streets, but they are to turn at the intersections and corners of said streets to avoid the inconveniences that might arise, such as wounding children or other persons and interfering with other passers-by along the road.[5]

As today, these regulations failed to cure all the ills they were meant to remedy.

In +XIII, a number of new towns were laid out by English, French, and German rulers. Examples are Monpazier, Carcassonne, Winchelsea, and Neubrandenburg. All were built on the gridiron plan. Usually the marketplace occupied one whole central block, with colonnades around it and the church adjacent. Town planning and urban renewal became active in the Renaissance, notably in the rebuilding of shabby old Rome by the popes.

A fine opportunity for city planning was offered after the great fire of London in 1666. The ashes had not cooled when John Evelyn, the diarist and civil servant, and Christopher Wren the architect rushed around to show Charles II their plans for rebuilding the burned area. They arrived almost at the same time and discovered that their plans had many features in common. Either plan would have been of the greatest advantage to London.

The king passed the plans on to Parliament, which debated them. But it would have taken a lot of effort and money to settle all the claims of the property owners in the district. Moreover, the burnt-out shop-keepers screamed that they would starve unless allowed to reopen for business at once at the same old place in any old shack. While Parliament dithered, the burned area was rebuilt on the old plan (except for some street widening and a new law against wooden houses) and the East End of London has remained a medieval tangle of alleys ever since.

Paving was long unknown in medieval cities. The stroller was blinded by clouds of dust in dry weather and sank to his ankles in muck during wet spells. It was said of anything particularly gluesome that *"Il tient comme le boue de Paris"*—"It sticks like the mud of Paris."[6]

In 1184, King Louis Philippe, tiring of the dust and mud, ordered Paris paved. The result, if not so finished a job as the paving of ancient Memphis and Babylon, was a step in the right direction. The paving was done with stones about 7 inches thick and of irregular shape, 20 to 52 inches long. If these stones did not help the drainage problem, they at least kept vehicles from sinking up to the hubs.

For several centuries thereafter, rulers and officials tried to clean up, pave, and modernize the streets of European cities. Their efforts were not very successful. Medieval Europe had fallen so far behind Roman standards of municipal service that it did not surpass the Romans in this regard until the last two centuries. Rome of +100 was better off for street maintenance, sewers, police, and fire protection than London or Paris of 1700.

One difficulty was caused by rulers themselves. Instead of collecting taxes—especially from the untaxed Church and nobility—and spending the money on improvements, they passed laws requiring every citizen to pave the street before his house, to keep it clean, to refrain from encroaching on the public right of way, and to refrain from throwing his garbage and sewage into the street. Dumps were provided outside the town for these purposes. Medieval Nantes even had a law forbidding people to throw dead cats into the water supply.

Most citizens, however, preferred to spend their time and money otherwise than in keeping up their cities, and the nobility simply ignored the rules. After a ruler had issued some resounding decree, there might be sporadic improvement. But soon the law would be quietly forgotten, and the city would become as foul as ever. As if the streets were not narrow enough to begin with, citizens narrowed them still further by extending their own structures out into the right of way.

Attempts to make citizens pave the streets before their houses proved ineffectual. When hauled before the courts, the householders wailed that they had no money to pay the pavers, and that anyway that stretch of street had never been paved. When paving was laid, it was highly irregular, since each property owner had it done as he saw fit.

Moreover, guilds of pavers grew up as certain stonemasons specialized in this trade. These men dug up good paving to compel the householders to hire them to replace it. This racket continued even after such destruction of pavement had been made a capital crime.

Most of us are familiar with the general appearance of medieval European private houses. Wood and plaster half-timbered construction was typical, and the second story often extended forward over the street below. The roofs were high-peaked and covered with thatch or shingle or tile. Domestic chimneys, adapted from those already used in bake ovens and smelting furnaces, appeared in +XIII.

The High Middle Ages, however, saw advances in structural forms. One advance was a greater variety of arches. In addition to the semicircular arch of the Romans, medieval builders learned from Islam the use of the pointed arch. The pointed arch had the advantage of exerting on its piers a thrust that was more nearly straight down; and, since there was less outward thrust of the lower ends of the arch, there was less tendency to push the piers apart. Therefore, with the pointed arch, the supporting piers or walls did not have to be so massive or so heavily buttressed.

Still another advance was the truss. We have seen the first steps toward the truss in Trajan's Danube bridge. In the Middle Ages, trusses of timber were used to hold up roofs. However, as nobody could analyze the forces in a truss, medieval roof trusses were often cluttered with extra members that added nothing to their strength. Really efficient trusses, in which every member counted, were developed by the Renaissance architect Palladio (+XVI).

Besides these changes in structure, the methods of designing buildings also changed. The profession of architect reappeared. In the Dark Ages, buildings were planned by master masons or by monks with a taste for design. In the High Middle Ages, professional architects once more evolved from these masons.

Not much is known about these men beyond the bare fact that they existed. We know of Villard de Honnecourt solely because we have his notebook. But these nameless architects were competent men who, if innocent of higher mathematics, nevertheless had a good practical knowledge of weights, strengths, and forces.

Their most signal accomplishment was the development of the Gothic cathedral. While in Italy the Romanesque church with its thick walls and round-topped doors and windows continued to be built right down to the Renaissance, a very different style of church appeared in northern France in the latter half of +XII. Although the Italians called it "Gothic," meaning "barbarous," it spread over most of Europe.

The most obvious features of the Gothic church were its pointed arches and its large windows, adapted to the dim light of the gray northern skies. The large windows were achieved by a basic change in structure. The weight of the roof was carried, not by a whole wall as in earlier structures, but by piers, which were incorporated in the outer wall. These piers functioned somewhat like the colonnades that had held up the roofs of Greek temples.

The space between the piers was filled at first by thin walls and later by large stained-glass windows. The fully developed Gothic cathedral anticipated the modern skyscraper with its steel skeleton holding up separate panels of wall. The skeleton does the work while the wall merely keeps out the weather.

In addition, parts of the inner structure were often held up by rows of pillars. But these columns differed from those of antiquity. The capital, which had become the impost block of Byzantine times, dwindled away altogether, so that the ribs of the vaulting sprang directly from the top of the column like branches from the trunk of a tree. The column itself became a pier built up of small stones with a ribbed pattern resembling a cluster of small columns, not unlike the ribbed columns of King Joṣer's Egypt.

Vaulting was developed to a new pitch. Medieval architects learned to roof their churches with vaults of stone and tile, because wooden roofs were always catching fire from sparks or lightning. As the people had no means of getting water up to such a height, such a fire always destroyed the whole church.

Vaults were made with a pointed peak like that of a Gothic arch. As in the walls, stresses were concentrated along stone ribs that soared up from one pier to the zenith of the ceiling, often over 100 feet from the floor, and down on the other side. At least, that was the effect, although the architects probably had no such clear engineering concepts in mind.

These heavy roofs naturally posed the problem of outward-spreading thrust. Pointed arches took care of some of the horizontal thrust. To cope with the rest, medieval architects developed the flying buttress. This was a structure of stone, like a small section of an arch, leaning against the piers from outside the building, propping the wall and counterbalancing the thrust of the roof vault.

Another characteristic of the Gothic church was its ornateness. Sections of the stained-glass windows were divided up not only by partitions of iron and lead but also by delicate curlicues of stone, called "tracery." The European architect, like the Indian temple builder, liked to fill up with statuary every available space on his building. The roof was embellished with scores of spiky spires, often surmounted by statues. Although some stone saints and angels were works of high art, a good deal of medieval sculpture was fairly crude; but quantity made up for quality.

Lacking scientific principles, medieval architects developed their methods by guess and by trial. Some of their churches fell down on the heads of the faithful, and some surviving churches contain mistaken, illogical features. The roof of Beauvais Cathedral, whose ceiling was the highest of any–154 feet–collapsed twice in +XIII and had to be rebuilt with stronger supports. The spire, not built until +XVI, also toppled with a crash.

During the High Middle Ages, much of the economic surplus of Europe went into the building of cathedrals, as that of Egypt had once gone into pyramids. Cities vied to build the biggest and most sumptuous churches. Contributions of money and voluntary work brought into being an amazing array of towering Gothic fanes. Regional styles developed. For example, the French liked very tall towers, while the Germans preferred smaller towers but lofty side walls.

In fact, many cities undertook projects larger than they could carry out. Sometimes these churches took three or four centuries to complete, during which time the plans were several times changed. Some have not been completed to this day; thus the original nave planned for Beauvais Cathedral has never been built.

As a result of these vicissitudes, many Gothic churches have bizarre asymmetries. For instance, one of the pair of towers at the west front of Amiens Cathedral is a whole story taller than the other.

The rage for building giant Gothic cathedrals died out in +XV, although work continued fitfully on churches already begun. The Renaissance produced other building styles, with a strong flavor of classical Rome.

Still, from time to time, since the Middle Ages, architects have reverted to the Gothic form for churches. The original churches built in this style were such remarkable *tours de force,* they were so huge and splendid and numerous, and their heaven-piercing spires made such a deep impression on the minds of men, that ever since then they have become in many European minds the very symbol of the Christian religion.

After the Gothic cathedral, the most characteristic large building of the High Middle Ages was the castle. The castle or fortress-dwelling was an ancient structure known to antiquity, and found from prehistoric Sardinia to the Far East. But the loose feudal government of medieval Europe, under which any local lord might wage private war on any other, encouraged castle building on an unprecedented scale. These castles ranged from the little family dwelling towers of Scotland and Germany–one-family houses built like miniature castles–to huge fortresses like the citadel at Carcassonne, which was practically a fortified city.

The landscape of Europe and the northern Mediterranean lands was dominated by these castles and their surrounding farms, carved out, as it were, from the primeval forests and worked by miserable serfs. Paths, rather than roads, wound past these castles and on toward the occasional walled town, which was prepared to hold its own against the warring sieurs.

When the Crusaders conquered the Byzantine Empire in 1204, they carved the Empire up into feudal domains and dotted the mountain tops of Hellas with castles, whose ruins still rise jaggedly hither and yon against the clear blue sky. One of the world's largest and best preserved castles, Kerak des Chevaliers, was built in Syria for the Knights Hospitallers of St. John of Jerusalem in the late +XII. It is curious that the best example of medieval European castle architecture should stand today in a Muslim land, while the finest medieval Muslim palace, the Alhambra at Granada, is in Christian Spain.

The art of pre-gunpowder fortification reached its peak in medieval Europe. Although most of the elements of the European castle–the round tower, the crenelated parapet, the moat, and the portcullis–were old, medieval castle-builders combined them with greater skill and care than their predecessors.

In fact, medieval castles reached such a pitch of perfection that few, before the coming of cannon, were ever captured save by surprise or treachery. As in the First World War, defense had a strong if temporary advantage over offense. Hence most wars took the form of indecisive sieges. The besieger tried to starve out the besieged. But, if the latter had prudently prepared for the siege, the attacker might well have to give up first when his men's short enlistments ran out and they trickled away.

Along with the other arts and sciences, fortification had declined in Europe during the Dark Ages. But then it began to revive. At the time of the Norman conquest of England (1066) the typical castle was no more than a large house of stone or wood, surrounded by a wooden stockade, on a mound surrounded by a ditch.

However, some larger and stouter castles had also been built, with turrets and crenelated parapets on the old Roman model. When the Crusaders saw the vast walls of Constantinople, they brought back to western Europe ideas for still more stubborn strongholds.

The medieval castle threupon developed two main features: a tall, thick, crenelated stone wall, the enceinte, surrounding the whole; and a large thick-walled round tower, the keep or donjon, for a last stand. The biggest keep of all was that of Château Coucy, near Soissons, 100 feet in diameter, 180 feet high, and made of walls 18 feet thick. In the First World War, General Erich Ludendorff ordered it blown up, not for any military reason but for the pleasure of destroying things.

The living quarters of early medieval castles, for the most part extremely crude and uncomfortable, were tucked away in odd corners of the structure. Provisions for warmth, cleanliness, and privacy were rudimentary.

The enceinte comprised the main defense. But, if it were overrun, the defenders retired into the keep. As the keep had only one small door and was of such massive masonry that it could not be battered down, the defenders could hold out there as long as their provisions lasted.

Fig. 22. The castle of Coucy, near Soissons (restored), showing the keep-and-enceinte plan, fortified gatehouses, machicolation, and conical roofs on the turrets.

On the other hand, the attackers could just as easily prevent the defenders from coming out.

As a result of crusading experience, castle builders changed from the simple enceinte-and-keep plan to one of concentric walls. Such a structure was really two or three castles, one inside the other. As one line of walls was taken, the garrison retired to the next. Because the inner walls were taller than the outer, missile troops on an inner wall could prevent the enemy from using an outer wall to attack an inner one.

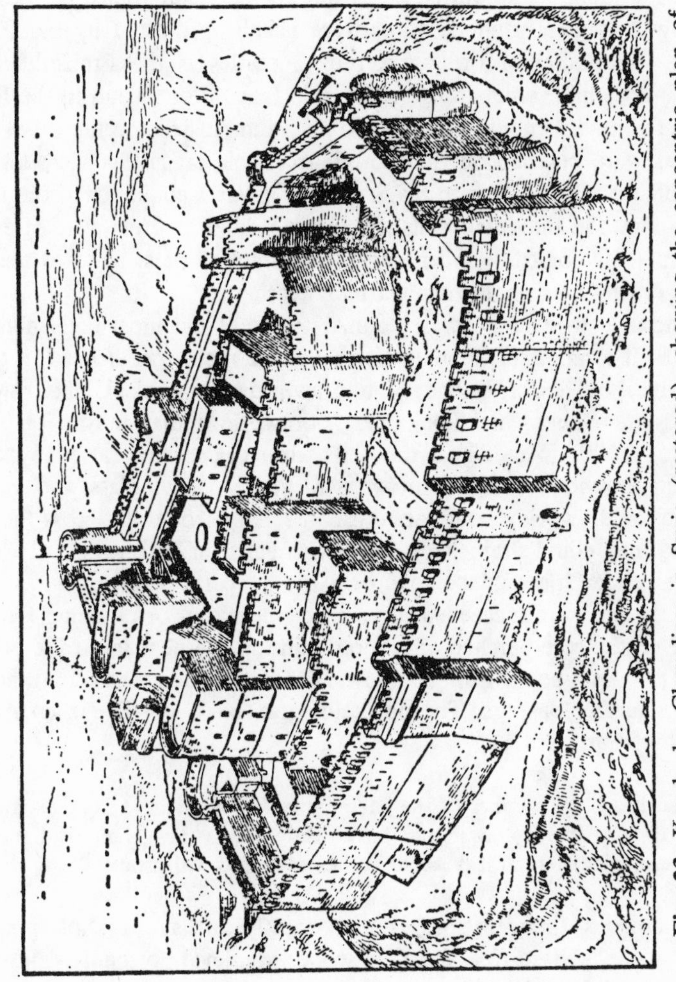

Fig. 23. Kerak des Chevaliers in Syria (restored), showing the concentric plan of castle building.

In the later Middle Ages, the keep was revived, but its purpose was now different. The soldiers of the castle owner were likely to be mercenaries instead of feudal retainers. The main purpose of the keep was now to enable the lord to defend himself against his own men-at-arms in case they took it into their heads to cut his throat and share his wealth.

Castle architecture varied according to local conditions. Castles built in lowlands near running water were usually protected by moats. The most remarkable of these waterborne castles is the Pfalzgrafenstein, which rises like some strange stone ship from a low island in the Rhine.

In the mountainous parts of Germany and Austria, on the other hand, castles were perched on the most inaccessible crags. Such castles were so difficult for attackers to reach that the owners could afford the luxury of windows in the outer walls. In northern lands, turrets were often surmounted by conical wooden roofs, like witches' hats, to ward the warriors on the battlements against bad weather.

Entrance to a castle was planned to daunt the most determined attacker. If a moat surrounded the castle, the moat was crossed in peaceful times by a drawbridge. The drawbridge is ancient; the Egyptian fortress at Buhen, in Nubia, had a drawbridge that moved on rollers about –2000. In medieval times the drawbridge was raised by chains wound around drums, which were turned by a windlass and gearing. When the drawbridge was raised, the assailants could not cross the moat unless they could drain it or fill it with earth or brushwood. To try to tunnel under it would be to drown the sappers.

If the besieger succeeded in crossing the moat–or climbing the crag, if the castle were perched on a mountain top–he next faced the portcullis. This was a heavy iron gate, lowered from a slot over the entranceway to bar further passage. The portcullis goes back to –IV, when Aineias the Tactician wrote:

> And if a large number of the enemy come in after these fugitives and you wish to stop them, you should have ready above the center of the gate a portcullis of the stoutest possible timbers overlaid with iron.[7]

In Aineias' time, iron was still too costly to make a whole portcullis of it, so he advised iron-sheathed wood. Medieval portcullises, however, were of solid iron.

Besides the drawbridge and portcullis, entrance to a castle might also be strengthened by strong outer gate towers that constituted a small fortress in themselves. Mastiffs or bears might be chained at the outer gate to discourage unwanted visitors.

If an attacker overcame the defenses at the outer gate, he often found that he had to follow a spiral path around the castle to the main gate. The spiral went clockwise so that the attackers' shields, on their left arms, were on the side away from the inner wall and so did the assailants no good.

The cannon, invented in China or Germany, put an end to the castle-building art, or at least transformed it beyond recognition. During its first century, the cannon was so feeble that it had little effect on fortifications. The chronicler of the city of Ulm briefly noted in his entry for 1380: "A knight came and besieged the town and shot at it with thunder guns. It did no harm." A little siege now and then was the sort of thing you had to expect in those days.

However, the gun improved along with all the other devices of the time. By 1414, cannon were formidable enough so that the Elector of Brandenburg and Prussia was able in two days to demolish the castle of a rebellious noble. Soon castle walls were seen to be of no more avail against heavy artillery than magical spells. Lords converted their castles to more or less comfortable mansions and palaces. After the Turkish sultan, Muḥammad II, took Constantinople in 1453 with his mighty artillery, it looked as though no stronghold could withstand the new weapon.

After the Turks conquered the Balkan peninsula and Hungary, they besieged Vienna in 1529. Vienna was then under the rule of Emperor Charles V, who reigned over a patchwork empire that included nearly all of western Europe save France and the British Isles. The wall of Vienna was a crumbling little thing only six feet thick, and there was no time to build a proper wall. So Count Salm, the Emperor's general, made the Viennese build thick earthen embankments instead.

Although the Turks outnumbered the defenders by ten to one and were deemed the world's bravest and best-disciplined soldiers, the defenders nevertheless beat off their fiercest attacks, while the sultan's cannon balls buried themselves harmlessly in the embankments. After a month of trying, the Turks gave up and marched back home.

The lesson of Vienna was soon learned. Engineers, notably Michele Sanmichele, invented a fortress of a new kind, with a polygonal or star-shaped plan, a low ground-hugging profile, and huge earthen ditches and embankments.

Machinery also advanced in medieval Europe. Europeans used catapults and crossbows much like those of classical times, save that for throwing heavy stones the counterweighted trebuchet took the place of the classical onager and the two-armed stone thrower.

In the late Middle Ages, for sieges and shipboard fighting, Europeans developed a heavy crossbow as powerful as the much larger dart throwers of classical times. This weapon had a steel bow. Much too stiff to be bent by hand, it was cocked by clamping a windlass with a rack and pinion gear over its butt and cranking the windlass to draw back the string.

Use of these powerful crossbows in open warfare, however, did not prove successful because of their slow rate of fire. For example, the English longbowmen routed the Genoese crossbowmen at Crécy (1346). Although the Genoese weapons were probably more powerful and accurate, the English archers shot so fast that they pincushioned the Genoese before the latter could make much impression, and a volley from three small English cannon completed the rout of the mercenaries.

Among the other machines of medieval Europe, we have seen how clocks evolved from the water clocks and geared astronomical computers of antiquity, with the addition of the escapement from China. Other mechanical advances were the combination of the crank and the foot treadle, applied to turning the lathe and the grindstone.

The advance of machinery was hindered by the guilds. The main purpose of the guilds was to make life easy for the master craftsmen who dominated them. The guilds kept out competition, restricted entry into the guild, and opposed innovations that might upset their business. They enjoyed political power through their representation on the councils of self-governing cities and thus were able to legislate against innovations.

An example of guild activity is the case of Hans Spaichl, a coppersmith of Nuremberg. In 1561, Hans invented an improved lathe slide rest. Hearing of this, the agitated Council showered Hans with commands: he should make no more such lathes until a committee had examined his first one and reported whether it might harm the city; he should not sell lathes to anybody outside his own craft; he should not leave town without permission. The Council offered Hans 100 florins if he would agree to let them destroy his lathe, which had cost him 300 florins to build, and promise to make no more like it.

After the case had dragged on for years, Hans built another lathe and sold it to a goldsmith, but the Council seized and destroyed it. When, some years later, another coppersmith built and sold one of the improved lathes, the Council decreed that "he shall be imprisoned in a barred dungeon for eight days to teach him not to do it again."[9]

Despite such handicaps, the design of machinery forged ahead. The most notable advances occurred in the millwright's art. From the medieval millwright, the mechanical engineer of later times evolved.

Medieval Europe inherited three types of water wheel from the classical world: the undershot, the overshot, and the horizontal. These continued in use with small improvements; see for instance the reversible overshot wheel depicted in Agricola's mining treatise (Pl. XXI). By moving the spout at the top, one could direct the flow of water into either of two sets of buckets around the rim of this wheel. One set of buckets turned the wheel one way and the other set the other way.

A sore point in the conflict between the medieval social classes was the feudal lords' assertion of a monopoly on grinding grain within their demesnes. All yeomen, tenants, and serfs were supposed to bring their grain to their lords' mills to have it ground at a price set by the lord. They were not even supposed to grind their own grain with querns in their own houses. The peasants paid no more heed to these rules than they were compelled to; hence much of the flour milled in medieval Europe was bootleg flour ground in illegal private mills.

In 1274, for example, St. Albans Abbey at Cirencester, England, asserted a feudal monopoly of the milling in Cirencester and demanded that the townsfolk surrender their querns. Fifty years later, the townspeople attacked the Abbey with arms and extorted from the Abbot the right to own their own querns. A few years later, the Abbot swooped on the town with his bully boys, searched the houses, and broke all the querns he found save a few that he carried off to pave the parlor floor of the abbey. Quarrels over milling rights dragged on for centuries, and some of these medieval monopolies were not finally done away with until +XIX.

The biggest medieval stride in the use of water power was not so much in the wheels themselves as in the uses to which they were put. In ancient times, water power was used only for milling grain and raising water, save for Ausonius' solitary mention of a water-powered sawmill.

The next mention of a water-powered sawmill comes eight centuries later. It is a sketch in Villard de Honnecourt's notebook (+XIII). The saw is hung by one end from a sapling braced at an angle and attached to the ground by a wooden linkage at its lower end. Four spokes on the mill-wheel axle push down the linkage as the wheel turns, and the spring of the sapling pulls the saw up again each time. This looks like a crude preliminary design that would not work in practice.

All we can say for certain about the origin of the water-powered sawmill is that it may have existed in some form from the time of Ausonius to that of Villard, but we know nothing of how these mills worked. In the century after Villard, however, allusions to water-powered sawmills become common in France and Germany. Nevertheless, as late as +XVII and +XVIII, when some enterprising Englishmen tried to set up such mills in England, the mills were wrecked by mobs of hand sawyers, who feared that their bulging muscles would become obsolete.

During the Middle Ages, water power was also applied to the bellows of smelting furnaces, to trip hammers for crushing ore in smelteries and bark in tanneries, to fulling mills, and to grinding and polishing armor and other metal wares.

Horses and mules also powered these machines by an apparatus called a horse whim. The animals were hitched to the ends of booms 10 to 15 feet long, or to the rim of a wheel of 10- to 15-foot radius. They could now work much more efficiently than in ancient times. For one thing, with the horse collar they could pull four times as hard. For another, they were allowed to walk around a big enough circle so that they could make full use of their strength.

The biggest novelty in European power machinery, however, was the windmill. The origin of the European windmill is another mystery. In +I, Heron of Alexandria proposed to power a hydraulic organ by means of a little windmill or pinwheel. Nine centuries later, the Iranians of Seistan built windmills to mill their grain.

Now, Heron's pinwheel had a horizontal shaft, like windmills of the familiar European type; whereas the Persian windmill, as we have seen, had a vertical shaft like a revolving door. That was all very well in Seistan, where the wind blows for months from one direction. But it would not be practical in Europe, where the wind blows every which way. Therefore European windmills had to be pivoted, so that they could turn to face the wind.

The earliest trustworthy accounts of windmills in Europe dated from +XII. These windmills are of a type called the post mill. On a pedestal is mounted a small room, roughly cubical, so that it can be turned in any direction by means of a long wooden boom. The room contains the millstones, driven through gearing by a shaft that projects through the wall of the room and carries the sails. The sails may be of wood or a combination of wood and cloth.

The number of sails varied. There were usually four sails, but sometimes there were six or eight. They were usually adjustable for different wind velocities.

Along the northern shore of the Mediterranean, a type of mill still in use has ten cloth sails, like the jib sails of a ship. When the wind freshens, the miller takes in sail by stopping the mill, unhitching one of his sails, winding it around its boom to reduce its exposed area, and tying it fast again.

Whence came the European windmill? Is it an enlargement of Heron's pinwheel? Was it invented in response to a rumor about the windmills of Iran? Nobody knows, although the basic difference of principle between horizontal-shaft and vertical-shaft windmills argues for Heron and against Iran.

Post mills persisted through the High Middle Ages with gradual improvements, such as brakes and means of adjusting the clearance of the millstones. In +XV, however, windmills increased in size in response to the demand for more power. At this time, for instance, the Dutch began to use windmills to drain the many lakes and marshes of their country.

As post mills were enlarged, they became unwieldy, since the whole mill had to turn. They were also likely to be wrecked in gales. So the millwrights developed a mill of another type, called the turret, tower, smock, or cap mill. In the turret mill, the main body of the mill is a fixed, solid structure of wood, brick, or stone. On top of this tower is a revolving turret bearing the sails. The first mill of this kind may have been one built by Leonardo da Vinci at Cesena in 1502. The great Florentine dreamer designed such a mill, though it is not certain that it was actually built. Half a century later, the Dutch were using turret mills for drainage.

With this improvement, mills could be made much larger and more efficient. Millwrights soon were turning out turret windmills that developed as much power as three water wheels, or twenty-five horses, or three hundred men.

There are various ways of turning the turret of a smock mill. In Greek mills of the jib-sail type, the turret is turned by hand. The miller levers it around with a crowbar, which engages pegs on the top of the tower wall and recesses in the turret.

The final refinement in windmill construction was the British invention (+XVIII) of the fantail or fly. This was a little windmill mounted on the rear of the turret, with its sails at right angles to those of the main mill. By means of gearing, the fantail kept the mill facing the wind. This was one of the first self-regulating machines.

Renaissance engineers also designed a number of windmills with vertical shafts like those of Iran. They devised systems of guiding the wind by vanes and feathering the sails so that the mill should turn regardless of the wind direction. But none of these designs came into general use. The fact is that the horizontal-shaft windmill is more efficient than the other kind, because the whole area of its sails is in use all the time, whereas in a vertical-shaft mill the sails transmit power only during half of each revolution.

Medieval improvements in iron- and steel-making led to advances in armor. During the Dark Ages a man was considered armed if he had a shield on one arm and a sword, spear, or ax in his other hand. If he were well off, he might also wear a simple helmet and a vest of hardened leather. If he were really rich, he might have a shirt of chain or scale mail.

The Byzantines continued the late-Roman practice of armoring mounted men from head to foot. As conditions improved in western Europe, so did the armorer's trade. By the time of the Norman conquest of England, the horseman wore a whole chain-mail suit, with mail sleeves to the elbow and mail breeches to the knee. His helmet came to a point on top and had a vertical bar called a nasal in front of his face and perhaps a neck guard, either of plate or of mail, behind.

Thus armored, the first Crusaders set out for the Holy Land. During the next century, the suit of mail was extended over the whole body, including hands and feet. A mail hood covered all the head and neck except the face. Over the hood the knight might wear a barrel-shaped closed helmet with narrow slits in front to see through.

Chain mail had many advantages. It was flexible, allowed free movement, and was easy to tailor to the wearer's size. But it also had shortcomings. For one thing, the knight had to wear thick padded clothing under it, lest a heavy blow break his bones even though it did not pierce his mail. This padding got terribly hot in summer, especially in the Mediterranean lands. For another, to raise either arm meant dragging up the mail on that whole side of the body. This extra effort helped to wear out the warrior.

The mailed knights of the High Middle Ages proved very effective against the Turks and Arabs of the Near East in the earlier Crusades, until the Saracens learned to deploy armored cavalry, too. But these knights proved helpless against the lightly armored Mongol horsemen, who easily slaughtered whole armies of mailed knights in Poland and Hungary (1241). The secret of Mongol success, however, lay not in particular weapons or armor but in their formidable organization, discipline, and tactics, all of which were in short supply in medieval Europe.

Early in +XIV, just when the gun was appearing, armorers began adding plate to chain mail. At first they strapped a few iron plates on here and there over the mail. By 1400, mounted men wore complete suits of plate armor over mail. Thereafter, chain mail dwindled in area until it only covered a few gaps in the plate, as under the armpits.

People often think of the knight in full plate armor as living in the High Middle Ages. In fact, however, knights wore only chain mail through +XII and +XIII. The complete suit of plate, topped off by a crested helm with a movable visor, belongs not to the High Middle Ages but to the Renaissance, when feudalism was crumbling, knighthood had become a mere picturesque affectation, and the gun was beginning to dominate the battlefield. In fact, the ironclad Renaissance horseman, in addition to all his other hardware, often carried one or two enormous horse pistols in saddle holsters.

Armor for infantry evolved along similar lines, except that a footman, because of the weight, did not wear so complete a suit. While a 75-pound suit of armor might be no great burden to a strong mounted man, it was impractical on foot.

Thus at Agincourt (1415) the French, having experienced disasters with mounted charges against English longbowmen, decided to advance on foot. But a heavy rain had made the battlefield into a bog, into which the full-armored French knights sank to their calves and were slaughtered. During the plate-armor period, more than one knight died of heart failure without striking a blow, simply from the strain of bearing his armor.

In its heyday, however, the full suit of plate armor was a marvelous thing. Its weight was so well distributed, and the joints were so craftily contrived, that as long as his heart held out the wearer could get around almost as actively as he could without armor. Its protection was so complete that in the important battle of Zagonara (1423), when the Milanese beat the Florentines in a place that had been flooded by heavy rains, "no death occurred except those of Ludovico degli Obizi and two of his people, who having fallen from their horses were drowned in the morass" by the weight of their armor. When in 1440 the Florentines in their turn beat the Milanese at Anghiari, "only one man died, and he, not from wounds inflicted by hostile weapons . . . but, having fallen from his horse, was trampled to death."[10] If men will make war on one another, that, it seems to me, is the right way to do it.

Such a suit was also very costly. It had to be fitted to the wearer, and tailoring sheet iron is not so easy as cutting and stitching cloth. During +XV and +XVI, when plate armor was at its height, the weapons of footmen–bills, halberds, guisarmes, and fauchards–resembled nothing so much as enormous can openers on poles.

While plate armor grew up at the same time as the gun, in the long run the latter prevailed. Early handguns were no more effective than longbows and crossbows. Any of these weapons could pierce armor at close range with a square hit. But at a distance, or with a glancing blow, good armor shed all these missiles.

As guns improved in +XVI and +XVII, armorers made armor thicker to keep out musket balls. Then the armor became so heavy that the wearer could hardly move. As a result, soldiers began giving up armor, piece by piece, starting with the lower legs. Three-quarter suits (down to the knees only) with visored helms were still worn in the English Civil War (1642–46). But thereafter armor was quickly abandoned and had almost disappeared by 1700, although a few units of heavy cavalry continued to wear cuirasses and helmets down through the Franco-Prussian War (1870–71).

A startling revival of fighting in armor almost occurred as lately as +XIX. Medieval Europe inherited from the Germanic tribes the barbarous custom of trial by battle, rationalized on the dubious grounds that God would grant the victory to the right. In 1817, when everybody thought the custom had been safely embalmed in the history books, an Englishman named Ashford accused another man, Thornton, of murdering Ashford's sister. Thornton challenged Ashford to present himself in the lists, armed *de cap à pié,* for the wager of battle with lance and sword. When Ashford failed to appear, armored or otherwise, Thornton claimed he had won his case. The lawyers found to their amazement that indeed Thornton had, because Parliament had never gotten around to abolishing trial by battle–an oversight hastily corrected at the next session.

After the burning of Charlemagne's wooden bridge over the Rhine, few bridges were built in Europe for several centuries. These few were wooden bridges, too. Some Roman bridges also survived the tooth of time. Otherwise, those who wished to cross rivers went back to using fords and ferries.

Stone bridges revived in +XII. The first bridgebuilders seem to have been the monks of certain monasteries. The story of the *fratres pontifices,* however, has become so overlaid with legend that it is hard to recover any solid facts about the bridgebuilding brothers.

The most celebrated bridge of this era was the bridge at Avignon, about which a well-known folksong was woven. It is supposed to have been built by a nebulous Saint Bénezèt, 1178–88. The original design is doubtful, because the bridge was rebuilt in +XIV.

However, the Pont d'Avignon probably consisted of two bridges in tandem, meeting at an island in the river and forming a blunt angle. Each section consisted of eight spans, more or less, with some more spans across the island. The arches were probably of the semicircular type, copied from those of the Roman aqueduct bridge called the Pont du Gard, twenty-odd miles away. A chapel stood on one of the piers.

At the same time that the Pont d'Avignon was built, an English cleric, Peter Colechurch, collected funds from everybody from king to peasant to build a stone bridge at London. Old London Bridge (1176–1209) showed how backward England was at that time compared to France. It was a crude and crazy affair with nineteen arches, all of different sizes, and a draw span. The arches were of the pointed Gothic type, though for bridges such an arch has no advantage over the semicircular arch.

The piers were erected on starlings or artificial islands. These were boat-shaped structures of piles and stone, all of different sizes and spacings and taking up more than two-thirds of the width of the river. As the Thames at London is a tidal estuary, a strong current runs back and forth. The bridge so choked this flow that at times the water on one side of the bridge was several feet higher than on the other. The water raced between the piers at dizzy speed, and only the most foolhardy boatmen "shot the bridge" at these times.

Further to complicate matters, people began erecting houses on the piers, straddling the roadway, until there were about a hundred such structures, many of three or four stories. Between the overloading of the bridge above and scouring by the river below, one part or another of the bridge was always crumbling. Hence the song: "London Bridge is falling down, falling down. . . ." In the 1820s the picturesque monstrosity was finally demolished to make way for a new bridge.

Medieval bridges differed in some ways from Roman bridges. One difference was that the builders deliberately made their roadways narrow so that they should be easier to defend. The roadway of the Pont d'Avignon varied from 6½ to 16 feet. Many bridges were fortified. The fortification of Southwark Gate on Old London Bridge helped to break at least two rebellions. The handsome Pont Valentré at Cahors, France, has three 130-foot fortified towers bestriding its roadway, one at each end and one in the middle.

Moreover, medieval piers were provided with pointed cutwaters downstream as well as up. The Romans had furnished bridges with such cutwaters on the upstream side only. As the engineer Alberti (+XV) correctly explained:

> . . . the water is much more dangerous to the stern than to the head of the piers, which appears from this, that at the stern the water is in a more violent motion than at the head, and forms eddies which turn up the ground at the bottom, while the head stands firm and safe, being guarded and defended by the banks of sand thrown up before it by the channel.[11]

From +XIV on, bridgebuilders experimented with a variety of arches. One was the segmental arch, beginning with Taddeo Gaddi's Ponte Vecchio or "Old Bridge" at Florence (1345). This arch is an arc of a circle less than a full semicircle. Although its outward thrust is greater than that of the semicircular arch, this did not much matter when the bridge was braced at each end against solid river banks. And the piers, being fewer in proportion to the size of the bridge, interfered less with the flow of the river.

In the Renaissance, other curves were tried. Sometimes, as in the Pont Neuf at Toulouse (1540–1603), the arches were semi-ellipses. That is, they were the upper halves of ellipses lying on their sides. Once the standard types of stone arched bridge had been established in the Renaissance, nearly all large bridges were built this way with little change down to the development of steel bridges in +XIX.

Although the magnificent Roman roads continued to carry a dwindling traffic for several centuries after the fall of the Western Empire, by the High Middle Ages they had largely disappeared. People had pried up their stones for local use, and the feudal system could not support any such elaborate road system. At this time the road net of Europe was a mixture of Roman roads, with or without their original paving; roads, sometimes lightly paved, built under orders of kings and nobles; and supplementary tracks across the land, wide enough only for a single man or beast.

In wet weather, unpaved roads became elongated quagmires.This was no great disadvantage so long as the traffic consisted entirely of men afoot or mounted. Women and old men of the ruling class, unsuited to long mounted journeys, traveled in horse litters. A horse litter was a sedan chair slung between two horses, one fore and the other aft.

During the High Middle Ages, wheeled traffic, which had almost vanished, increased. The roads of Europe then bore a brisk traffic not only of pedestrians, mounted men, pack animals, and litters, but also of wagons, carts, and handcarts.

Technical improvements added to the efficiency of four-wheeled wagons, which had never been very popular in ancient times. One advance was the discovery that horses could be hitched in tandem. This system probably came from China, where it appeared in +II. Thus the load could be increased without requiring a wider road to move on.

Another improvement was making the front wheels small and cutting away the body in front, underneath, so that, as the front axle turned on its pivot, either wheel could roll in under the body without striking it. In this way the wagon could make sharper turns than had been possible in ancient times.

Throughout the High Middle Ages, a few persistent souls put tops on four-wheeled wagons and board seats inside for passengers. Lacking brakes, these vehicles were kept from running away downhill by chaining one wheel fast so that it could not turn. Because of the cost, discomfort, and lack of suitable roads, these springless vehicles never became really popular. Most of them were state carriages in which royalty rode from time to time to awe their subjects.

In 1457 a clever but nameless mechanic of Kocs, Hungary, built a four-wheeled passenger wagon, nicely appointed and upholstered. Although it was not much of an improvement on the ancient Persian *harmanaxa* or the Roman *rheda,* it was ahead of anything seen in Europe for many centuries. It probably had the body hung by leather straps from posts that rose from the corners of the bed frame. This primitive springing system substituted a sloshing motion for the hard jouncing of the earlier, springless carriages.

The new vehicle spread. In early +XVI, scores of them rattled the roads of Italy, when there were only three in Paris and one in England. "Coach" is simply a phonetic spelling of Kocs. Although the development of the new art of coachmaking was hindered by a struggle between the wheelwrights' and saddlers' guilds as to which should have the right to make these vehicles, by 1534 Ferrara had a busy firm that did nothing but make coaches.

With straps for springs and leather curtains for windows, these carriages were hardly up to Detroit's standards of comfort. In late +XVII, however, coach design leaped ahead. Brakes, glass windows, and steel springs appeared. In the following century came such luxurious accessories as built-in lanterns and a sword-and-pistol case for coping with highwaymen.

In +XVIII, omnibuses carrying as many as sixty passengers began to rattle across Europe on regular schedules. For the first time in history, a traveler on land could go whither he wished by simply paying his fare and taking his seat, without having to organize his own safari.

Another revival of ancient usages was the postal system. Beginning in later +XIII, European universities, led by the University of Paris, organized their own letter-carrying systems. In +XV, French and German rulers set up systems of couriers to carry official mail, like the imperial riders of Darius and Hadrian. The kings, seeing that the university postal systems were not only popular but even profitable, first regulated them and then took them over, merging them with their own governmental systems. They also wanted to be able to censor everybody's letters. By the end of +XVI, public postal systems not unlike those of our times existed, although the cost of sending a letter was still high.

All this increase in traffic, especially wheeled traffic, brought an increasing demand for better roads. For some centuries, kings sought to improve their roads. But they had little success, because they repeated all the errors of their long-dead predecessors. They commanded landowners to maintain roads that ran past their property; they sent agents to round up peasants for forced road work. But, as fast as the roads were slightly improved, the increase of traffic tore them to pieces again. As the kings were never willing or able to build really heavy paved roads of the Roman type, their roads began to dissolve into lines of mudholes as soon as they were completed. This was not entirely the fault of the kings; for, in feudal Europe, the power of a king to make his subjects do what he wished was much less than that of an ancient watershed emperor or a modern commissar.

The story of municipal waterworks in the Middle Ages is one of slowly regaining the ground lost at the fall of Rome. During the High Middle Ages, sporadic efforts were made either to put the ruined Roman aqueducts back into service or to build new ones on the Roman model. In Paris, for instance, the Roman aqueduct built under the emperor Julian was destroyed by Norse invaders in +IX, and for three centuries Parisians depended wholly on wells and on the Seine.

In +XII, two abbeys on the north side of the Seine, finding the river inconveniently far away, built their own aqueducts to lead water by masonry conduits and lead pipes from springs in the hinterland to basins in the monasteries. People living in the neighborhood were allowed to share the water. After 1200, as Paris grew, Parisians demanded more fountains. Not deeming this their proper business, the abbeys turned their waterworks over to the city.

Then, just as had happened in Rome over a thousand years before, lords and royal officers began asking for permits to tap public lines and run private pipes to their houses. At first there seemed no harm in this, as only a little water was diverted and the petitioners were privileged persons. By 1392, the number of taps was so large that the public fountains, being the last delivery points on the lines, often ran dry.

So Charles IV issued a resounding proclamation to reform this abuse: ". . . and We recall, cancel, annul, and revoke all privileges, grants, licenses, rights, duties, permissions, or suffrances . . . excepting only," he weakly finished, "so far as it affects Us and Our uncles and brother."[12]

The result was what one would expect. All the other nabobs in Paris said: if the King and the dukes can have private running water, why cannot I? And private water supplies they proceeded to get, by fair means or foul.

But, as nobody provided additional sources of water, the scramble became more acute than ever. Paris went through another spasm of reform in 1553. This, like the previous one, proved ineffectual because Henri II, like Charles IV, insisted on excepting his royal relatives.

This dismal contest for the available water continued until Henri IV, the French Perikles, firmly grasped the problem in 1600. After another abortive attempt to cut off all the private connections, he sensibly decreed that the pipes might remain, but that the users must pay fees large enough to support the water system.

Furthermore, he authorized a Flemish engineer, Jean Lintlaer, to install a pumping station under one of the arches of the Pont Neuf. Here an undershot water wheel worked four pumps, which raised river water high enough so that gravity could speed it to the royal palaces. Henri turned the surplus water over to the public, and Paris was on the way to acquiring a modern water system.

As for sewers, the people of medieval Europe not only started far behind the Romans, but also have only recently begun to catch up. Again Paris may serve as an example.

Medieval Paris had open ditches, intended for storm drains only. People were not supposed to dump household waste into them, but people did. As Paris seldom receives downpours heavy enough to flush out such ditches, the result was noisome in the extreme.

Around 1400, Hugues Aubriot, Provost of Paris under Charles VI and builder of the Bastille, roofed over a stretch of one of these sewers. After Aubriot's time, the rest were gradually covered.

Paris still lacked a proper sewer system, even though by +XVI the connection between filth and disease was dimly realized. Around 1550, Henri II repeatedly tried to get the Parliament of Paris to build new sewers. But, as the king did not propose to pay for this work himself, Parliament evaded the king's recommendations.

Later proposals met a similar fate. Arguments about stench and disease made no impression on those who would have had to pay for this improvement. The man of the medieval and Renaissance city, having grown up with the stink of ordure in his nostrils, saw nothing very terrible about it, while the prospect of parting with money made him scream with anguish.

In +XVII, under Louis XIII, a beginning was made towards building a few real sewers. But the great and celebrated web of modern Parisian sewers is only about a century old, having been built under Napoleon III.

In another branch of hydraulic engineering, however, Europeans soon advanced beyond the Romans. This was the building of canals. Small irrigation canals appeared in the Po Valley in the early Middle Ages, as part of the general upsurge of technology that took place in North Italy.

In 1161 the Holy Roman Emperor, Frederick Barbarossa, captured Milan and razed it to the ground. When they rebuilt their city, the Milanese decided to surround it not only with an impregnable wall but also by a moat. To furnish water for this moat, they dug a canal for sixteen miles–an unheard-of distance then–to the river Ticino, which carries the waters of Lake Maggiore to the Po. This canal they called the Ticinello, or Little Ticino.

At first used only to fill the moat and to irrigate fields along its path, the canal was soon enlarged to navigable size and renamed the Naviglio Grande, or Great Canal. Its story then became the common one of the struggle between farmers, who wanted to lead off enough water to irrigate all their lands and to clutter the channel with water wheels and fish weirs, and boatmen and merchants, who wanted the channel kept full of water and open for navigation.

Historians argue as to when and where the first canal lock was built. Some believe it to have been at Vreeswijk in the Netherlands in 1373, or at Spaarndam a little earlier. Others find the origin of the lock in Italy around 1400.

In any case, about 1400 the Bishop of Milan was building a cathedral. To make it easier to get stone to the site, he persuaded the government of the city to extend branch canals into the city by way of the moat. These branches were also used by merchants for delivery of goods. A single gate controlled the flow of water into the branch canals, which were several feet lower than the moat. As this system did not work very well, two engineers[13] installed a lock in 1438, and later a second lock.

The first clear written description of a canal lock is by the versatile Alberti, who wrote:

> Also, if you wish, you can make two gates cutting the river in two places at such a distance one from the other that a boat can lie for its full length between the two; and if the said boat desires to ascend when it arrives at the place, close the lower barrier and open the upper one, and conversely, when it is descending, close the upper and open the lower one. Thus the said boat shall have enough water to float it easily to the main canal, because the closing of the upper gate restrains the water from pushing too violently, with fear of grounding.[14]

In the 1450s, the engineer Bertola da Novate put these ideas into practice. The dukes and republics of North Italy kept Bertola, the ablest canal engineer of his time, busy all his life digging canals for them. Sometimes they quarreled over who should have priority on his services. His only trouble was that his workmen sometimes could not understand his advanced concepts.

All these early canal locks seem to have been of the portcullis type, sliding up and down like a window sash. Later, the hinged gate, swinging like a door, was found more practical.

Besides Milan, other cities of Lombardy also built canals, some as early as the 1180s. At the same time, Mantua began reclaiming marshland. Land reclamation—*la bonifica*—became a major Italian interest. Whereas the ancient Mesopotamians preserved marshes in which to grow reeds for house-building, and whereas we sometimes preserve marshes to provide refuges for wild life—today's most persecuted minority—medieval Italians wanted to dry up the marshes for the extra farmland. Besides, they had a suspicion that marshes were connected with malaria, even if they did not know how the disease was transmitted. So they energetically dug, ditched, and drained.

Meanwhile, across Europe on the shores of the North Sea, another vast land-reclamation project was taking shape. The Netherlands is a flatland as innocent of stone as Babylonia. Much of it is below sea level, at least at high tide. The coastal region was originally a tangle of marshes, shallow lakes, dunes, tidal flats, and low islands half submerged at high tide. Through this perpetual wetland, the rivers Rhine, Meuse, and Schelde, with many local changes of name, lazily wound their way.

Men long ago began tinkering with this ameboid mixture of land

and water. In −12 the Roman general Nero Claudius Drusus is said to have joined the largest of the lakes, the Flevo Lacus, with the Rhine by way of the Yssel.[15]

In Drusus' time, the people of the Low Countries were Germanic tribes, some of whose names–Batavi, Frisii–still appear in one form or another on the map. When the Western Empire was crumbling, the Netherlanders were called Franks. Dreaded for their ferocity and treachery, they overran Gaul. Around +500, their fierce and crafty king, Clovis, smashed the remnants of Roman and Gothic power and founded the kingdom of Frank-land, or France.

Many Franks remained in the Low Countries to become the peaceful and progressive Dutch of today. In the early Middle Ages, the Dutch began building dykes to keep within bounds the bodies of water that cut up their country.

During +XIII, a series of floods occurred. In 1277, these culminated in a great storm, which broke through the dunes that formed a natural dyke between the North Sea and the Flevo Lacus. The sea poured in, drowning thousands and enlarging this lake to the Zuider Zee of later times.

Little by little the Dutch once more began dyking, draining, and pumping to turn water back into land. Over the centuries they worked out very efficient methods of doing this, using local materials: sand, clay, straw, seaweed, reeds, brush, and pilings.

Dutch land-reclamation became especially active after 1400, when the Dutch applied windmill-driven pumps to the task. One patch of sub-sea land after another was surrounded by dykes and pumped out, creating what the Dutch call "polders." The Zuider Zee itself is now being liquidated, piece by piece.

And by the bye, the story of the little Dutch boy who stuck his finger in a hole in a dyke could only have been invented by somebody who had never seen a dyke. The story implies that a dyke is some sort of wall. Actually it is a long, low ridge whose slopes are so gentle that you might not even notice them. A typical dyke is about 10 to 16 feet high and 15 to 20 feet wide at the top–but at least 100 feet wide at the bottom.

The remaining aspect of medieval European technology is shipbuilding. During the Dark and early Middle Ages, the ships that plied the Mediterranean and the North Sea were not very different from those of classical times. Galleys were used for war and tubby sailing ships for commerce.

In addition, there grew up a class of merchant galleys, with the same number and arrangement of oars as regular war galleys but with larger hulls, which could hold cargo as well as rowers. These carried goods of large value and small bulk, such as gold, silk, and spices, and took pilgrims to the Holy Land. The pilgrims preferred galleys to roundships because, whereas the latter sailed directly from some European port to the Levant, the galleys, having little sea-keeping capacity, crept along the coast, stopping at famous cities of antiquity. The pilgrims, like any other tourists, wanted to see all there was to be seen. The merchant galley disappeared in +XVI as a result of the high cost of running it, the improvement of the sailing ship, and the development of competing Atlantic routes to the Orient.

There had been some changes in galley design since ancient times. A Mediterranean galley now carried its ram above the waterline, like a thickened bowsprit. In +XIV and +XV, the typical Mediterranean galley was a trireme, with twenty-five to thirty benches on a side and the rowers arranged as shown in Fig. 24. The Byzantine arrangement of rowers, vertically staggered on two levels, had given way to one in which all three rowers of each group sat on one bench but pulled separate oars.

In the 1520s Vettor Fausto, a professor of Greek in Venice, persuaded the Venetian government to let him build a quinquireme he had designed. It should be, he said, like those of the ancient Greeks and Romans, of which he had read in his classical studies.

What Fausto did was to enlarge the ordinary Mediterranean galley so that five men could sit on each bench instead of three. Each, however, still pulled a separate oar. Although this ship beat a standard trireme in a race, the design was not repeated because she proved hard on the rowers, who were excessively crowded and exposed to the weather. Nevertheless, Fausto led a successful career as a shipwright, converting some Venetian merchant galleys to quadriremes and devising other improvements.

Later in the century a Venetian pupil of Fausto, Giovanni di Zaneto, developed another type of ship. This design, called a *galeazza* or galeass, was based partly upon the merchant galley, then disappearing, and partly on Fausto's quinquireme. Giovanni, with some mathematical help from Galileo, went back to the Hellenistic system of mounting large oars in a single bank, with five to ten rowers pulling each oar.

For several centuries, Venice was the leading Mediterranean naval power. It was an oligarchic republic, governed by a merchant aristocracy like that of ancient Carthage and ruling a large overseas dominion.

Most of the Venetian empire consisted of Greek islands taken from the Byzantines when the Crusaders temporarily overthrew the Byzantine Empire in 1204.

Venetian shipwrights led the world not only in the design of ships but also in methods of manufacture. For instance, the Arsenal of Venice made serious efforts to standardize the sternposts and other parts of galleys, thus anticipating Eli Whitney and others who, in the early 1800s, developed the principle of interchangeable parts. The Venetian shipyards even employed a kind of assembly-line manufacture, as an admiring Spanish visitor noted in 1436:

> And as one enters the gate there is a great street on either hand with the sea in the middle, and on one side are windows opening out of the house of the arsenal, and the same on the other side, and out came a galley towed by a boat, and from the windows they handed out to them, from one the cordage, from another the bread, from another the arms, and from another the balistas and mortars, and so from all sides everything which was required, and when the galley had reached the end of the street all the men required were on board, together with the complement of oars, and she was equipped from end to end. In this manner there came out ten galleys, fully armed, between the hours of three and nine.[16]

Venetian sea power slowly declined from +XVI on. The reasons for this decline were the conquest of the Venetian overseas possessions by the Turks, the withering of the Mediterranean trade routes as new routes opened up in the Age of Exploration, and the failure of Venetian supplies of timber. In particular oak, used for the frames and hull planking, became scarce, despite intelligent efforts by the Venetian government at conservation and reforestation. The Venetian nation disappeared in the Napoleonic wars, coming first under French and then under Austrian rule.

Another change in maritime methods was that rowers, instead of being free workers as in ancient times, became captives chained to their benches. Use of slaves and prisoners first became common in +XV. No doubt this change was fostered by the fact that multitudes of people were captured in the constant wars and piratical raids of the Christian and Muslim powers of the Mediterranean against each other. The Turks, like the Romans of the late Republic, were given to slave-raiding and slave-owning on a huge scale; while, during the Reformation, the French and Spanish kings found the galleys a useful way to dispose of Protestants.

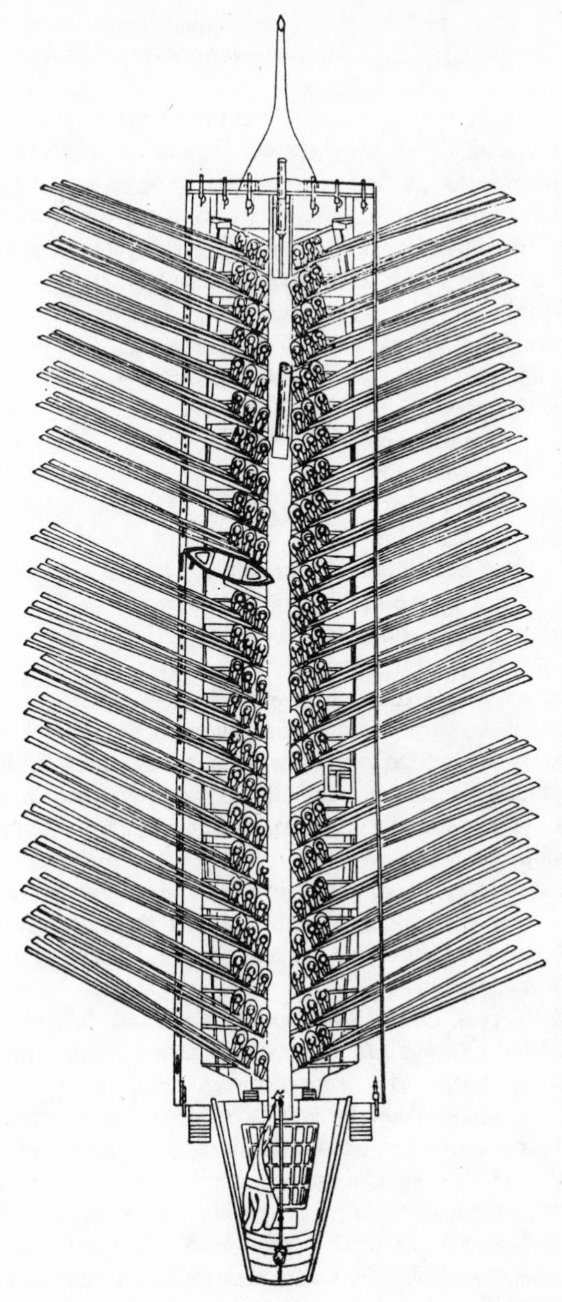

Fig. 24. Plan of a 16th-century Venetian galley, showing rowers in groups of three on the same bench, pulling separate one-man oars (*from E. A. d'Albertis:* Le Construzioni Navali . . . , *1893*).

The Venetian navy stuck to free rowers, who could fight in a pinch, as long as it could. But at last even Venice was forced to use prisoners, because employment of such persons had made free workers unwilling to serve as rowers.

A danger of using slaves and prisoners was that, since the rowers comprised a large majority of all the people on board, they might seize the ship, toss the free men overboard, and sail away to become pirates. To prevent this, captive rowers were permanently chained to one place. As a rower had to eat, sleep, and do everything else in that one spot, galleys became so foul that their officers went about their tasks with handkerchiefs sprinkled with musk pressed to their noses. The rowers sickened and died like flies, while those who lived must have been in such poor condition that they cannot have been very efficient oarsmen. In this respect, at least, they did things better in ancient times.

As for sailing rig, large galleys now bore lateen sails on one, two, or even three masts. Some merchantmen also flew lateen sails, although many still plodded along under one square sail.

The ships of the North Sea differed in small ways from those of the Mediterranean. They stuck to the square sail, and they carried but one steering oar or quarter rudder, on the starboard ("steer-board") side. Norse galleys lacked the ram but possessed extremely sleek, graceful, efficient lines, like those of an Adirondack guide boat.

Hulls also differed in construction. In the "carvel-built" Mediterranean hulls the planks were set edge to edge, while in the northern "clinker-built" hulls the planks overlapped. The latter system, adapted to the heavy Atlantic seas, gave a stronger hull but, because this hull was less smooth, a slower ship.

Soon, however, a series of revolutions overtook ship construction. One change was the central rudder, which probably reached Europe from China. Several pictures from about the middle of +XIII, on municipal seals, church windows, and so forth, show this feature.

In the High Middle Ages, naval powers often reinforced their galley fleets with converted merchantmen or *nefs*. Although these ships could carry large numbers of troops, they were not very successful because, save in heavy weather, galleys could always dodge the wallowing roundships. In late +XIV and early +XV, however, changes occurred, which drastically altered the rôles of these ships.

One was the development of a larger roundship with a large mainmast in the center and two smaller masts, one at each end. The foremast normally carried a square sail and the aftermost or mizzen mast a lateen sail. Although these small sails did not make the ship go much faster, they greatly helped in controlling its direction. The ship could be steered by its sails alone. The lateen mizzen was especially useful in holding the ship on a close-hauled course into the wind.

As ships got larger, their rig was divided into more and more sails until the full-rigged ship of the last two centuries evolved. The Age of Exploration (+XV to +XVIII) took place as soon as European ships and navigating instruments had developed to the point where voyages across oceans were practical. During the Age of Exploration, the Dutch were the most active people in the invention of new rigs and sails, such as the gaff-headed sail and the jib or forestay sail.

The other change was placing guns on shipboard. A few guns were mounted along the sides of nefs and in the forward deckhouses of galleys. Galley guns fired straight forward only, so that one aimed the guns by aiming the ship. For over a century, however, not enough guns were mounted on ships to affect the outcome of any battle.

In 1538, however, a Venetian and Genoese fleet set out to attack the mighty Turkish navy. The Genoese admiral, Doria, had 166 galleys, 64 nefs, and a new experimental ship, the *Galleon of Venice*. The *Galleon* was a large, solidly built sailing ship bearing the unheard-of number of 128 guns.

The Italians found a smaller Turkish fleet at Preveza. While the Christians were patrolling in front of the harbor, the *Galleon* fell behind the rest, becalmed. Out rushed the 122 Turkish galleys like wolves to pounce upon the straggler, while Doria timidly watched from a distance.

The Venetians, unable to move, received the onset with a terrific blast of cannon and arquebus fire. The galleys lost way as the fire slaughtered their rowers; one was sunk by a single shot. Others stove in their own bows in ramming the *Galleon*. All afternoon, the Turks kept attacking in divisions of twenty ships at a time. At nightfall the Venetians, with fifty-three dead and wounded and their deck littered with wreckage, were still holding out. The equally battered Turks withdrew, and the Italians towed the *Galleon* away.

Although this battle achieved nothing politically, it showed the shape of things to come. At the great galley battle of Lepanto in 1570, which loosened the Turkish grip on the Mediterranean, guns played an important part. Here the combined Italian and Spanish fleets included six Venetian galeasses with seventy guns apiece, the larger cannon being mounted in the deckhouses at the ends and the smaller ones along the sides between the oars.

The weight of the cannon made these ships so heavy that their rowers could scarcely move them, and they had to be towed into action. Nevertheless, standing out in front of the allied line, they fearfully strafed the advancing Turkish galleys with gunfire:

> Don John's hunting, and his hounds have bayed–
> Booms away past Italy the rumor of his raid.
> Gun upon gun, ha! ha!
> Gun upon gun, hurrah!
> Don John of Austria
> Has loosed the cannonade.[17]

This bombardment weakened the Turks and so contributed to the Christian victory.

Thereafter galleys swiftly declined. Although because of their rowers they could move in any direction regardless of the wind, galleys were so flimsy and feebly armed, compared to the full-rigged galleon, that even overwhelming numbers and a flat calm did not assure the victory.

Although Mediterranean powers built a few galleys as late as +XVIII, except for a few with the Spanish Armada they never fought a major battle after Lepanto. They lingered on to the dawn of steam in early +XIX, because they could still do one thing better than sailing warships. They could pursue pirates, who also used galleys.

With the battle of the Armada in 1588, naval warfare, which for over 3,000 years had consisted mainly of ramming and boarding contests, turned into a series of artillery duels. So it remained right down to +XX, although ramming and boarding continued to play some small part, even in the two World Wars. Therefore, despite the steam-and-steel revolution in ship construction of +XIX, naval warfare became modern, in an important sense, between Lepanto and the Armada.

The name "Renaissance," literally "rebirth," is applied to the period, roughly +XV and +XVI, when many medieval things like feudalism, Gothic architecture, and the religious monopoly of the Catholic Church were passing away and many modern things like vernacular literature, centralized national governments, political parties, and experimental science were coming into being.

In the narrowest sense, "Renaissance" refers to the revival of learning that took place, mainly in Italy, with the rediscovery of most of the remains of classical literature that we possess today. Some classical writings had of course never disappeared from circulation, but many more were now exhumed from monastic libraries or imported from the crumbling Byzantine Empire.

Hence there arose a class of scholars who devoted their lives to the discovery, publication, and explanation of classical texts. The study of Greek was revived. Fashionable people acquired at least a veneer of this scholarship; they proved their gentility by lacing their speech with quotations from Plato and Cicero.

The period +XV and +XVI, however, involved other revolutionary movements besides the revival of learning. These included the Reformation, the Age of Exploration, and the downfall of the old astronomy, which put the earth at the center of things. And it included the first patent systems for the encouragement of inventions.

It also involved an expansion of engineering. The new engineering, except in the field of architecture, had but little connection with the revival of classical learning. There was, in fact, a good deal of hostility between the humanists or "ancients," who cared for nothing but the revival of classical literature, and the "moderns," who wanted to expand and exploit the new discoveries in the arts and sciences.

In architecture, the recovery of the treatise of Vitruvius led to a passion for imitating classical temple design. This vogue has lasted right down to the present, as a glance at the relief of the Lincoln Memorial on a new cent will show you.

For the most part, however, Renaissance engineering grew out of the experience of the Middle Ages. In some ways it lagged behind the examples of antiquity, while in others it leaped ahead of them.

The main change in engineering in the Renaissance was that it waxed much greater in scale. Engineering again became a respected profession. Engineers became famous and even, sometimes, well paid. They were no longer anonymous craftsmen, humbly serving their temporal and spiritual lords. In line with the Renaissance tendency towards uninhibited self-assertion, they promoted themselves, grasped for personal fame, and told off their rivals and employers when they thought themselves wronged.

Moreover, they took direct action against those they considered their enemies. When about 1575 Henri II of France sent the engineer Adam de Crapone to inspect some fortifications at Nantes, which the king suspected of having been faultily built, the builders at first tried to flatter Crapone into giving them a clean bill of health. When that did not work, they invited him to a banquet and poisoned him.

One of the earliest Renaissance engineers was Filippo Brunelleschi (1379–1446). The son of a notary of Florence–the Athens of the Renaissance–Brunelleschi was, like most youths of artistic leanings at the time, apprenticed to a goldsmith. Here he learned clockmaking. Later he put this knowledge to use in building machines.

Like many Renaissance engineers, Brunelleschi spent most of his time on art and architecture, though he was also active in other fields. He mastered the new science of perspective and applied it to his architecture.

Failing to win a competition for the design of a pair of gates for a church in Florence, Brunelleschi went to Rome with the sculptor Donatello. There he studied for five years. He particularly studied the dome of the Pantheon, as rebuilt by Hadrian.

In Florence stood a cathedral, Santa Maria del Fiore, which had been under construction for over a century. No roof had been built over the crossing of the nave and the transepts. The covering of this part would have to be something unusual, as the space to be roofed was about 138 feet across.

In 1407, hearing that a conference of architects was meeting in Florence to consider this problem, Brunelleschi hurried home to present his proposals. Nothing came of this action and Brunelleschi returned to Rome. Eleven years later he was back in Florence, urging his plan for a huge octagonal dome. His plan seemed so daring that once the committee had him thrown bodily out of its meeting. But he returned with a model and a host of arguments, until he finally won over the committee.

In 1419, Brunelleschi at last received his order to build. The committee, however, had given him an unwanted collaborator, the sculptor Ghiberti, who had won the contest for the church gates. Since Ghiberti knew nothing much about architecture, Brunelleschi determined to get rid of him. By shamming sickness he tricked Ghiberti into undertaking the design for the chain to hold the dome against bursting stress.

There is, you see, an outward, tensile component of stress in the sides of a dome, a short distance up from the base. If too much load is piled on top of the dome, the dome will fail at this point by bursting outward. Although these forces were not mathematically known in Brunelleschi's time, the Florentine still felt that such a stress existed. To counteract it, he proposed to wrap a girdle or chain around the dome.

Ghiberti accordingly produced a design for a chain, which Brunelleschi easily proved to be utterly inadequate. So poor Ghiberti was forced to resign. Then Brunelleschi installed his own chain, made of lengths of oaken timbers bolted together. Modern study shows that Brunelleschi's chain, likewise, was too weak to reinforce the dome appreciably; but the dome was stiff enough to do without it. It is still there despite earthquakes and aerial bombs.

As work progressed, Brunelleschi discovered that his workmen were losing too many hours climbing up and down the lofty ladders of the scaffolding at mealtime. So the resourceful architect installed a canteen on the scaffold.

Brunelleschi finished his task in 1436, although the 600-ton lantern—the name given the stone ornament atop a church dome—was not completed until after his death, and the cathedral itself did not receive its finishing touches until 1888. The dome so dominated the city that the cathedral has come to be called simply *il Duomo*—"the Dome."

This dome is made of two shells, one inside the other. It measures 143 feet across and 105 feet high. The total height of the cathedral, from pavement to lantern, is 351 feet, the height of a thirty-five-story building. Although Brunelleschi was a leader in the revival of classical architecture, his dome was neither Roman nor Gothic; it was something brand new. Its octagonal form, emphasized by thick ribs, was intermediate between the true dome and the square dome or cloister vault.

Brunelleschi would not have been a true man of the Renaissance if he had not been working on a multitude of other jobs at the same time as the dome. He acted as architect for several buildings, including the huge Pitti Palace for the Medici family, and two churches. He also invented construction machinery. In 1421 the Republic of Florence gave him the first known patent. This was for a canal boat equipped with cranes for handling heavy cargo.

After Brunelleschi's time, in 1474, the Republic of Venice adopted the first formal patent law and in 1594 issued to Galileo a patent on a system for raising water. A patent is a temporary legal monopoly on an invention: a license to stop all others from making, selling, or using specimens of the invention without the patent owner's permission. The purpose of patents is to encourage invention by giving the inventor a profitable monopoly, for a limited time, in return for his making and disclosing the invention, so that it becomes public property after the patent expires. If any one change could have caused our modern technological revolution to begin in classical times instead of 1,500 to 2,000 years later, it would have been a good patent law.

It is not likely, however, that any of the early Renaissance patentees made money from his patent. The territory covered by the Italian city-states was too small, and the requirements of a sound patent law were not worked out until +XIX. Yet the beginning of patents and patent laws in Renaissance Italy was one of those portentous developments that, like the revolution in reading, aroused little comment at the time but were bound in the long run to bring about a vast overturn in human affairs.

Of course, Brunelleschi had to serve as a military as well as a civil engineer. In 1430 he accompanied the Florentine army to the siege of Lucca, with Ghiberti, Donatello, and two other artists as assistants. Brunelleschi advised diverting the river Serchio by a canal, so that its water should surround Lucca and cut off the town from aid. It worked, though not quite as Brunelleschi had planned. By providing Lucca with a ready-made moat, the scheme kept the Florentines from attacking. So it was they who had to withdraw, to the joy of the Luccans.

A younger friend of Brunelleschi was Leon Battista Alberti (1404–72), born in Florence but reared in Venice because his family was banished during a political struggle.

If the man of the Renaissance was versatile, Alberti was the ideal man of the Renaissance. Painter, poet, philosopher, musician, architect, and engineer, as a youth he faked a Latin comedy in verse so well as to convince some scholars that it was a genuine ancient Roman work. He played the organ and experimented with the camera obscura.

The camera obscura, from which we get our word "camera," was a closed room, facing the street, with a small aperture for light. It could be made by covering the window with a sheet of thick paper, having a small hole. A screen of sheeting was hung opposite the aperture, and those in the room amused themselves by watching the upside-down images of passers-by on the screen. It was the Renaissance equivalent of television.

Most of his life, Alberti worked for successive popes. Nicholas V employed him to restore the papal palace and to build the Acqua Vergine, one of the new aqueducts which the Renaissance popes constructed to replace the ruined Roman waterworks.

At the behest of Cardinal Colonna, Alberti investigated the Roman ships lying at the bottom of Lake Nemi. He tried to raise one of these ships by windlasses mounted on a raft buoyed by barrels, but all he succeeded in doing was to tear off a piece of the bow.

In addition to all the palaces and churches he designed, Alberti found time to write on many subjects: political philosophy, horse breeding, family life, surveying, sculpture, Greek mythology, and architecture. His main work was *De re aedificatoria* (or *I dieci libri dell' Architettura*), a treatise on building. Although Alberti could not yet apply mathematical measures to the strength of materials, he described the properties of many different kinds of stone and wood for construction. He gave rules of thumb for the proportions of structures, such as this one for bridges:

> And there should not be a single stone in the arch but what is in thickness at least one tenth part of the chord of that arch; nor should the chord itself be longer than six times the thickness of the pier, nor shorter than four times.[18]

Alberti wrote this work in Latin about 1452. During the following years, copies circulated among his friends in manuscript. Some hitch in his arrangements delayed publication until 1485, after the author's death. When it did appear in printed form, it became very popular. The work was translated into Italian, French, Spanish, and English and led a long and honorable career.

The leading architect after Alberti was Bramante (1444–1514)—born Donato d'Agnolo in Urbino and at some point in his life nicknamed Bramante, "he who wants." After the wandering life of a Renaissance technician, he arrived in Rome in his fifties to work for the Pope. The list of all his minor architectural works would be tedious. However, he devised a screw press for stamping coins and rediscovered the old Roman art of pouring liquid concrete into wooden forms.

Pope Julius II chose Bramante to rebuild St. Peter's cathedral, on the site of a ruined early-Christian basilica. Bramante designed a huge building in the form of a cross with four stubby arms and a huge dome over the middle. The four main piers had been erected when the aged Bramante died. A succession of popes and architects wrestled with the task, making little changes here and there, which reduced the whole project to confusion. Moreover, it transpired that much of Bramante's work had been hasty and unsound.

In 1546, Paul III gave the job to the one man with the force to push the project through despite any obstacles. This was the Florentine Michelagniolo Buonarroti (1475–1564), known to us as Michelangelo. Most of his work was in the pure arts, but he was also an able engineer. At Florence and again at Rome, he was called upon to build fortifications for an expected siege. He built them and then, correctly divining that the city was bound to fall through the defenders' incompetence, fled through the enemy's lines.

Michelangelo was one of the most peculiar, difficult, and cantankerous geniuses of the Renaissance world, wherein ruthless individualism was deemed right and normal. A compulsive worker, a compulsive penny pincher, and rigidly honest in a corrupt age, Michelangelo also had an extraordinary talent for making enemies. While he was still a youth, a taunt flung at a fellow sculptor bought him a broken nose, which disfigured him for life.

Bramante could not stand him either. Michelangelo believed that Bramante had persuaded Julius II to commission him to decorate the Sistine chapel with frescoes in the hope that he would fail and fall into disgrace. But that may have been Michelangelo's persecution complex.

During the last few years of his life, Michelangelo was employed on St. Peter's. He simplified the fussiness of Bramante's design and raised the dome higher. He died at eighty-nine leaving a tidy fortune in golden ducats, although he had loudly complained all his life that he was poor to the point of starvation. The dome was finally built a quarter-century after Michelangelo's death, and the whole cathedral was completed in 1626.

Another of Michelangelo's enemies was a man who ordinarily had no time for enemies and no desire to make them. This was the most famous of all Renaissance geniuses, Leonardo da Vinci (1452–1519).

All educated people have heard of Leonardo as a painter; many have also heard that he was a scientist and an engineer, although they might have trouble naming any particular discovery or invention of his. He is one of history's most celebrated but most paradoxical characters. He has often been justly acclaimed as one of the greatest of all creative geniuses. Historians of Renaissance science and engineering often give him a whole chapter.

But, when we look into his story, we find that he had hardly any influence at all on the science and engineering of his time. In this regard he has been as much overrated as Roger Bacon.

Was he persecuted? Not at all. Although a religious skeptic, he kept his doubts in his notebooks and so was never bothered by the Church. His lack of impact in the technical and scientific fields was due to his own peculiar nature.

Leonardo was the illegitimate son of a notary of Vinci, a village near Florence. His parents were married almost at once, but not to each other, and Leonardo's father brought him up. He was apprenticed to the artist and goldsmith Verrocchio in Florence, where he knew Toscanelli, the leading physicist of the age. In his late twenties he painted pictures, some for Lorenzo de' Medici, political boss of the Republic of Florence.

In 1483 Leonardo removed to Milan, then a bigger city than Paris. Here he met the Duke, Lodovico Sforza, and submitted the following employment résumé:

Having, My Most Illustrious Lord, seen and now sufficiently considered the proofs of those who consider themselves masters and designers of instruments of war and that the design and operation of said instruments is not different from those in common use, I will endeavor without injury to anyone to make myself understood by your Excellency, making known my own secrets and offering thereafter at your pleasure, and at the proper time, to put into effect all those things which for brevity are in part noted below—and many more, according to the exigencies of the different cases.

I can construct bridges very light and strong, and capable of easy transportation, and with them pursue or on occasion flee from the enemy, and still others safe and capable of resisting fire and attack, and easy and convenient to place and remove; and have methods of burning and destroying those of the enemy.

I know how, in a place under siege, to remove the water from the moats and make infinite bridges, trellis work, ladders, and other instruments suitable to the said purposes.

Also, if on account of the height of the ditches, or of the strength of the position and the situation, it is impossible in the siege to make use of bombardment, I have means of destroying every fortress or other fortification if it be not built of stone.

I have also means of making cannon easy and convenient to carry, and with them throw out stones similar to a tempest; and with the smoke from them cause great fear to the enemy, to his grave damage and confusion.

And if it should happen at sea, I have the means of constructing many instruments capable of offense and defense and vessels which will offer resistance to the attack of the largest cannon, powder, and fumes.

Also, I have means by tunnels and secret and tortuous passages, made without any noise, to reach a certain and designated point; even if it be necessary to pass under ditches or some river.

Also, I will make covered wagons, secure and indestructible, which, entering with their artillery among the enemy, will break up the largest body of armed men. And behind these can follow infantry unharmed and without any opposition.

Also, if the necessity occurs, I will make cannon, mortars, and field pieces of beautiful and useful shapes, different from those in common use.

Where cannon cannot be used, I will contrive mangonels, dart throwers, and machines for throwing fire, and other instruments of admirable efficiency not in common use; and in short, according as the case may be, I will contrive various and infinite apparatus for offense and defense.

In times of peace I believe that I can give satisfaction equal to any other in architecture, in designing public and private edifices, and in conducting water from one place to another.

Also, I can undertake sculpture in marble, in bronze, or in terra cotta; similarly in painting, that which it is possible to do I can do as well as any other, whoever he may be.

Furthermore, it will be possible to start work on the bronze horse, which will be to the immortal glory and eternal honor of the happy memory of your father, My Lord, and of the illustrious House of Sforza.

And if to anyone the above-mentioned things seem impossible or impracticable, I offer myself in readiness to make a trial of them in your park or in such place as may please your Excellency; to whom as humbly as I possibly can, I commend myself.[19]

Here, evidently, was a young man bursting with ideas. But he could not say that he had actually done all these remarkable things, only that he could do them if given a chance.

As Lodovico was not much impressed, Leonardo had a painful struggle to live, until he joined forces with another artist who had better connections. Leonardo scrimped, studied, and at last was commissioned to paint Lodovico's mistress. After that, Sforza hired him from time to time for various jobs, artistic and technical. Since the Duke had a way of paying the promised fees late or not at all, Leonardo once quit altogether but reconsidered.

During this time, Leonardo planned the entertainments and celebrations of which the ducal court was fond, with decorations, parades, and fireworks. He indulged his fondness for practical jokes and mechanical toys and read the ducal family stories that he had composed. He began a colossal equestrian statue of Lodovico's father, a self-made duke. However, when the clay model of the statue was completed, the Duke ordered the bronze for the statue cast into cannon to fight his many foes.

Leonardo stayed in Milan for sixteen years. When at last the French drove Lodovico out of his dukedom, Leonardo went to Venice. There he spent some months as a military engineer, casting cannon and strengthening fortifications. After a period of free-lance painting in Florence, he worked briefly for the sinister Cesare Borgia, the Pope's son and the most accomplished betrayer and murderer in Italy. Then Pope Alexander VI died, the realm that Cesare had conquered crumbled away, and Leonardo went back to Florence.

During the next two or three years, Leonardo, now moderately prosperous, free-lanced in Florence. He painted, studied, and filled notebooks with endless sketches and notes. For a time he was associated with Niccolò Machiavelli, the political philosopher. While Machiavelli preferred republics to princedoms, he thought a strongman like Borgia was needed to unite Italy against the French, German, and Spanish armies then devastating the peninsula. He startled readers by coolly heading a chapter in one of his books: "Description of the Methods Adopted by [Cesare Borgia] When Murdering Vitellozzo Vitelli . . ."

and three other notables whose continued existence he found inconvenient.[20]

During this sojourn in Florence occurred Leonardo's brush with Michelangelo. Some of Leonardo's acquaintances were arguing over a passage in Dante in the piazza as Leonardo passed by. They called to him and asked his interpretation. Just then Michelangelo stalked by on the other side of the square, head down in thought. Leonardo was in his fifties and very dignified of mien, tall and well-built, with robe and long red beard. Michelangelo, on the other hand, was about thirty, and an ugly little man who wore the same clothes day and night until they rotted off him. Indulgently Leonardo said:

"You had better ask Michelagniolo; he will expound it to you."[21]

Michelangelo looked up, startled. "Explain it yourself!" he shouted, assuming that sarcasm was meant. He burst into a tirade, berating Leonardo for not having finished the Sforza statue. Leonardo flushed angrily but walked off in silence.

During the next few years, Leonardo traveled between Florence and Milan, working at the latter place for Louis XII of France. In 1512 he went to Rome, hoping for work from the Pope, who belonged to the family of his old patrons the Medici; for his colleagues Bramante, the latter's protegé Raphael, and Michelangelo were all hard at work for the papal court.

But this Pope was the fat, indolent, and worldly Leo X, who said: "Since God has given us the Papacy, let us enjoy it," and "What profit has not that fable of Christ brought us!"[22]

As a humanist, Leo had no use for Leonardo's science. Moreover, after Bramante's death he found the amiable Raphael more to his taste as his artist-general. He gave Leonardo only small jobs, such as making mirrors, and dismissed him when told that Leonardo had been dissecting bodies to study their anatomy.

Now well into his sixties and prematurely aged, Leonardo at last found a generous patron: François I, the clever, enthusiastic, hearty, frivolous, and extravagant French king. Although crippled in one arm by a stroke, Leonardo spent three years at Amboise, where the king set him up in a villa. As of old, Leonardo planned his patron's entertainments, with mechanical lions and other wonders. He tried to bring his vast mass of notes into order for publication, but with no more success than before. After making meticulous plans for his own funeral, he peacefully died.

Leonardo never married or displayed any interest in women. Many have thought him to have been a homosexual, which is quite possible though not definitely proved. When he was living in Florence in his twenties, an anonymous informer accused a youth of the city of homosexual relationships with several men, including Leonardo. The morals police investigated but soon dropped the charge. Whatever his tendencies, Leonardo was one of those who, like his contemporary Michelangelo and like the physicists Henry Cavendish and Willard Gibbs, is so driven by his urge to discover and create that he has little time left for human relationships.

Leonardo was a frugal and abstemious vegetarian. Although he could turn on the charm when he chose, his usual façade was aloof and reserved to the point of secretiveness. He combined boundless curiosity to know and discover and meticulous perfectionism with limitless ambition for fame and accomplishment.

These features united to form one of the most unfortunate, self-defeating characteristics that a fame-seeker can have, namely: inability to complete the tasks he has started. Either Leonardo's plans were too grandiose and hence could never be realized, or Leonardo himself lost interest before the end.

He was a man of the type who sometimes appears among modern scientists, driving laboratory directors mad. The scientist does brilliant work but, when a task is nine-tenths done, he is seized by a passion for some new idea and goes haring off on another line of investigation. Because report-writing bores him, he never reports on his previous work, which remains in the form of scattered notes and, as far as the laboratory is concerned, might as well have never been done at all.

Leonardo's many abortive projects included two colossal equestrian statues, the construction of a canal to Pisa, urban renewal at Florence, completion of the Milan Cathedral, and drainage of the Pomptine Marshes. It was not always his fault that the project failed, but he seldom pushed a task to completion even when he could. He was an incorrigible dabbler and dilettante.

He also tried to master all the sciences. His studies included astronomy, anatomy, aeronautics, botany, geology, geography, phonetics, and physics. The physics included optics, ballistics, component forces, friction, lubrication, stress, levers, pulleys, loaded beams, and strength of materials. He often dropped his work and forgot his obligations for months at a time while he buried himself in another science.

In many scientific fields, Leonardo went as far as any man of his time. Sometimes he anticipated the discoveries of later scientists like Galileo or Stevin. But, as he never published, nothing came of all his studies.

Likewise, he worked on mechanical problems covering an extraordinary range. He designed armored battle cars, a steam gun, a wheel-lock gun, many-barreled cannon, a city built on two levels, a mechanical turnspit, an oil press, a clockwork automobile with a differential gear, mechanical musical instruments, an improved bawdyhouse with a private entrance for clients of quality, a polygonal fortress like that of his friend Martini, textile machinery, a rolling mill, a grinding machine, a rope-making machine, a diving suit, a submarine, a turret windmill, a brake with curved shoes, flying machines, a parachute, a printing press, a screw propellor, a screw-cutting machine, a ribbon-drawing machine, valves, pumps, surveying instruments, poison gas, catapults, a file-cutting machine, sprocket chains, spiral gears, a needle-making machine, a coining machine, a lathe, a truss bridge, a well-boring machine, a road scraper, a crane, a dredge, excavating machines, and a mitered canal lock with a wicket gate.

A mitered lock is one with tongues and grooves along the edges of the valves, which interlock when the gates close. Each pair of valves form a blunt angle pointing upstream, so that water pressure forces them all the more tightly together. A wicket gate is a small portal in the main gate, set low down, to equalize water levels before the main gate is opened. Otherwise the surge of water tosses boats about like chips.

But, of all these gadgets, only a few—the canal lock, and perhaps the screw-cutting machine and the turret windmill—were actually reduced to practice. Sometimes the idea was not workable.

One of Leonardo's battle cars, for example, was a turtle-shaped armored vehicle, loopholed for guns. It was supposed to be moved from within by hand-operated cranks. Leonardo soon realized that it would be far too heavy to be thus propelled. Another of Leonardo's designs was for an unmanned two-horse chariot with whirling scythe blades, like that with which the unknown fourth-century author of *De rebus bellicis* had wished to mechanize the Roman army.

Several inventors of the time devised such proto-tanks, but all these war-wagons failed for want of adequate power. Sometimes they were designed to be pulled or pushed by horses or oxen. If, however, the animals were left outside, they would soon be killed, while there was not enough room for them inside. The Spaniards actually tried 200 battle cars against the French at Ravenna in 1512, but the French won anyway.

It was the same with Leonardo's flying machines. He drafted several designs in which batlike wings were to be flapped by human muscle. He may even have tried one out. Convinced that this was not the right way, he designed a helicopter with a big spiral screw-shaped rotor on a vertical shaft. This screw was to be turned by men working a treadmill below.

Not until 1680 was Leonardo's basic error pointed out. Then the physicist Borelli showed that man has no muscles anywhere nearly so massive and powerful, in proportion to his weight, as the flying muscles of birds. Therefore, flight by human muscles might as well be forgotten.

What did Leonardo actually accomplish? He painted several immortal pictures, such as "The Last Supper" and "Mona Lisa" (*La Gioconda*). He dug some canals, cast some cannon, staged charades for kings and dukes, and made countless notes and sketches.

As he studied, speculated, and designed, Leonardo entered his notes and drawings in notebooks. He wrote from right to left, backwards, either for secrecy or because, being left-handed, he found it easier. He meant, he said, some day to put this mass of material in order and make one or more publishable treatises out of it. But "some day" never came.

When Leonardo died, thirty-odd volumes of notes passed to his friend and pupil Melzi. The complete edition of Leonardo's notes, without the pictures, makes a 1,200-page book. Melzi's heirs sold the collection to various purchasers, so that it became scattered over Europe. The buyers were more interested in the notebooks as collector's items than as technical works, for by then Leonardo's fame as a painter had quite eclipsed his renown as an engineer. Some volumes were lost, but the rest have been collected into various European libraries.

For several centuries, Leonardo's only published work was a treatise on painting extracted from his notebooks and printed in 1551. Nobody even began to publish the rest of the material until the 1880s. By that time the mechanical arts had advanced so far beyond Leonardo's time that his designs were only historical curiosities.

During his life, Leonardo had some influence by personal contacts with other engineers. Some historians of technology claim to find traces of his influence in later sixteenth-century machines. A few men, during the centuries following Leonardo's death, mined the manuscripts for ideas. One such was the mathematician Girolamo Cardano, whose father had been a friend of Leonardo. We cannot tell for sure how much influence Leonardo exerted in this way. But certain it is that his influence would have been many, many times greater if he had published his ideas.

This brings us to the final paradox of Leonardo. Despite his eagerness for fame and his wide knowledge, he failed to grasp the importance of an invention that would have enabled him to achieve his goal: the printing press. He knew all about the printing press, too. In fact, he sketched mechanical improvements for it.

But Leonardo was content to reach for fame by word of mouth only, as most scientists and engineers had been doing from Imḥotep on down. He did not realize that even a small printed book would multiply his voice thousands of times over and expand his influence accordingly; or that this influence, applied in this way, might have advanced the mechanical arts by decades. Or, if he did realize these things, he never acted upon his knowledge. And for these reasons I class him, not as the first of the modern engineers, but as the last of the ancient ones.

After Leonardo, the changes leading to the modern era in engineering came thick and fast.

One change was the growth of printed technical literature. In 1482, Valturio brought out *De re militari,* surveying the state of military engineering. Three years later Alberti summarized construction with his book on building. In early +XVI, many minor technical books–handbooks and how-to books–appeared, such as Bayfius' *De re navali,* on shipbuilding (1536).

In 1540 Biringuccio's great treatise on metallurgy, *Pirotecnia,* was published posthumously. Biringuccio had been influenced by his German colleague Agricola,[23] who in turn borrowed passages from Biringuccio's book for his own masterpiece on mining, *De re metallica* (1556). In 1912 Herbert Hoover, then a young mining engineer, and his wife translated this book into English. Agricola also wrote a perfectly serious book for mining engineers on gnomes and how to get rid of them. However, some statesmen and scholars of more recent times have believed in queerer things than gnomes; not many years ago the Prime Minister of Canada was an ardent Spiritualist.

Later in the century, Palladio, who perfected the bridge truss, covered the subject of architecture in *I quattro libri dell' architettura* (1581). Ramelli and Verantius wrote on machinery. Fontana told just how he moved the second biggest of Rome's obelisks from Nero's Circus to a position in front of St. Peter's, in case anyone wanted to move another obelisk. And many other, less famous technical books appeared in print.

Another change was the advance of those pure sciences destined most to affect engineering, especially statics, mechanics, kinematics, and hydraulics. In the Netherlands, Simon Stevin (1548–1620) discovered the triangle of forces. This discovery enabled men to calculate the actual loads on the members of cranes, trusses, and other simple structures.

Stevin also speeded up the process of calculation by inventing the decimal system. The effect of this discovery may be compared with that of the modern introduction of computers. In addition, Stevin devised the plan of flooding the Dutch polders as a means of defending the Netherlands. In 1672, this defense stopped the invincible armies of Louis XIV in their tracks.

Stevin's younger contemporary Galileo Galilei (1564–1642) solved the problems of accelerated movement and began the analysis of stresses in beams. He did this in his old age, after his trouble with the Inquisition. The Inquisition had ordered him not to expound the hypothesis of Copernicus, which put the sun, instead of the earth, at the center of the solar system, as Aristarchos had done long before. Galileo, a peppery man who did not suffer fools gladly, published a book arguing this hypothesis anyway.[24] The Inquisitors forced Galileo by threats of torture to recant and put him under house arrest for the rest of his life. Nevertheless, and despite failing eyesight, he managed to complete his *Dialogue on Two New Sciences* (1636), giving the science of mechanics its greatest impetus since Heron. Although his analysis of the stresses in a beam contained a basic error, others soon corrected this.

After Galileo, Italian science and engineering did not fare so well. For one thing, during +XVI the great powers–France, Spain, and the "Holy Roman" or German Empire–used Italy as their favorite battleground. The divided and mutually hostile Italian city-states were as helpless to resist these mighty, marauding armies as the Greek city-states had been to resist the Macedonians and the Romans many centuries before.

Furthermore Italy, where the Renaissance had started, received the brunt of the Counter-Reformation of the Catholic Church. One of the more puritanical popes sent workmen through St. Peter's to remove the genitalia of Michelangelo's cherubs with hammer and chisel. The counter-reformers cleaned up not only the morals of the clergy but also any scientists suspected of heresy. One of the latter, Domenico Oliva, escaped the tortures of the Inquisition by jumping out a window to his death.

Still, technical progress continued elsewhere, ever faster. Technical men began to organize; the Society of Lynxes, to which Galileo belonged, was the first of many such organizations. The first research institute was founded in 1560, and the first industrial exposition was held at Nuremberg in 1568.

Engineering schools appeared in France in +XVIII. At the same time, specialization within the engineering profession took place. John Smeaton, who had to go to France in the 1750s to round off his technical education, called himself a "civil engineer," meaning a non-military one. Simultaneously, the experimental vacuum mine pumps of +XVII were evolving into the steam engine.

So, after Leonardo, engineering became modern. We should not be fooled by the gaudy medievalistic costumes of these engineers—the feathered hats, flowing hair, ruffs, laces, and swords. The thoughts in these plumed and periwigged heads were modern thoughts.

Nor was the transition to modern times a matter of any particular invention, such as the steam engine. It was, rather, the coming of a time when technical development became so rapid that men could see the world about them change in their own lifetimes. It is no accident that about +XVI, men began to write of the "wonderful discoveries of this modern age" and to wonder aloud what would happen next.

For that matter, we are still wondering. The pace of change has speeded up until change itself has become one of the most pressing human problems. Older readers can remember streets lit by gas, wagons and carriages drawn by horses, stoves fired by coal or wood, and outside plumbing. Even the younger ones may recall the days before self-service markets and transoceanic air travel.

This constant and rapid change affects those who live under it. As Russell has pointed out: "Few men's unconscious feels at home except in conditions very similar to those which prevailed when they were children."[25] Hence people who live in an era of rapid change tend to suffer from a vague but persistent feeling of unease. Something is wrong, though they do not know just what. Some react to this discomfort by wildly irrational fanaticism, as by asserting that fluoridation of drinking water is part of a Communist plot to poison the nation. Moreover, it can be shown that speeding up the rate of change probably fosters crime.

Some blame engineers for these difficulties, but that is futile. The whole process began back in the days of Imḥotep. Once started, technology was bound to develop much as it has, although the timing of this development might have varied if political history had been different.

Nay rather, technology started back in the early Pleistocene, when our apish ancestors, instead of evolving fangs, horns, or claws to enable them to make their way in the world, developed brains for the purpose. Once that portentous step had been taken, everything else followed naturally. And now there is no more stopping or turning back than there is on a ski slide, once the skier has pushed off.

Others blame the technical men because they have not succeeded in making everybody kind, honest, and peaceful. Therefore, they say, technical progress has not "civilized" men at all.

But it is a mistake to confuse these virtues with civilization. Civilization is a matter of power over the world of nature and skill in exploiting this world. It has nothing to do with kindness, honesty, or peacefulness. These virtues are found scattered–rather thinly, alas–through the entire human species, although they occur in some people more than in others and are encouraged in some cultures more than in others. No doubt it would be a good thing if they were universal, but the engineer is not the man to ask this of. He can heat your house, dam your river, or build your space ship, but it is hardly fair to expect him also to make you love your fellow man. Priests and philosophers have tried for thousands of years to accomplish this, but with indifferent success.

An engineer is merely a man who, by taking thought, tries to solve human problems involving matter and energy. Since the Mesopotamians tamed their first animal and planted their first seed, engineers have solved a multitude of such problems. In so doing, they have created the teeming, complex, gadget-filled world of today.

But problems are like the Hydra of the myth of Herakles; cut off one head and two more sprout. Scientists and engineers have, by using their intelligence, solved some human problems. But the solutions themselves have given rise to still greater problems–nationalism, nuclear war, population explosion, and degenerative mutation pressure, to name but a few. If civilization is to last, and the problems are not to grow beyond all coping, the rest of mankind, and not just the scientists and engineers, will have to use intelligence also, and more than they have so far.

Otherwise, in place of vanished man, the remote future may see the specialized descendant of some present-day monkey, rat, or lizard beginning to dig his species' first irrigation ditch–and so starting the whole process over again.

NOTES

CHAPTER ONE

1. Basham, pp. 85 f.

CHAPTER TWO

1. Also spelled Djoser, Zoser, or Zeser. Because of the ambiguities of ancient Egyptian writing, the pronunciation of early Egyptian words is very uncertain. Hence, Hurry (pp. 191 f.) lists 34 ways to spell the name of Imḥotep: Aiemhetp, I-em-hotep, Iḥotep, Yemhatpe, &c.

2. Manetho (Loeb Classics), pp. 41–45. Joṣer's name is also spelled Sesorthos or Sosorthus in other quotations from Manetho, whose name in Greek was Manethôs or Manethôn.

3. Greek, Imouthês. The ḥ represents a guttural fricative between our *h* and the *ch* sound of German *ach*. For other Egyptian sounds without close English equivalents, see any Arabic grammar.

4. *Maṣṭaba* is an Arabic word meaning "bench." Tombs of this type are so called because they were shaped like the benches of mud brick that Egyptian peasants build in front of their huts.

5. Greek, Troia; modern Ṭura.

6. Latin, Aesculapius.

7. These are estimates, based upon what we can see of the pyramid. There is no way to ascertain these figures exactly short of taking the pyramid apart and counting and weighing the stones–especially as a ledge of rock rises into the interior, no man knows how far.

8. *A General Summary,* in *Departmental Ditties* (1912), pp. 11 f.

9. Greek, Chephrên or Kephrên.

10. Greek, Mykerinos or Mencherês; Latin, Mycerinus.

11. Greek, Amenemmês.

12. Or Aaḥmes; Greek, Amasis. Edwards, p. 196; Fakhry, pp. 234 f.

13. Glanville, p. 355.

14. Herodotus, II, 125.

15. Deḥutiḥotep, or Tehutihotep, Jehutihetep, &c. For the argument about rollers, see Davison; Engelbach, pp. 58 ff.; Neuburger, p. 210.

16. Because of the precession of the earth's axis, the north celestial pole moves in a circle in the sky, 47° in diameter, making one complete circuit every 25,800 years.

17. Dunham.

18. Shear legs are a pair of long timbers arranged like an inverted V, with a rope passing over the apex for hoisting.

19. Millet.

20. "Ozymandias" is Shelley's corruption of Diodorus' "Osymandyas" (I, xlvii) which in turn is a corruption of User-ma-Ra, one of Rameses' many names or titles. The obscure phrasing of lines 6–8 is understood if you remember that "survive" is here a transitive verb. Shelley is saying: "those passions . . . survive . . . the [sculptor's] hand . . . and the [king's] heart."

21. Herodotus, II, 108.

22. Greek, Nechô, Nekôs, or Nechaô.

23. Persian, Kambujiya.

24. Greek, Dareios or Dareiaios; Persian, Darayavahush. Diodorus (I, xxxiii) says that Darius gave up the project when erroneously warned that the Red Sea was higher than Egypt and would flood that land. This is wrong and Herodotus, who says (II, 158) that Darius completed the canal, is right, because Darius left monuments commemorating his completion. Some think the canal was begun before Nikau's time, perhaps as far back as –2000.

CHAPTER THREE

1. Forbes: *Studies,* II, p. 20.
2. Pritchard (1950–55), p. 197.
3. King, III, p. 14.
4. Pritchard, op. cit., p. 270.
5. Or Iddû; Greek, Is; modern Hît.
6. The biblical Shushan; modern Shush, Iran.
7. Onesikritos, quoted by Strabo, XV, iii, 10.
8. Wittfogel, p. 155.
9. Arabic *tall,* pl. *tulûl.*
10. The biblical Erech, modern Warka.
11. Akkadian, *ziqquratu,* "peak" or "high place."
12. Pritchard (1950–55), pp. 275–95; Champdor, p. 90.
13. Akkadian, Sinakherîba.
14. Akkadian, Kaldu, corresponding roughly to ancient Sumeria.
15. Akkadian, Ninua; Greek, Ninos.
16. Called Dûr Sharrukîn (Home of Sargon); modern Khorsabad.
17. The modern Khosr.
18. Jacobsen & Lloyd, p. 35.

19. Akkadian, Arbaila; modern Erbil.
20. Jacobsen & Lloyd, p. 38.
21. Finegan, p. 215.
22. Nahum iii, 1, 7, 18.
23. Akkadian, Nabu-apal-uṣur.
24. Or Nebuchadnezzar; Greek, Naboukodonosoros; Akkadian, Nabu-kudurri-uṣur. "Nebuchadrezzar," the spelling used by Jeremiah and Ezekiel, is nearer the original than the "Nebuchadnezzar" used elsewhere in the Bible.
25. Formerly Boğazköy or Boghaz Keui; now Boğazkale, Turkey.
26. Ingraham, p. 18.
27. Greek, *kremastos;* Latin, *pensilis.*
28. Koldewey, p. 91. The Arabic word mentioned seems to be *dulâb,* meaning also "cupboard."
29. Herodotus, I, 186; Diodorus, II, viii, 1–3.
30. Persian, Kurush; Greek, Kyros.
31. Greek, Nabonidos.
32. Persian, Khshayarsha.
33. Arrian: *Anabasis,* III, xvi; VII, xvii; Strabo, XVI, i, 5.
34. Persian, Hakhamanish.
35. Sabaea was the biblical Sheba and the Arabic Saba'; Ma'rib was the classical Mariaba.
36. Luqmân ibn-'Âd.
37. Esther viii, 10; Herzfeld, I, p. 228.
38. Persian, Bubranda and Artahaya.
39. Herodotus, VII, 23.
40. Greek, Ekbatana, Ag-; modern Hamadan, Iran.
41. Persian, Artakhshathra.
42. Polybius, X, xxvii.
43. Medieval Khuwaresm; modern Khurassan.
44. Greek, Arados; modern Ruâd.
45. Greek, *diêrê,* "two-er"; Latin, *biremis,* "two-oar-er."
46. A knot = 1 nautical mile (= $1\frac{1}{7}$ land miles) per hour. "Knots per hour" is incorrect unless it refers, not to speed, but to acceleration.
47. The modern Nahr Na'mein.
48. Canaanite: Ṣiduna, Ṣôr, Qarth-hadasht, Beroth; Greek: Sidôn, Tyros, Karchêdôn, Berytos; modern Ṣaïda, Ṣûr, Tunis, Beirût.

CHAPTER FOUR

1. Or Harpalos.
2. Herodotus says (VII, 184) that Xerxes had 1,800,000 fighting men, but this is a preposterous exaggeration. See Munro's analysis of these figures in *Cambridge Ancient History,* IV, Ch. ix. Few ancient armies could exceed

50,000 combatants for long, because of the difficulty of collecting enough food in one place to keep them from hunger.

3. Originally called the New Hekatompedon ("Hundred-footer") of Athena Polias; later known as the Parthenon ("Place of the Maidens") from the name of the treasure chamber at the west end.

4. Keats: *Ode to a Nightingale.*

5. "Athena of the City," later called Athena Promachos, "Front-rank Fighter."

6. Architects use the Latin term, *guttae.*

7. This member was called an *echinos* or "sea-urchin."

8. Vitruvius, IV, i, 9–10.

9. Vitruvius, V, v. Vitruvius tells of bronze vases. These have all disappeared, for obvious reasons, but fragments of the cheaper pottery vases have been found at Greek theaters.

10. Aristotle: *Politics,* II, viii (1268b).

11. Diogenes Laërtius, II, iii, 8.

12. Cicero: *The Republic,* I, x (15).

13. His real name was Aristokles. *Platôn,* from *platys,* "broad," was a nickname.

14. Plato: *The Republic,* VII, xi. Some defenders of Plato aver that this passage is not really anti-scientific, but is a plea for the use of mathematical calculations and hypotheses in astronomy.

15. Vitruvius, V, iii, 1.

16. Roman, Agrigentum; mod. Italian, Agrigento; Sicilian, Girgenti.

17. Modern Selinunte. Diogenes Laërtius, VIII, 70.

18. Herodotus, III, 60.

19. Thucydides, IV, 100. For a similar device used by the Greeks of Ambrakia against the Romans in –189, see Polybius, XXI, 28.

20. Diodorus, XIV, xli, 3–4.

21. Diodorus, loc. cit., & xlii, 2; Pliny sr., VII, lvi (207); Tarn, pp. 130 f.

22. Diodorus, XIV, l. 4.

23. Plutarch: *Sayings of Kings and Commanders: Archidamos son of Agesilaos;* Aeneas Tacticus, xxxii, 8.

24. Ramelli, pp. 322–32; Biringuccio, p. 437; Leonardo da Vinci, *passim.*

25. Or Sun-dz ("Philosopher Sun"), Sun Tzŭ. See Sun Tzu Wu, Ch. v (ed. Phillips, p. 56).

26. Or Chih-chih. See Duyvendak. We shall learn much more about Chinese catapults when Dr. Joseph Needham completes publication of the fourth volume of his monumental *Science and Civilisation in China.*

27. Shakespeare: *The First Part of King Henry IV,* Act I, Scene iii, ll. 60–63.

CHAPTER FIVE

1. Greek, Akadêmeia, Lykaion.
2. Diogenes Laërtius, V, i, 20–21.
3. Aristotle: *Politics,* III, v (1278a).
4. Sarton (1927–48), I, p. 132.
5. Aristotle: *Mechanics,* Introd. (847a).
6. Ibid., (848a).
7. Philon Byz.: *Pneumatics,* lxiii (ed. Carra de Vaux, p. 205); Heron Alex.: *Pneumatics,* xxxi (ed. Woodcroft, p. 49).
8. A. G. Drachmann has suggested (personal communication) that the priests, wishing their bronze magic-machine to look brightly polished all over, made it in the form of a wheel so that each worshiper should grasp it in a different place.
9. Aristotle: op. cit., i (849b).
10. Ibid., vii (851b).
11. Greek, *periagôgê,* literally "turning about" or "guiding roundabout."
12. Snorri: *King Olaf Trygvesson's Saga,* Ch. 85; Achilles Tatius, III, 1.
13. Aristotle: op. cit., xxi (854a).
14. Dionysius of Byzantium (+II), quoted by Adams, p. 14.
15. His name was really 'Abû 'Abdallah Muḥammad ibn-Muḥammad ibn-Abdallah ibn-Idrîs. "Idrîs" (or Edrîs) was the name of his great-grandfather.
16. Idrisi, III, iv (ed. Dozy, pp. 166 f.). The height given by Idrisi, 100 fathoms = 645 to 755 feet, is probably much exaggerated; cf. ibn-ash-Shaykh's more careful measurements.
17. *Hê Alexandreia têi pros Aigyptôi.*
18. Polybius, XII, 13.
19. Pliny sr., XIII, xxiii (74–77).
20. Landa, pp. 16 ff.
21. "Such are the crimes to which religion leads" (Lucretius, I, l. 101). See also E. A. Parsons, pp. 420 f.
22. Greek, *mouseion.*
23. Emma Lazarus: *The New Colossus.* The tradition seems to have originated with Blaise de Vigenère (Adams, p. 44).
24. Vitruvius, X, i, 1–3.
25. Waldemar Kaempffert: *A Popular History of American Invention* (N.Y.: 1924), I, p. 481.
26. Athenaeus Nauc., IV, 174c.
27. Vitruvius, IX, viii, 4.
28. Philon Byz.: *Pneumatics,* lxi, lxv.
29. Pliny sr., XXXIV, xlii (148); Ausonius: *Mosella,* ll. 311–17.
30. Plutarch: *Marcellus,* xvii, 3–4 (307).

31. "I have found!" Not *eureka* as often written; Archimedes as a western Greek would not have dropped his h's. The latter spelling probably originated in a copyist's error. Archimedes' nudity would not have surprised his fellow Syracusans, because classical Greeks thought nothing of appearing naked in public.

32. Vitruvius, IX, Pref., 9–12. For later versions of this story, see Clagett (1959), pp. 57, 86 f.

33. Diodorus, XXVI, xviii; Pappus, VIII (ed. Ver Eecke, pp. 836 f.). For a slightly different version, see Plutarch: *Marcellus,* xiv, 7 (306).

34. Athenaeus Nauc., V, 204c–204d.

35. Plutarch: *Demetrius,* xliii.

36. Greek, *sphairopoiia.*

37. Diodorus, XXVI, xviii, *apud* Tzetzes. For the siege of Syracuse, see also Polybius, VIII, 5–9; Livy, XXIV, xxxiii–xxxiv; Plutarch: *Marcellus,* xv–xvii.

CHAPTER SIX

1. Frontinus: *The Aqueducts of Rome,* I, 16.
2. Eric T. Bell: *Numerology* (Baltimore, 1933), p. 57.
3. Tacitus: *Annals,* III, 37; Vitruvius, I, i, 3; VI, Pref., 7.
4. To be fair, a small part of Arcadia, in the extreme west along the valley of the Alpheios, does look like the fictional Arcadia of Theokritos and his colleagues.
5. Strabo, V, iii, 7; Seneca: *De ira,* III, xxxv, 5; *De beneficiis,* VI, xv, 7; Juvenal, III, 190–202.
6. Pliny sr., XXXVI (lxvi), cxcv; Petronius, li.
7. Suetonius: *Vespasian,* xviii.
8. Pliny sr., IX, lxxix (168–69); Valerius Maximus, IX, 1. See also Cicero: *De finibus,* II, xxii (71); *De officiis,* III, xvi (67); Columella: *On Agriculture,* VIII, xvi, 5.
9. Modern *pozzuolana,* Pozzuoli.
10. G. Giovanni (in Bailey, p. 437) explains *opus incertum* as meaning random-width ashlar masonry, while all my other sources explain it as concrete studded with a facing of irregular stones. Some class random-width ashlar masonry as a kind of *opus quadratum.*
11. Pliny sr., XXXVI, iii (7).
12. Suetonius: *Augustus,* xxviii.
13. Modern Spljet, Yugoslavia.
14. Modern Timgad, Algeria.
15. *Panem et circenses;* Juvenal, X, 80.
16. They were actually begun by Caracalla's father Septimius Severus and finished some years after Caracalla's reign by Severus Alexander (+III).
17. Frontinus: *Aqueducts,* I, v; Livy, IX, xxix, 5–9.

18. Greek Epidamnos, modern Italian Durazzo, Albanian Durrës.

19. Statius: *Silvae,* IV, iii: *The Domitian Road.*

20. *"Cedo alteram";* Tacitus: *Annals,* I, 23.

21. Modern Mitrovica, Yugoslavia. Tacitus: *Annals,* I, 20; XI, 20; Dio Cassius, LXI, xxx, 6; Eutropius, IX, 17.

22. Caesar: *The Gallic War,* IV, xvii.

23. Thomas Babington Macaulay: *Horatius;* Livy, II, ix–x; Polybius, VI, 55. Porsena's name is also spelled Porsenna or Porsinna.

24. From north to south, the ancient bridges of Rome, *selon* Gest (Ch. vii), were as follows. First comes the century in which the bridge was begun, then the names by which it has been known in more or less chronological order. Those starred still stand:

–III Mulvius = Milvio or Molle*
+II Aelius = Hadriani or Adrianus = Sancti Petri = Elio or S. Angelo*
+I Neronianus
–I Agrippae
+II Aurelius or Antoninus = Valentinianus = Sisto*
–I Cestius = Gratiani = Cesti or S. Bartolomeo*
–I Fabricius = Judaeorum = Fabricio or Quattro Capi*
–II Aemilius = Lepidi = Senatorum
–VI Sublicius
+III Probi = Theodosii = Marmoreus

The Pontes Cestius and Fabricius are in tandem, not parallel as might be thought from their listing. The broken-down bridges have been called *pons ruptus* or *ponte rotto* at various times, but the term does not mean any particular bridge. In recent centuries a number of new bridges have been built across the Tiber, and more are planned.

25. Hart (1928), p. 4.

26. Vitruvius, VIII, vi, 10–11.

27. Also called a modulus or adjutage.

28. Tacitus: *Annals,* XII, 57; Suetonius: *Claudius,* xxi; Dio Cassius, LX, xxxiii.

29. Meaning Tiberius, Gaius Caligula, Claudius, and Nero respectively. Quoted from an unidentified source by Basil Davenport, in *The Roman Reader* (N.Y.: 1951–59), p. 479.

30. Horace: *Satires,* I, v. The marsh was part of the Pomptine Marshes, already described, while "Feronia" was a shrine of the goddess of that name near Tarracina.

31. Landström, p. 213.

CHAPTER SEVEN

1. The ending *-ianus* meant that, before joining the clan of the Iulii, he had belonged to that of the Octavii.
2. Seneca: *On Mercy,* I, xxiv, 1.
3. Wright, pp. 140 f.
4. Suetonius: *Nero,* xli, xlix.
5. Steinman & Watson, p. 48.
6. Dio Cassius, LXIX, iv; Spartianus: *Hadrian,* xix.
7. Specifically, Hadrian's sending the *plans* implies that the temple was not yet built, in which case the faults of the design could have been easily remedied. If on the other hand the temple was already built, Hadrian would more likely have asked Apollodoros' opinion of the completed structure, not of the plans.
8. Pliny sr., XXXI (iii), xxiii.
9. Pliny jr.: *Epistles,* VI, xvi.
10. Apuleius: *Metamorphoses,* or *The Golden Ass,* IX, 12.
11. Pliny sr., XVIII (xi), xxviii (107).
12. Strabo, XII, iii, 31.
13. *The Greek Anthology,* IX, 418.
14. Vitruvius, X, v, 2.
15. Ausonius, X, *Mosella,* ll. 359–64.
16. Procopius: *History,* V, xix.
17. Aristotle: *On Marvelous Things Heard,* 62.
18. Ley (1954).
19. Thorndike, I, p. 197.
20. Plutarch: *Crassus,* ii, 4.
21. Heron Alex.: *Pneumatics,* II, xi (50) (ed. Woodcroft, p. 72); Cohen & Drabkin, pp. 254 f.
22. From *aeoli pila,* "ball (or cap) of Aiolos," the wind god; used by Vitruvius (I, vi, 2) for some sort of ornamental kettle, probably shaped like a statuette of the god, with a steam spout at the top.
23. St. Gregory (ed. Gardner), pp. 17 f.
24. Xenophon: *Anabasis,* III, ii, 18–19.
25. Falko Daim: "The Avars," *Archaelogy,* Mar.-Apr. 1984, p. 33.
26. Vegetius, X; Anonymus, Pref., 4 (ed. Thompson, pp. 106 f.). Cf. the invention of a lightweight ram by men of the Sabeiroi, a Caucasian nation, at the siege of Lazica in +550 (Procopius: *History,* VIII, xi, 27–31).
27. Frontinus: *Stratagems,* III, Introd.
28. Pliny sr., XXXVI, xxvi (199); Seneca: *Quaestiones naturales,* I, vi, 5; see also Aristophanes: *The Clouds,* ll. 766 ff.
29. Seneca: op. cit., VII, xxv, 2–4.
30. Ibid., VII, xxix, 4.

31. Strabo, I, ii, 8.
32. B. Russell: *The ABC of Relativity* (N.Y.: 1926), p. 166.

CHAPTER EIGHT

1. Diener, p. 167.
2. Ioannes or John Philiponus: *Commentary on Aristotle's Physics,* quoted by Cohen & Drabkin, p. 220, & by Clagett (1955), p. 173. "Philiponus" means that he belonged to a particular sub-sect of the Monophysites.
3. A. D. White, I, pp. 91 f.; Lactantius: *Divine Institutes,* III, xxiv.
4. Encyclopaedia Britannica, s.v. *Vladimir, St.*
5. Miller, p. 109; Gibbon, II, p. 1355.
6. Procopius: *Buildings,* II, iii, 16–21 (B219).
7. Aeneas Tacticus, xxxv.
8. Heron Byz., xxii (ed. Barocius, p. 39b).
9. Merckel, pp. 311 f. Where I have written "Mazdai," Merckel has "Masmaeus." I think this is an error for Mazdai (a common Old Persian name) and its Latin equivalent "Masdaeus," arising from the fact that the Arabic letters *ḍâd* and *mîm* look much alike in medial position. Ardeshir (Old Persian, Artakhshathra; Greek, Artaxerxes) was the founder of the Sassanid Empire and the father of Shapur I.
10. Lloyd (1942–45), p. 14.
11. Ibid., p. 15.
12. "Caste" is loosely used for two Indian social groupings: the main groups, properly called *varṇa* or color, and *jâti,* a subdivision of the varṇa. Ever since Aryan times there have been four varṇas: priests, warriors, merchants, and workers. But the number of jâtis runs into the thousands. In recent decades the system has been losing its rigidity.
13. Homer: *Odyssey,* XII, ll. 109–10.
14. Better spelled "Aśoka," because the Indian languages have the combination *s* + *h* as in "gashouse" as well as the sound of *sh* as in "fish."
15. Respectively called *ardhamaṇḍapa, maṇḍapa, antarala,* and *vimâna.*
16. Or *śikhara.* The shrine room is the *garbhagriha.*
17. King Kumaragupta.
18. Known to medieval Europeans as Rhazes, Albiruni, Avicenna, and Alhazen.
19. Lîsharḥ ibn-Yaḥṣub.
20. Properly Ṣâlaḥ-ad-Dîn Yûsuf ibn-Ayyûb.
21. Gibbon, II, p. 1334. Twelve palms would be about four feet, but a 600-pound ball would be only about two feet in diameter.
22. Arnold, p. 337.
23. Or Huang-ti. Of the several ways of transliterating Chinese names, I use a system like that of Yale and Gardner which, if it does not fit the language exactly, comes closer to it without special directions than the others,

especially the familiar but misleading Wade-Giles system with its apostrophes. Wade-Giles spellings are given as alternatives in the notes. To approximate the Chinese, render *ê* by the vowel of "up"; *ï* by a sound like that of the word "err"; *ü* as in German; *o* as in "horse"; *ou* as in "soul"; and *hs* like *ch* in German *ich*. Well-known Chinese place-names are given in their conventional forms regardless of systems. When I first wrote this book, the now official Pinyin system was not available."

24. Or Yao.

25. Or Chou.

26. Or Chu-ko Liang. Needham (1954), p. 118; Forti, pp. 25, 104; Agricola, pp. 154 f.

27. Or K'ung Fu-tze, -tzŭ.

28. Or Lao-tze, -tzŭ; a nickname meaning "old philosopher." His real name was Li Êr.

29. Because the correct reading of the Chinese characters of this name is doubtful, it has several possible forms: Hsüan Chuang, Yuan Chwan, Huan Tsang, Hiuen Tsiang, &c.

30. Or Ts'in, Chin, Ch'in Chêng, &c.

31. Or Shih Huang Ti.

32. Or Li-sze, -ssŭ.

33. Geil, p. 204. This is a paraphrase or abridgment of the original, which in turn is probably a fictitious speech like those which classical historians put in the mouths of Perikles and other notables. But it may give the true reasons motivating Tsin.

34. Or Jin, Kin, Chin.

35. Or Ch'in Erh Shih, &c.

36. Or Liu Pang.

37. Swei Wên-di and Swei Yang-di, or Sui Wên Ti and Sui Yang Ti.

38. Polo, II, liii.

39. Or Ts'ai Lun.

40. Or *chih*.

41. Carter, p. 3.

42. Loc. cit.

43. Or Chieh; Carter, p. 41.

44. Or Pi Shêng.

45. It is uncertain whether "Coster" was the man's surname or merely a sobriquet meaning "the sacristan."

46. It is not certain that this job was finished while Gutenberg was working with Fust or whether it was completed by Fust and his later partner Schöffer.

47. Or Liang Ling-tsan.

48. Lyou Hsü: *Jyou Tang Shu,* xxxv, pp. la ff., quoted in Needham, Wang, & Price, pp. 78 f.

49. Hsüan Dzung or Tsung.

50. Needham, Wang, & Price, p. 79.

51. Or Chang Ssŭ Hsün, Sze Hsün.
52. Or Tsung.
53. Or K'itan, Ch'itan, &c.; "Cathay" comes from "Kitan." The Kitan dynasty was called the Lyau or Liao.
54. Or K'ai-fêng-fu; *fu* = "city."
55. Shun-di or Shun Ti.
56. Forti, p. 272.
57. Ley (1942), pp. 61 f.; Partington, Ch. ii.
58. Biringuccio, p. 433. Biringuccio's "cubit" is the Florentine *bracchio* of about 23 inches.
59. Lu Gwei-Djen *et al.:* "The Oldest Representation of a Bombard," *Technology and Culture,* Jul. 1988, p. 594.
60. Biringuccio, p. 319.
61. Or Chêng Ho.
62. J. & D. Needham, pp. 256 f. Marxist Needham hastily adds that he "does not mean that capitalist economy is necessarily the best today or the best for the future."

CHAPTER NINE

1. Theophilus, III, xlviii.
2. Bacon: *Epistola de secretis operibus artis et naturae,* quoted in Klemm, p. 95.
3. Bacon: *Opus majus,* II, pp. 574, 582.
4. Singer, III, p. 230.
5. W. B. Parsons, pp. 277 f.
6. Ibid., p. 271.
7. Aeneas Tacticus, xxxix, 3.
8. Ley (1955), p. 92.
9. Klemm, p. 159.
10. Machiavelli (1960), IV, i (p. 164); V, vii (p. 253).
11. L. B. Alberti: *Architectura* (Leoni transl., London: 1755), quoted by Klemm, p. 119.
12. W. B. Parsons, p. 241.
13. Filippo da Modena and Fioravante da Bologna.
14. W. B. Parsons, p. 375.
15. Or IJssel. Suetonius: *Claudius,* i; Tacitus: *Annals,* II, 8; XIII, 53.
16. Lane, p. 172; quoting *Pero Tarfur, Travels and Adventures, 1435–1439,* (N.Y.: 1926), p. 170.
17. G. K. Chesterton: *Lepanto.*
18. Klemm, p. 121.
19. W. B. Parsons, pp. 16 f. All of Leonardo's many biographers print this memorandum; *e.g.,* Hart (1925), pp. 43 ff.; Vallentin, pp. 75–85, &c.
20. Machiavelli (1908–28), p. 217. Actually a separate short article (*Del modo tenuto dal duca Valentino nell'ammazzare Vitellozzo Vitelli, Oliverotto*

da Fermo, il signor Pagolo, ed il duca di Gravini Orsini, 1502) but sometimes printed as an appendix to *Il Principe* (*The Prince*).

21. Vallentin, p. 349.

22. Ibid., p. 462; Encyclopaedia Britannica, *s.v. Renaissance.*

23. Born Georg Bauer, he latinized his name to Georgius Agricola for literary purposes, as they did at that time. *Bauer* and *agricola* both mean "farmer." For that matter, his given name comes from the Greek *georgios,* also meaning "farmer."

24. G. Galilei: *Dialogo . . . dei Due Massimi Sistemi del Mondo Tolemaico, e Copernicano* (Florence: 1632).

25. B. Russell: *The Impact of Science on Society* (N.Y.: 1953), p. 108.

BIBLIOGRAPHY

Abbreviations: EL = Everyman's Library (London, Toronto, & N.Y.); LC = Loeb Classical Library (London & Cambridge, Mass.); ML = Modern Library (N.Y.); PB = Pelican-Penguin Books (Harmondsworth, England); GH = *The Greek Historians* (N.Y.: Random House, 1942).

ACHILLES TATIUS: *Clitophon and Leucippe*, LC.

ADAMS, W. H. DAVENPORT: *Lighthouses and Lightships*, Lon.: T. Nelson & Sons, 1870.

ADCOCK, F. E.: *The Greek and Macedonian Art of War*, Berkeley: Univ. of Calif. Pr., 1957.

AENEAS TACTICUS: *On the Defence of Fortified Positions*, in *Aeneas Tacticus, Asclepiodotus, Onasander*, LC. See also ROCHAS.

AGRICOLA, GEORGIUS: *De Re Metallica*, N.Y.: Dover Publications, Inc., 1912–50.

ALBA, DUKE OF: "The Pharos of Alexandria (Summary of an Essay in Spanish by Don Miguel de Asin with Architectural Commentary by Don M. Lopez Otero)," in *Proceedings of the British Academy*, XIX (1933), pp. 277–92.

AMMIANUS MARCELLINUS: *Roman History*, Lon.: Henry G. Bohn, 1862, & LC.

ANONYMUS: *A Roman Reformer and Inventor* (*Being a New Text of the Treatise* De Rebus Bellicis *with a Translation and Introduction by E. A. Thompson . . .*), Oxf.: Clarendon Pr., 1952.

APPIAN (APPIANOS OF ALEXANDRIA): *Roman History*, LC.

APULEIUS, LUCIUS: *Metamorphoses*, or *The Golden Ass*, Lon.: Henry G. Bohn, 1853, & LC.

ARISTEAS: *Aristeas to Philocrates* (*Letter of Aristeas*), N.Y.: Harper & Bros., 1951.

ARISTOPHANES: *The Eleven Comedies*, N.Y.: Liveright Publ. Corp., 1912–23, & LC.

ARISTOTLE: *Mechanics, On Marvelous Things Heard, & Politics*, in *The Works of Aristotle Translated into English*, Oxf.: Clarendon Pr., 1913; *The Basic Works of Aristotle*, N.Y.: Random House, 1941, & LC.

ARNOLD, Sir THOMAS (ed.): *The Legacy of Islam,* Oxf.: Oxford Univ. Pr., 1931–43.

ARRIAN (FLAVIUS ARRIANUS): *The Anabasis of Alexander,* GH & LC.

ASHBY, THOMAS: *The Aqueducts of Rome,* Oxf.: Clarendon Pr., 1935.

ATHENAEUS MECHANICUS: *Traduction du Traité des Machines d'Athénée, par M. de Rochas d'Aiglun, c.* 1880.

ATHENAEUS OF NAUCRATIS: *Deipnosphistae,* LC.

AUSONIUS, DECIMUS MAGNUS: LC.

BACON, ROGER: *The Opus Majus of Roger Bacon,* Phila.: Univ. of Penna. Pr., 1928, 2 vols.

BAEDEKER, KARL: *Egypt and the Sûdân,* Leipzig: Baedeker, 1929.

——: *Palestine and Syria* (*with Routes Through Mesopotamia and Babylonia and the Island of Cyprus*), Leipzig: Baedeker, 1912.

BAILEY, CYRIL (ed.): *The Legacy of Rome,* Oxf.: Clarendon Pr., 1923.

BASHAM, A. L.: *The Wonder That Was India* (*A Survey of the Culture of the Indian Sub-Continent Before the Coming of the Muslims*), N.Y.: Macmillan Co., 1954.

BELL, H. IDRIS: *Egypt from Alexander the Great to the Arab Conquest* (*A Study in the Diffusion and Decay of Hellenism*), Oxf.: Clarendon Pr., 1948.

BERTHELOT, M.: *Introduction à l'Étude de la Chimie des Anciens et du Moyen Âge,* Par.: Steinheil, 1889.

BEVAN, EDWYN: *A History of Egypt under the Ptolemaic Dynasty,* Lon.: Methuen, 1927.

BIRINGUCCIO, VANNOCCIO: *The Pirotecnia of Vannoccio Biringuccio* . . . , N.Y.: Basic Books, 1942–59.

BITON: *Bitons Bau von Belagerungsmaschinen und Geschützen,* in *Abhandlungen der Bayerische Akademie der Wissenschaften, Philosophisch-Historische Abteilung, Neue Folge 2,* 1929.

BORCHARDT, LUDWIG: *Gegen die Zahlenmystik an der grossen Pyramide bei Gize,* Berlin: Verlag von Behrend & Co., 1922.

BREASTED, JAMES HENRY: *Ancient Records of Egypt* (*Historical Records* . . .) Chi.: Univ. of Chicago Pr., 1906–07, 5 vols.

BROWN, PERCY: *Indian Architecture* (*Buddhist and Hindu Periods*), Bombay: D. B. Taraporevala Sons & Co., Ltd., 1942.

BURY, J. B., *et al.: The Hellenistic Age,* Camb.: Cambridge Univ. Pr., 1925.

CAESAR, GAIUS JULIUS: *The Gallic War,* LC.

CALZA, G., & BECATTI, G.: *Ostia,* Rome: Instituto Poligrafico dello Stato, 1958.

Cambridge Ancient History, Camb.: Cambridge Univ. Pr., 1923–56.

CARCOPINO, JÉRÔME: *Daily Life in Ancient Rome,* New Haven: Yale Univ. Pr., 1940.

CARTER, THOMAS FRANCIS: *The Invention of Printing in China* (*and Its Spread Westward*), N.Y.: Columbia Univ. Pr., 1925.

CASSIODORUS SENATOR, M. A.: *The Letters of Cassiodorus* (*Being a

Condensed Translation of the Variae Epistolae *of Magnus Aurelius Cassiodorus Senator*), Lon.: Henry Frowde, 1886.

CASSON, LIONEL: *The Ancient Mariners,* N.Y.: Macmillan Co., 1959.

———: "The Sprit-Rig in the Ancient World," in *The Mariner's Mirror,* XLVI, 4 (Nov. 1960), p. 241.

CHAMPDOR, ALBERT: *Babylon,* Lon.: Elek Books, 1958.

CICERO, MARCUS TULLIUS: *Brutus, De Finibus, De Natura Deorum, De Officiis, The Republic, & Tusculan Disputations,* LC.

CLAGETT, MARSHALL: *Greek Science in Antiquity,* N.Y.: Abelard-Schuman, Inc., 1955.

———: *The Science of Mechanics in the Middle Ages,* Madison: Univ. of Wis. Pr., 1959.

CLAUDIAN (CLAUDIUS CLAUDIANUS): LC.

COHEN, MORRIS R., & DRABKIN, I. E.: *A Source Book in Greek Science,* Cambridge, Mass.: Harvard Univ. Pr., 1958.

COLLAS, PERICLES: *A Concise Guide to Athens,* Athens: C. Calcoulides.

———: *A Concise Guide to the Acropolis of Athens,* Athens: C. Calcoulides.

COLUMELLA, LUCIUS JUNIUS MODERATUS: *On Agriculture,* LC.

CONTENAU, GEORGES: *Everyday Life in Babylon and Assyria,* Lon.: Edward Arnold, 1954.

CORY, ISAAC PRESTON: *Cory's Ancient Fragments* (*of the Phoenician, Carthaginian, Babylonian, Egyptian and Other Authors*), Lon.: Reeves & Turner, 1826–76.

COTTRELL, LEONARD: *Lost Cities,* N.Y.: Rinehart & Co., 1957.

———: *The Mountains of Pharaoh,* N.Y.: Rinehart, 1956.

———: *Wonders of the World,* N.Y.: Rinehart, 1956.

CRUMP, C. G., & JACOB, E. F. (eds.): *The Legacy of the Middle Ages,* Oxf.: Clarendon Pr., 1926–43.

CURTIUS, QUINTUS: *History of Alexander,* LC, 2 vols.

DAREMBERG, C., & SAGLIO, E.: *Dictionnaire des Antiquités Grecques et Romaines,* Paris: Hachette, 1904–07.

DAVIS, WILLIAM STEARNS: *A Day in Old Rome,* N.Y.: Bilbo & Tannen, 1959.

DAVISON, C. St. C.: "Transporting Sixty-Ton Statues in Early Assyria and Egypt," in *Technology & Culture,* II, 1 (Winter 1961), pp. 11–16.

DE BORHEGYI, SUZANNE: *Ships, Shoals, and Amphoras,* N.Y.: Holt, Rinehart & Winston, 1961.

DE CAMP, L. SPRAGUE: "Before Stirrups," in *Isis,* LII, Part 2, 164 (Jun. 1960), pp. 159 f.

———: "Master Gunner Apollonios," in *Technology & Culture,* II, 3 (Summer 1961), pp. 240–44.

———: "Sailing Close-Hauled," in *Isis,* L, Part 1, 159 (Mar. 1959), pp. 61 ff.

DEMOSTHENES: *Orations,* LC.

DIBNER, BERN: *Moving the Obelisks,* Norwalk, Conn.: Burndy Library, 1950.

DIELS, HERMANN: *Antike Technik* (*Sieben Vortrage*), Leipzig: B. G. Teubner, 1930.

DIENER, BERTHA: *Imperial Byzantium,* Boston: Little, Brown & Co., 1938.

DIO CASSIUS COCCEIANUS: *Dio's Roman History,* LC.

DIODORUS SICULUS: *The Library of History,* Lon.: Davis, 1814, 2 vols., & LC.

DIOGENES LAERTIUS: *Lives of Eminent Philosophers,* LC, 2 vols.

DIONISI, FRANCESCO: *Le Navi Sacre di Claudio nel Lago di Nemi,* Rome: Casa Editrice Galilei, 1956.

DIONYSIUS OF HALICARNASSUS: *Roman Antiquities,* LC.

DRACHMANN, AAGE G.: *Ktesibios, Philon, and Heron* (*A Study in Ancient Pneumatics*), in *Acta Historica Scientiarum Naturalium et Medicinalium* IV, Copenhagen: Munksgaard, 1948.

——: *The Mechanical Technology of Greek and Roman Antiquity* (*A Study of the Literary Sources*), Copenhagen: Munksgaard, 1963.

DUNHAM, DOWS: "The Biographical Inscriptions of Nekhebu in Boston and Cairo," in *Journal of Egyptian Archaeology,* XXIV (1938), pp. 1–8.

DUYVENDAK, J. J. L.: "An Illustrated Battle-Account in the History of the Former Han Dynasty," in *T'oung Pao,* 2d ser., XXXIV (1938–39), pp. 249–64.

EDWARDS, I. E. S.: *The Pyramids of Egypt,* PB.

ENGELBACH, R.: *The Problem of the Obelisks* (*From a Study of the Unfinished Obelisk at Aswan*), N.Y.: George H. Doran Co., 1923.

EUTROPIUS: see JUSTIN.

FAIRSERVIS, WALTER A., Jr.: *The Origins of Oriental Civilization,* N.Y.: Mentor Books, 1959.

FAKHRY, AHMED: *The Pyramids,* Chi.: Univ. of Chicago Pr., 1961.

FEDDEN, ROBIN: *Syria* (*An Historical Appreciation*), Lon.: Robert Hale, Ltd., 1946–56.

FINCH, JAMES KIP: *The Story of Engineering,* Garden City, N.Y.: Doubleday & Co., Inc., 1960.

FINEGAN, JACK: *Light from the Ancient Past* (*The Archeological Background of Judaism and Christianity*), Princeton: Princeton Univ. Pr., 1946–59.

FORBES, R. J.: *Man the Maker* (*A History of Technology and Engineering*), N.Y.: Abelard-Schuman Ltd., 1958.

——: *Studies in Ancient Technology,* Leiden: E. J. Brill, 1955–58, 6 vols.

FORTI, UMBERTO: *Storia della Tecnica dal Medioevo al Rinascimento,* Florence: Sansoni Editore, 1959.

FRONTINUS, SEXTUS JULIUS: *The Aqueducts of Rome & Stratagems,* LC.

GARDNER, ERNEST ARTHUR: *A Handbook of Greek Sculpture,* Lon.: Macmillan & Co., 1896–1924.

GEIL, WILLIAM EDGAR: *The Great Wall of China,* N.Y.: Sturgis & Walton, 1909.

GELLIUS, AULUS: *Attic Nights,* LC.

GESSI, LEONE: *Rome, the City and its Environs,* Rome: Libraria dello Stato, 1949.

GEST, ALEXANDER PURVIS: *Engineering,* N.Y.: Longmans, Green & Co., 1930.

GIBBON, EDWARD: *The Decline and Fall of the Roman Empire,* ML, 2 vols.

GIBSON, CHARLES E.: *The Story of the Ship,* N.Y.: Henry Schuman, 1948.

GILFILLAN, S. C.: *Inventing the Ship* (*A Study of the Inventions Made in Her History Between Floating Log and Rotorship*), Chi.: Follett Publ. Co., 1935.

GLANVILLE, S. R. K. (ed.): *The Legacy of Egypt,* Oxf.: Clarendon Pr., 1942.

The Greek Anthology, LC.

GREGORY, Pope, I: *The Dialogues of St. Gregory the Great* (*ed. by Edmund G. Gardner*), Lon.: Philip Lee Warner, 1911.

GRIFFO, PIETRO: *Agrigento* (*Guide to the Monuments and Excavations*), Agrigento: 1956.

GUIDO, MARGARET: *Syracuse* (*A Handbook to Its History and Principal Monuments*), Lon.: Max Parrish & Co., Ltd., 1958.

HADDAD, GEORGE: *History of Baalbak* (*Heliopolis*), Damascus: Karam, 1960.

HAMLIN, TALBOT: *Architecture Through the Ages,* N.Y.: G. P. Putnam's Sons, 1940.

HARRIZ, MICHEL A.: *Baalbek,* Lebanon: 1955.

HART, IVOR B.: *The Great Engineers,* Lon.: Methuen & Co., 1928.

——: *The Mechanical Investigations of Leonardo da Vinci,* Lon.: Chapman & Hall, 1925.

HERODOTUS, *History,* GH & LC.

HERON OF ALEXANDRIA: *Herons Belopoiika,* in *Abhandlungen der Königlich Preussichen Akademie der Wissenschaften, Jahrgang 1918, Philosophisch-Historische Klasse, Nr. 2.*

——: *Les Méchaniques ou l'Élévateur de Héron d'Alexandrie, Publiées pour la Première Fois sur la Version Arabe de Qostâ ibn-Lûqâ, et Traduites en Français par M. le Baron Carra de Vaux,* in *Journal Asiatique, ou Recueil de Mémoires d'Extraits et de Notices Relatifs à l'Histoire, à la Philosophie, aux Langues et à la Littérature des Peuples Orientaux, Neuvième Série, Tome Premier,* Par.: Imprimerie Nationale, 1893.

——: *The Pneumatics of Hero of Alexandria* (*From the Original Greek, Translated and Edited by Bennett Woodcroft . . .*), Lon.: Taylor Walton & Maberly, 1951.

HERON OF BYZANTIUM: *Heronis Mechanici Liber des Machinis, Necnon*

Liber de Geodaesia, Venice: Franciscus Barocius (Barozzi), 1572. See also ROCHAS.

HERZFELD, ERNST: *Zoroaster and His World,* Princeton: Princeton Univ. Pr., 1947, 2 vols.

HIRTH, FRIEDRICH: *The Ancient History of China* (*to the End of the Chou Dynasty*), N.Y.: Columbia Univ. Pr., 1908.

HITTI, PHILIP K.: *History of Syria* (*Including Lebanon and Palestine*), Lon.: Macmillan & Co., 1951.

——: *History of the Arabs* (*from the Earliest Times to the Present*), Lon.: Macmillan & Co., 1957.

——: *Lebanon in History* (*from the Earliest Times to the Present*), Lon.: Macmillan & Co., 1957.

HOPKINS, ARTHUR JOHN: *Alchemy, Child of Greek Philosophy,* N.Y.: Columbia Univ. Pr., 1934.

HORACE (QUINTUS HORATIUS FLACCUS): *The Complete Works of Horace,* ML & LC.

HURRY, JAMIESON B.: *Imhotep* (*The Visier and Physician of King Zoser and Afterwards the Egyptian God of Medicine*), Oxf.: Oxford Univ. Pr., 1928.

IAMBLICHUS OF CHALCIS: *Iamblichus' Life of Pythagoras,* Lon.: Thomas Taylor, 1915.

IDRISI: *Description de l'Afrique et de l'Espagne, par Edrîsî . . . ,* Leyden: E. J. Brill, 1866.

INGRAHAM, JOSEPH C.: *Modern Traffic Control,* N.Y.: Funk & Wagnalls, 1954.

JACOBSEN, THORKILD, & LLOYD, SETON: *Sennacherib's Aqueduct at Jerwan,* Chi.: Univ. of Chicago Pr., 1935.

JOSEPHUS, FLAVIUS: *The Works of Flavius Josephus . . . ,* Phila.: Grigg, Elliot & Co., 1847, & LC.

JUSTIN, CORNELIUS NEPOS, AND EUTROPIUS, Lon.: Henry G. Bohn, 1853.

JUVENAL (DECIMUS IUNIUS IUVENALIS): *Satires,* LC.

KING, L. W.: *The Letters and Inscriptions of Hammurabi, King of Babylon About* B.C. *2200 . . . ,* Lon.: Luzac & Co., 1900, 3 vols.

KIRBY, RICHARD SHELTON, *et al.*: *Engineering in History,* N.Y.: McGraw-Hill Book Co., Inc., 1956.

KLEMM, FRIEDRICH: *A History of Western Technology,* N.Y.: Charles Scribner's Sons, 1959.

KOLDEWEY, ROBERT: *The Excavations at Babylon,* Lon.: Macmillan & Co., 1914.

KRAMER, SAMUEL NOAH: *From the Tablets of Sumer,* Indian Hills, Colo.: Falcon's Wing Pr., 1956.

LACTANTIUS, LUCIUS CAECILIUS FIRMIANUS: *The Works of Lactantius,* in *The Ante-Nicene Christian Library,* Edinburgh: T. & T. Clark, 1867 *sqq.*, Vol. 21.

LANDA, DIEGO DE: *Landa's Relación de las Cosas de Yucatan,* Cambridge, Mass.: Peabody Museum, 1941.

LANDSTRÖM, BJÖRN: *The Ship (An Illustrated History),* Garden City, N.Y.: Doubleday & Co., Inc., 1961.

LANE, FREDERIC CHAPIN: *Venetian Ships and Shipbuilders of the Renaissance,* Baltimore: Johns Hopkins Pr., 1934.

LECKY, W. E. H.: *History of the Rise and Influence of the Spirit of Rationalism in Europe,* N.Y.: D. Appleton & Co., 1866, 2 vols.

LEONARDO DA VINCI: *Leonardo da Vinci's Notebooks . . . ,* N.Y.: Empire State Book Co., 1923.

LEY, WILLY: "The Elements of Khujut Rabu'a and Ctesiphon," in *Galaxy Science Fiction,* IX, 3 (Dec. 1954), pp. 44–51.

——: "Konstantin Anklitzen alias Friar Bertholdus," in *Galaxy Science Fiction,* X, 3 (Jun. 1955), pp. 92–103.

——: *Shells and Shooting,* N.Y.: Modern Age Books, 1942.

LIVINGSTONE, R. W. (ed.): *The Legacy of Greece,* Oxf.: Clarendon Pr., 1922.

LIVY (TITUS LIVIUS): *History of Rome,* Lon.: J. M. Dent & Sons, Ltd., 1912–24; & LC.

LLOYD, SETON: *Ruined Cities of Iraq,* Bombay: Oxford Univ. Pr. & Humphrey Milford, 1942–45. See also JACOBSEN.

LUCIAN (LOUKIANOS): *The Works of Lucian of Samosata,* Oxf.: Clarendon Pr., 1905–39, & LC.

McCARTNEY, EUGENE S.: *Warfare by Land and Sea,* Boston: Marshall Jones Co., 1923.

MACHIAVELLI, NICCOLÒ: *History of Florence (From the Earliest Times to the Death of Lorenzo the Magnificent),* N.Y.: Colonial Pr., 1901; Harper Bros., 1960.

——: *The Prince,* EL.

MAHAFFY, JOHN PENTLAND: *The Empire of the Ptolemies,* Lon.: Macmillan & Co., 1895.

——: *The Story of Alexander's Empire,* N.Y.: G. P. Putnam's Sons, 1894.

MAIURI, AMADEO: *Herculaneum,* Rome: Instituto Poligrafico dello Stato.

——: *The Phlegraean Fields,* Rome: Libraria dello Stato, 1953.

——: *Pompeii,* Novara: Instituto Geografico de Agostini, 1957.

MANETHO: LC.

MARTIAL (MARCUS VALERIUS MARTIALIS): *On Spectacles & Epigrams,* LC.

MATSCHOSS, C.: *Great Engineers,* Lon.: G. Bell & Sons Ltd., 1939.

MAYERS, W. F.: "Chinese Explorations of the Indian Ocean During the Fifteenth Century," in *China Review,* III (1874–75), Hong Kong, pp. 219–25, 321–31.

MEIGGS, RUSSELL: *Roman Ostia,* Oxf.: Clarendon Pr., 1960.

MERCKEL, CURT: *Die Ingenieurtechnik in Altertum,* Berlin: Julius Springer, 1899.

MIELI, ALDO: *La Science Arabe* (*et son Rôle dans l'Évolution Scientifique Mondiale*), Leiden: E. J. Brill, 1938.

MILLER, JOHN ANDERSON: *Master Builders of Sixty Centuries,* N.Y.: D. Appleton-Century Co., Inc., 1938.

MILLET, NICHOLAS: "An Obelisk is Raised," in *Archaeology,* XIV, 2 (Jun. 1961), pp. 138 f.

MINNS, ELLIS H.: *Scythians and Greeks* (*A Survey of Ancient History and Archaeology on the North Coast of the Euxine from the Danube to the Caucasus*), Camb.: Cambridge Univ. Pr., 1913.

MONTET, PIERRE: *Everyday Life in Egypt* (*in the Days of Ramesses the Great*), N.Y.: St. Martin's Pr., 1958.

MORET, ALEXANDRE: *The Nile and Egyptian Civilization,* N.Y.: Alfred A. Knopf, 1927.

MOULE, A. C.: "Boats Towed by a Swimming Buffalo in China," in *T'oung Pao,* 2d ser., XXXIII (1937), pp. 94 f.

NEEDHAM, JOSEPH: *The Development of Iron and Steel Technology in China,* Lon.: Newcomen Soc., 1958.

——: *Science and Civilisation in China,* Camb.: Cambridge Univ. Pr., 1954, Vol. I.

NEEDHAM, JOSEPH, & NEEDHAM, DOROTHY (eds.): *Science Outpost* (*Papers of the Sino-British Science Co-operation Office*), *1942–46,* Lon.: Pilot Pr., Ltd., 1948.

NEEDHAM, JOSEPH; WANG LING; & PRICE, DEREK J. DE SOLLA: *Heavenly Clockwork* (*The Great Astronomical Clocks of Medieval China*), Camb.: Cambridge Univ. Pr., 1960.

NEUBURGER, ALBERT: *The Technical Arts and Sciences of the Ancients,* N.Y.: Macmillan Co., 1930.

OLMSTEAD, A. T.: *History of the Persian Empire,* Chi.: Univ. of Chicago Pr., 1948.

OMAN, Sir CHARLES WILLIAM CHADWICK: *A History of the Art of War* (*The Middle Ages from the Fourth to the Fourteenth Century*), N.Y.: G. P. Putnam's Sons, 1898.

OVID (PUBLIUS OVIDIUS NASO): *The Art of Love and Other Love Books of Ovid,* N.Y.: Grosset & Dunlap, 1959, & LC.

Oxford Classical Dictionary, Oxf.: Clarendon Pr., 1949–57.

PALLOTTINO, M.: *The Etruscans,* PB.

PAPPUS: *Pappus d'Alexandrie* (*La Collection Mathématique . . . avec une Introduction et des Notes par Paul Ver Eecke*), Bruges: Desclée de Brouwer et Cie., 1933, 2 vols.

PARROT, ANDRÉ: *Babylon and the Old Testament,* N.Y.: Philosophical Library, 1958.

——: *Nineveh and the Old Testament,* N.Y.: Philosophical Library, 1955.

——: *The Tower of Babel,* N.Y.: Philosophical Library, 1955.

PARSONS, EDWARD ALEXANDER: *The Alexandrian Library* (*Glory of the Hellenic World*), Lon.: Cleaver-Hume Pr., 1952.

PARSONS, WILLIAM BARCLAY: *Engineers and Engineering in the Renaissance,* Baltimore: Williams & Wilkins Co., 1939.

PARTINGTON, J. R.: *A History of Greek Fire and Gunpowder,* Camb.: W. Heffer & Sons, Ltd., 1960.

PAUSANIAS: *Description of Greece,* Lon.: George Bell & Sons, 1886, 2 vols., & LC.

PAYNE, ROBERT: *The Canal Builders,* N.Y.: Macmillan Co., 1959.

PAYNE-GALLWEY, Sir RALPH: *A Summary of the History, Construction and Effects in Warfare of the Projectile-Throwing Engines of the Ancients . . . ,* Lon.: Longmans, Green & Co., 1907.

PELLIOT, PAUL: "Les Grands Voyages Maritimes Chinois au Début du XVe Siècle," in *T'oung Pao,* 2d ser., XXX (1933), pp. 237–455.

PENDEAS, EV.: *A Visitor's Guide to Mycenae, Argos-Tiryns-Nauplia, and Epidaurus,* Athens: P. Patsilniakos.

PETRIE, W. M. FLINDERS: *Memphis I,* Lon.: British School of Archaeology in Egypt, 1909.

PETRONIUS (GAIUS PETRONIUS ARBITER): *Satyricon,* N.Y.: Book Collectors Assn., 1934; New American Library, 1960; & LC.

PHILON OF BYZANTIUM: *Philons Belopoiika (Viertes Buch der Mechanik),* in *Abhandlungen der Preussischen Akademie der Wissenschaften, Jahrgang 1918, Philosophisch-Historische Klasse, Nr. 16.*

——: *Le Livre des Appareils Pneumatiques et des Machines Hydrauliques, par Philon de Byzance,* in *Notices et Extraits des Manuscrits de la Bibliothèque Nationale,* Tome 38, 1903.

——: *De Septem Orbis Spectaculis,* Rome: Leo Allatius, 1640. See also ROCHAS.

PIGGOTT, STUART: *Prehistoric India (to 1000* B.C.*),* PB.

PLATO: *The Republic, Critias, & Epistles,* in *The Works of Plato,* N.Y.: Tudor Publ. Co., & LC.

PLINY THE ELDER (GAIUS PLINIUS SECUNDUS): *Natural History,* Lon.: Henry G. Bohn, 1855–57, 6 vols., & LC.

PLINY THE YOUNGER (GAIUS PLINIUS CAECILIUS SECUNDUS): *Epistles,* LC.

PLUTARCH (PLOUTARCHOS): *Lives of the Noble Grecians and Romans,* ML & LC.

——: *Moralia,* LC.

POLO, MARCO: *Delle Cose de' Tartari e dell' Indie Orientali,* Venice: 1609–1954.

——: *The Travels of Marco Polo,* EL.

POLYBIUS: *The Histories of Polybius,* Lon.: Macmillan & Co., 1889, 2 vols., & LC.

PRICE, DEREK J. DE SOLLA: "An Ancient Greek Computer," in *Scientific American,* CC, 6 (Jun. 1959), pp. 60–67.

——: *On the Origin of Clockwork, Perpetual Motion Devices, and the Compass,* Washington: U. S. Natl. Museum Bull. 218. See also NEEDHAM.

PRITCHARD, JAMES B.: *Ancient Near Eastern Texts* (*Relating to the Old Testament*), Princeton: Princeton Univ. Pr., 1950–55.

——: *The Ancient Near East in Pictures,* Princeton: Princeton Univ. Pr., 1954.

PROCOPIUS: *History of the Wars & Buildings,* LC.

RAMELLI, AGOSTINO: *Le Diverse et Artificiose Machine,* Par.: 1588.

READ, JOHN: *Prelude to Chemistry* (*An Outline of Alchemy, Its Literature and Relationships*), N.Y.: Macmillan Co., 1937.

RICE, TAMARA TALBOT: *The Scythians,* N.Y.: Frederick A. Praeger, 1957.

RICHARDSON, R. C., & RICHARDSON, ROMOLA: *The Sailing Ship* (*Six Thousand Years of History*), N.Y.: Robert M. McBride & Co., 1947.

RIVOIRA, G. T.: *Roman Architecture* (*Its Principles of Construction Under the Empire . . .*), Oxf.: Clarendon Pr., 1925.

ROCHAS D'AIGLUN, ALBERT DE: *Poliorcétique des Grecs* (*Traité de Fortification, d'Attaque et de Défense des Places, par Philon de Byzance. . . . Accompagné de Fragments Explicatifs Tirés des Ingénieurs et Historiens Grecs*), in *Mémoires de la Société d'Émulation de Doubs, 1870–71;* Besançon: Imprimerie de Dodivers et Cie., *Quatrième Série, Sixième Volume.* See also ATHENAEUS MECHANICUS.

RODGERS, WILLIAM LEDYARD: *Greek and Roman Naval Warfare* (*A Study of Strategy, Tactics, and Ship Design from Salamis,* [*480* B.C.] *to Actium,* [*31* B.C.]), Annapolis: U. S. Naval Inst., 1937.

——: *Naval Warfare Under Oars* (*4th to 16th Centuries: A Study of Strategy, Tactics and Ship Design*), Annapolis: U. S. Naval Inst., 1939.

ROSTOVTSEFF, M.: *Iranians and Greeks in South Russia,* Oxf.: Clarendon Pr., 1922.

SANDYS, Sir JOHN EDWIN: *A Companion to Latin Studies,* Camb.: Cambridge Univ. Pr., 1910–43.

SANTANGELO, MARIA: *Selinunte,* Rome: 1953.

SARTON, GEORGE: *A History of Science,* Cambridge, Mass.: Harvard Univ. Pr., 1952–59, 2 vols.

——: *Introduction to the History of Science,* Baltimore: Williams & Wilkins Co., 1927–48, 3 vols. in 5.

SCHMIDT, ERICH F.: *Persepolis* (*I. Structures, Reliefs, Inscriptions*), Chi.: Univ. of Chicago Pr., 1953.

SCHREIBER, TH.: *Atlas of Classical Antiquities,* Lon.: Macmillan & Co., 1895.

Scriptores Historiae Augustae, LC.

SENECA, LUCIUS ANNAEUS: *De Ira, De Beneficiis, On Mercy, & Epistles,* LC.

——: *Physical Science in the Time of Nero* (*Being a Translation of the* Quaestiones Naturales *of Seneca*), Lon.: Macmillan & Co., 1910.

SESTIERI, PELLEGRINO CLAUDIO: "Greek Elea–Roman Velia," in *Archaeology,* X, 1 (Mar. 1957), pp. 2–10.

——: *Paestum (The City, the Prehistoric Necropolis in Contrada Gaudo, the Heraion at the Mouth of the Sele)*, Rome: Instituto Poligrafico della Stato.

SEXTUS EMPIRICUS: *Against the Logicians*, LC.

SINGER, CHARLES, *et al.* (eds.): *A History of Technology*, Oxf.: Oxford Univ. Pr., 1954–57, 5 vols.

SMITH, WILLIAM (ed.): *Dictionary of Greek and Roman Biography and Mythology*, Lon.: Taylor, Walton, & Maberly, & John Murray, 1849, 3 vols.

——: *Dictionary of Greek and Roman Geography*, Boston: Little, Brown & Co., 1854, 2 vols.

SPARGO, JOHN WEBSTER: *Virgil the Necromancer (Studies in Virgilian Legends)*, Cambridge, Mass.: Harvard Univ. Pr., 1934.

SPARTIANUS, AELIUS: see *Scriptores Historiae Augustae.*

SPAULDING, OLIVER LYMAN: *Pen and Sword in Greece and Rome*, Princeton: Princeton Univ. Pr., 1937.

SPEISER, E. A., *et al.*: *Everyday Life in Ancient Times*, Washington: National Geographic Soc., 1951–58.

STATIUS, PUBLIUS PAPINIUS: *Silvae*, LC.

STEICHEN, MICHEL: *Mémoire sur la Vie et les Travaux de Simon Stevin*, Brussels: A. van Dale, 1846.

STEINMAN, DAVID B., & WATSON, SARA RUTH: *Bridges and Their Builders*, N.Y.: Dover Publs., 1941–57.

STEVENS, WILLIAM OLIVER, & WESTCOTT, ALLEN: *A History of Sea Power*, N.Y.: Doubleday, Doran & Co., Inc., 1920–42, 2 vols.

STONE, GEORGE CAMERON: *A Glossary of the Construction, Decoration and Use of Arms and Armor . . .*, Portland, Me.: Southworth Pr., 1934.

STORCK, JOHN, & TEAGUE, WALTER DORWIN: *Flour for Man's Bread*, Minneapolis: Univ. of Minn. Pr., 1952.

STRABO: *Geography*, Lon.: Henry G. Bohn, 1854–57; Lon.: George Bell & Sons, 1881; & LC.

STRAUB, HANS: *A History of Civil Engineering (An Outline from Ancient to Modern Times)*, Lon.: Leonard Hill, Ltd., 1952.

SUETONIUS (GAIUS SUETONIUS TRANQUILLUS): *The Twelve Caesars & On Grammarians*, PB & LC.

SUN TZU WU: *The Art of War*, Harrisburg, Penna.: Military Service Publ. Co., 1943–57.

TACITUS, CORNELIUS: *History & Annals*, ML & LC.

TARN, W. W.: *Hellenistic Military and Naval Developments*, Camb.: Cambridge Univ. Pr., 1930.

TAYLOR, F. SHERWOOD: *The Alchemists (Founders of Modern Chemistry)*, N.Y.: Henry Schuman, 1949.

THEOPHILUS PRESBYTER: *The Treatise of Theophilus, Called Also Rugerus, upon Various Arts*, Lon.: John Murray, 1847.

THORNDIKE, LYNN: *A History of Magic and Experimental Science (During the First Thirteen Centuries of Our Era)*, N.Y.: Macmillan Co., 1923, 2 vols.
THUCYDIDES (THOUKYDIDES): *The Peloponnesian War,* GH & LC.
TOMPROPOULOS, MARIA: *A Concise Guide to Olympia,* Athens: C. Coulides.
TORR, CECIL: *Ancient Ships,* Camb.: Cambridge Univ. Pr., 1895.
——: *Rhodes in Ancient Times,* Camb.: Cambridge Univ. Pr., 1885.
TUNIS, EDWIN: *Wheels (A Pictorial History)*, Cleveland: World Publ. Co., 1955.
TUULSE, ARMIN: *Castles of the Western World,* Vienna: Thames & Hudson, 1958.
UCELLI, GUIDO: *Le Navi di Nemi,* Rome: Librario dello Stato, 1940–50.
USHER, ABBOTT PAYSON: *A History of Mechanical Inventions,* Boston: Beacon Pr., 1929–54.
VALERIUS MAXIMUS: *Faits et Paroles Memorables,* Par.: C. L. F. Panckoucke, 1828, 2 vols.
VALLENTIN, ANTONINA: *Leonardo da Vinci (The Tragic Pursuit of Perfection)*, N.Y.: Viking Pr., 1938.
VALTURIO, ROBERTO: *De Re Militari,* Verona: 1482.
VAUGHAN, AGNES CARR: *The House of the Double Axe,* N.Y.: Doubleday & Co., Inc., 1959.
VEGETIUS, FLAVIUS RENATUS: *Military Institutions of Vegetius,* Lon.: W. Griffin, 1797.
VITRUVIUS, MARCUS: *De architectura,* LC.
WELLER, CHARLES HEALD: *Athens and Its Monuments,* N.Y.: Macmillan Co., 1913.
WHIBLEY, LEONARD: *A Companion to Greek Studies,* Camb.: Cambridge Univ. Pr., 1906.
WHITE, ANDREW DICKSON: *A History of the Warfare of Science with Theology in Christendom,* N.Y.: Dover Publs., Inc., 1896–1960, 2 vols.
WHITE, LYNN, JR.: *Medieval Technology and Social Change,* Oxf.: Clarendon Pr., 1962.
WILSON, JOHN A.: *The Ancient Culture of Egypt,* Chi.: Univ. of Chicago Pr., 1951–59.
WINTER, H. J. J.: *Eastern Science (An Outline of Its Scope and Contribution)*, Lon.: John Murray, 1952.
WITTFOGEL, KARL A.: *Oriental Despotism (A Comparative Study of Total Power)*, New Haven: Yale Univ. Pr., 1957.
WOOLLEY, C. LEONARD: *Ur of the Chaldees (A Record of Seven Years of Excavation)*, N.Y.: Charles Scribner's Sons, 1930, & PB.
WRIGHT, F. A.: *Marcus Agrippa, Organizer of Victory,* N.Y.: Dutton, 1937.
XENOPHON: *Anabasis & On Horsemanship,* GH & LC.

INDEX

Italic numerals refer to illustrations in the text.

Big Five